Alloys: Structures, Properties and Applications

Alloys: Structures, Properties and Applications

Edited by **Keith Liverman**

Willford Press
New York

Published by Willford Press,
118-35 Queens Blvd., Suite 400,
Forest Hills, NY 11375, USA
www.willfordpress.com

Alloys: Structures, Properties and Applications
Edited by Keith Liverman

International Standard Book Number: 978-1-68285-231-6 (Hardback)

Printed in China.

Contents

Preface

This book has been a concerted effort by a group of academicians, researchers and scientists, who have contributed their research works for the realization of the book. This book has materialized in the wake of emerging advancements and innovations in this field. Therefore, the need of the hour was to compile all the required researches and disseminate the knowledge to a broad spectrum of people comprising of students, researchers and specialists of the field.

Alloys are widely used across different industrial and manufacturing sectors because of their strength, ductility and flexibility. They are primarily created by a combination of different metals or a mixture of metallic and non-metallic elements. This book elucidates the concepts around prospective developments with respect to synthetic metals, metalloids, metallurgy, evaluation and analysis of alloys created from different metals, etc. Those in search of information to further their knowledge will be greatly assisted by this book.

At the end of the preface, I would like to thank the authors for their brilliant chapters and the publisher for guiding us all-through the making of the book till its final stage. Also, I would like to thank my family for providing the support and encouragement throughout my academic career and research projects.

Editor

1

Mechanical Properties of Magnesium-Rare Earth Alloy Systems: A Review

Sravya Tekumalla [1], Sankaranarayanan Seetharaman [1], Abdulhakim Almajid [2] and Manoj Gupta [1,*]

[1] Department of Mechanical Engineering, National University of Singapore, 9 Engineering Drive 1, Singapore 117576, Singapore; E-Mails: tvrlsravya@nus.edu.sg (S.T.); seetharaman.s@nus.edu.sg (S.S.)
[2] Mechanical Engineering Department, College of Engineering, King Saud University, Riyadh 11421, Saudi Arabia; E-Mail: aalmajid@ksu.edu.sa

* Author to whom correspondence should be addressed; E-Mail: mpegm@nus.edu.sg

Academic Editor: Hugo F. Lopez

Abstract: Magnesium-rare earth based alloys are increasingly being investigated due to the formation of highly stable strengthening phases, activation of additional deformation modes and improvement in mechanical properties. Several investigations have been done to study the effect of rare earths when they are alloyed to pure magnesium and other Mg alloys. In this review, the mechanical properties of the previously investigated different magnesium-rare earth based binary alloys, ternary alloys and other higher alloys with more than three alloying elements are presented.

Keywords: magnesium; rare earth; binary alloys; ternary alloys; higher alloys and mechanical properties

1. Introduction

Magnesium is the sixth most abundant element in the earth's crust and is the lightest of all structural metals with a high specific stiffness. This is one of the prime reasons automobile manufacturers are in a quest to replace denser materials with magnesium (Mg) based materials. However, poor formability (ductility) and secondary processing induced crystallographic asymmetry (texture effects) due to the hexagonal closed pack (HCP) crystal structure represent the major limitations of Mg. It has limited slip systems and the activation of non-basal slip is difficult at room temperature, thereby limiting the ductility [1,2]. This limitation is being overcome with the development of new magnesium based alloys [3]. Amongst the several common alloying elements, rare earth (RE) addition has given promising results in terms of weakening the texture and improving the deformability of Mg [4,5]. Further, the hard eutectic phases formed as a result of RE addition also aid in increasing the strength of the alloy [6]. Thus, besides improving the ductility and formability, REs also act as effective strengtheners. Generally, the strengthening of Mg by the addition of RE is believed to be by a solid solution strengthening mechanism and a precipitation hardening mechanism [7].

The Mg-RE based alloys serve a useful purpose in automotive industry as superior light metal-alloys in cast or wrought condition [8,9]. Moreover, the Mg-RE alloys also gained prominence in biomedical applications as biodegradable implant materials that aid in healing of the tissues and leaving no implant residues e.g., bone implants, stents [10–12]. In order to serve such applications, the Mg-RE alloys are to be fabricated economically with minimum alloying compositions and simple processing. The Mg-RE phase diagrams suggest that each RE behaves uniquely when it is added as a dominant alloying element [13]. Hence, we study the effects of alloy compositions and processing on microstructure and texture and their effects on the tensile properties. This will in turn help to design alloy systems at the least cost to meet the requirements of the industry. The present article will give an in-detail review of the tensile and compressive properties of different magnesium-rare earth (Mg-RE) alloy systems.

2. Binary Systems

2.1. Yttrium

Addition of yttrium (Y) as an alloying element in Mg has been tested and tried by many researchers owing to the fact that there exists a large difference in the atomic radii of Mg (145 pm) and Y (212 pm) which allows strengthening of Mg by both solid solution strengthening and precipitation-strengthening (upon decomposition of the supersaturated solid solution). Zhao *et al.* [14] reported that Y with max solubility of 4.7 at.% (15.28 wt.%) in Mg can effectively strengthen Mg by solid solution strengthening. Similar observations regarding solution strengthening effects of Y in Mg have been made by Gao *et al.* [7]. In this study, cast Mg-Y alloys containing different Y concentrations ranging between 0.7 wt.% and 6.5 wt.% (below the solubility limit) were investigated. The properties of those Mg-Y alloys after heat treatment at 525 °C for 2–12 h are summarized in Table 1. It can be seen that the addition of Y resulted in the enhancement of strength properties, however, at the expense of ductility.

Table 1. Tensile properties of Mg-Y binary alloys produced by Gao *et al.* [7].

Alloy (wt.%)	Tensile Yield Strength (MPa)	Ultimate Tensile Strength (MPa)	Tensile Ductility (%)	Remarks Processing condition and Reference
Mg-0.7Y	39	150	14	As cast + heat treated at 525 °C for 2–12 h [7]
Mg-1.23Y	45	155	13	
Mg-3.1Y	70	160	11	
Mg-4.67Y	89	175	10	
Mg-6.5Y	110	230	8	

Xuenan *et al.* [15] investigated Mg-1%Y alloy in as cast and rolled conditions and reported that by rolling, the strength is improved and the elongation is decreased as compared to the as-cast Mg-1Y alloy. The authors also reported that Y has a negative effect on magnesium corrosion properties and Mg-1Y indicated no significant toxicity to osteoblasts and can be considered for biomedical alloy design. Zhou *et al.* [16] investigated the room temperature mechanical properties of as-extruded Mg-3%Y (extruded at 350 °C) and reported a tensile ductility as high as 33%. It is also interesting that the incremental tensile elongation has occurred without compromising the strength properties (Table 2) and the reason for such mechanical characteristics was reported to be due to the activation of multiple deformation modes. Similar observations of increases in both strength and ductility (Table 2) simultaneously were made by Edassiqi *et al.* [17] for Mg-2%Y and Sandlobes *et al.* [18] for Mg-3%Y where both the alloys were hot rolled and annealed. Contrarily, Wu *et al.* [19] reported increase in elongation to failure from 15% in pure Mg to 30% in Mg-4%Y alloy but this increase in elongation occured at the expense of strength (Table 2). This is reported to be due to the texture effects in case of Mg-2%Y and Mg-4%Y alloys extruded at 420 °C after heat treatment at 480 °C for 12 h. It has also been reported that the addition of Y to Mg as a solute in case of ultra-rapidly solidified and extruded Mg-alloys reduces critically resolved shear stress (CRSS) required to operate the pyramidal slip system. This activation of pyramidal slip system is believed to produce extensive plasticity. Hence, the ductility seems to increase in Mg-10%Y as compared to Mg-5%Y in reference [20]. The increase in strength from 5 wt%–10 wt% (Table 2), in this study, is reported to be due to the precipitation of β' phase from supersaturated solid solution [20].

Table 2. Tensile properties of Mg-Y binary alloys after secondary processing.

Alloy (wt.%)	Tensile Yield Strength (MPa)	Ultimate Tensile Strength (MPa)	Tensile Ductility (%)	Remarks Processing Condition and Reference
Mg-1Y	25	75	10	As-Cast [15]
Mg-1Y	148	200	9.3	Cast + Hot Rolled at 400 °C [15]
Mg-3Y	120	200	33	Cast + Extruded at 350 °C [16]
Mg-2Y	92	189	21	Cast + Heat treated at 480 °C for 12 h + Extruded at 420 °C [19]
Mg-4Y	87	177	30	Cast + Heat treated at 480 °C for 12 h + Extruded at 420 °C [19]
Mg-2Y	146	228	30.5	Cast + Hot Rolled at 450 °C + Annealed [17]
Mg-3Y	92	165	24	Gravity cast + Hot rolled + Annealed at 500 °C for 15 min [18]

Table 2. *Cont.*

Alloy (wt.%)	Tensile Yield Strength (MPa)	Ultimate Tensile Strength (MPa)	Tensile Ductility (%)	Remarks Processing Condition and Reference
Mg-5Y	-	350	7	Powder Metallurgy + Cold pressed (540 MPa) + Extruded at 420 °C [20]
Mg-10Y	-	440	9	Powder Metallurgy + Cold pressed (540 MPa) + Extruded at 420 °C [20]

2.2. Cerium

Cerium has almost negligible solid solubitlity in Mg at room temperature and is a eutectic forming solute (α-Mg + $Mg_{12}Ce$) [21]. Previous investigation [21] has shown that the addition of Ce as an alloying element not only produces the dispersion hardening effects but also contributes towards the intergranular percolation strengthening. Mishra *et al.* [22] investigated the microstructure, texture and mechanical properties of Mg-0.2 wt.%Ce alloy. In this study, the addition of 0.2%Ce to Mg has improved the tensile ductility of Mg from 9.1%–31% (Table 3). The increase in ductility however occurred alongside reduction in 0.2% offset yield strength. The reported mechanical properties were attributed to the favourable crystallographic orientation (rare earth assisted texture randomization) of Mg grains achieved upon dynamic recrystallization during extrusion. The ductility enhancement, in this study, was also accompanied with a little increase in the ultimate tensile strength as compared to pure Mg due to the reduction in the grain size upon Ce addition. Chino *et al.* [23] claimed that the improvement in ductility due to 0.2%Ce addition as compared to pure Mg is due to the increase in stacking fault energy unlike reduction in c/a ratio that led towards basal/non basal slip activation. Luo *et al.* [24] reported that the addition of 0.5% Ce to Mg resulted in more surface oxidation during extrusion and reduction in ductility and increase in strength. The authors also suggested that Ce concentration should not be higher than 0.5% in extruded alloys due to the excessive surface oxidation during extrusion. For Mg-1%Ce alloy [25], significant increase in ductility from 2.7%–11.9% was observed after annealing at 350 °C (Table 3). Chia *et al.* [6] have reported that when Cerium is added as an alloying element to Mg, an intermetallic $Mg_{12}Ce$ forms and the effect of increase in volume fraction of intermetallic is more than the effect of morphology of the intermetallic in increasing the strength and reducing the ductility of the alloy.

Table 3. Tensile properties of Mg-Ce binary alloys.

Alloy (wt.%)	Tensile Yield Strength (MPa)	Ultimate Tensile Strength (MPa)	Tensile Ductility (%)	Remarks Processing Condition and Reference
Mg-0.2Ce	68.6	170	31	Cast + Extruded at 400 °C [22]
Mg-0.2Ce	110–135	200–220	14–16	Cast + Extruded at 300 °C + Rolled at 400 °C [23]
Mg-0.2Ce	100	215	20.6	Cast+ Homogenized at 400 °C +
Mg-0.5Ce	130	230	8	Extrusion at 350 °C [24]
Mg-0.4Ce	140	160	20	As cast [21]
Mg-0.4Ce	90	120	29	Cast + Annealed at 520 °C for 1 h + water quenched [21]

Table 3. *Cont.*

Alloy (wt.%)	Tensile Yield Strength (MPa)	Ultimate Tensile Strength (MPa)	Tensile Ductility (%)	Remarks: Processing Condition and Reference
Mg-0.53Ce	80	140	5.5	High Pressure Die Cast [6]
Mg-0.93Ce	90	-	5	
Mg-1.48Ce	100	-	4	
Mg-2.87Ce	135	160	1.5	
Mg-4.78Ce	150	-	0.9–1	
Mg-1Ce	146 ± 5.5	168.5 ± 3	7.1	Cast + hot rolled at 400 °C [25]
Mg-1Ce	134 ± 2.5	205.5 ± 7	2.7	Cast + hot rolled at 400 °C and annealed for 1 h at 250 °C + Water Quenched [25]
Mg-1Ce	124.6 ± 1.5	212.7 ± 4.7	3.3	Cast + hot rolled at 400 °C and annealed for 1 h at 300 °C + Water Quenched [25]
Mg-1Ce	106 ± 4.7	197.6 ± 4.2	11.9	Cast + hot rolled at 400 °C and annealed for 1 h at 350 °C + Water Quenched [25]
Mg-1Ce	101.5 ± 1.6	203.1 ± 2.6	14.9	Cast + hot rolled at 400 °C and annealed for 1 h at 400 °C + Water Quenched [25]
Mg-1Ce	99 ± 2.1	203.3 ± 4.4	16.9	Cast + hot rolled at 400 °C and annealed for 1 h at 450 °C + Water Quenched [25]

2.3. Gadolinium

Gadolinium has a solublity of 23.49 wt.% at eutectic temperature [26] in Mg and thus contributes to solid solution strengthening when alloyed with Mg [12]. Hort *et al.* [12] reported that property profile of Mg-Gd alloys is similar to that of the cortical bone and can be adjusted over a wide range. They also reported that these alloys have better elongation to fracture compared to other metallic implant materials like stainless steels, *etc.* Gao *et al.* [27] studied the effects of Gd on the solid solution strengthening of Mg alloys. In this study, it was reported that Gd (due to size misfits and valency effects) is an effective solid solution strengthener in Mg as compared to Al and Zn. Peng *et al.* [28] reported that melt spun Mg-20Gd alloy contained mostly the supersaturated α-Mg solid solution while the as-cast Mg-20Gd alloy comprised of α-Mg + Mg_5Gd. In this study, the melt spun alloy exhibited fine grain morphology thus having higher strength when compared to that of the as-cast aloy containing metastable phases (Table 4).

Table 4. Tensile properties of Mg-Gd binary alloys.

Alloy (wt.%)	Tensile Yield Strength (MPa)	Ultimate Tensile Strength (MPa)	Tensile Ductility (%)	Remarks: Processing condition and Reference
Mg-2Gd	37.992	103.73	6.362	As-cast [12]
Mg-2Gd	33.430	87.002	4.928	As-cast + solutionized (T4) [12]
Mg-2Gd	41.274	101.368	5.686	As-cast + solutionized + artificially aged (T6) [12]
Mg-5Gd	54.752	128.468	6.620	As-cast [12]
Mg-5Gd	44.850	98.012	6.042	As-cast + solutioned (T4) [12]
Mg-5Gd	42.604	78.658	4.270	As-cast + solutionized + artificially aged (T6) [12]

Table 4. *Cont.*

Alloy (wt.%)	Tensile Yield Strength (MPa)	Ultimate Tensile Strength (MPa)	Tensile Ductility (%)	Remarks: Processing condition and Reference
Mg-10Gd	84.110	131.152	2.500	As-cast [12]
Mg-10Gd	69.120	111.650	3.152	Cast + solutioned(T4) [12]
Mg-10Gd	85.430	132.258	2.182	Cast + solutionized + artificially aged (T6) [12]
Mg-15Gd	127.646	175.220	0.950	As-Cast [12]
Mg-15Gd	118.052	186.844	2.440	Cast + solutionized(T4) [12]
Mg-15Gd	201.396	250.918	0.740	Cast + solutionized + artificial aged (T6) [12]
Mg-3.11Gd	60	160	13	Cast + Solution treatment at 535 °C/1.5 h [27]
Mg-5.73Gd	80	180	11	Cast + Solution treatment at 535 °C/4 h [27]
Mg-9.28Gd	100	190	9	Cast + Solution treatment at 535 °C/6.5 h [27]
Mg-14.2Gd	130	225	8	Cast + Solution treatment at 535 °C/9 h [27]
Mg-19.6Gd	150	255	7.5	Cast + Solution treatment at 540 °C/9.5 h [27]
Mg-20Gd	308	308	12	Melt Spun [28]
Mg-20Gd	254	254	13	As-cast [28]

Stanford *et al.* [29] investigated the microstructure-texture-mechanical property relationships in Mg-Gd alloys containing upto 4.5% Gd. They reported that the addition of Gd upto 1% was shown to significantly weaken the recrystallization texture. However, with further addition of Gd, it remained largely unchanged. Similarly, the strength values (Table 5) also increased until 1% Gd addition and remained unchanged thereafter. The authors stated that upon recrystallization annealing, the Gd solute locks the dislocation movement and causes matrix hardening. However, it is reported that such a mechanism did not affect the ducility of the developed Mg-Gd alloys. The same authors in [30] reported that Gd addition to Mg weakens the texture and produces the rare earth texture-component thereby resulting in extended plasticity when tested along the extrusion direction. They further reported that Mg-1.55Gd alloys exhibited high ductility when extruded at 450 °C as compared to 510 °C due to the suppression of formation of RE texture component at higher extrusion temperature. It is also interesting to note that the tensile ductility of Mg-1 wt.%Gd alloy was increased from 4.8%–30% upon annealing between 350 °C and 450 °C [25]. The properties of the Mg-Gd alloys available in literature are shown in Table 5.

Table 5. Tensile properties of Mg-Gd binary alloys after secondary processing.

Alloy (wt.%)	Tensile Yield Strength (MPa)	Ultimate Tensile Strength (MPa)	Tensile Ductility (%)	Remarks Processing Condition and Reference
Mg-0.22Gd	120	190	6	Cast + hot rolled at 400 °C and annealed for 1 h at 380 °C [29]
Mg-0.75Gd	145	210	12	Cast + hot rolled at 400 °C and annealed for 1 h at 380 °C [29]
Mg-2.75Gd	160	205	21	Cast + hot rolled at 400 °C and annealed for 1 h at 380 °C [29]
Mg-4.65Gd	165	210	26	Cast + hot rolled at 400 °C and annealed for 1 h at 380 °C [29]
Mg-1Gd	138.2 ± 1.7	173.3 ± 4	4.8	Cast + hot rolled at 400 °C [25]
Mg-1Gd	129.3 ± 4.9	191.2 ± 5.9	3.4	Cast + hot rolled at 400 °C and annealed for 1 h at 250 °C + Water Quenched [25]
Mg-1Gd	124.5 ± 1.4	225 ± 2.6	4.2	Cast + hot rolled at 400 °C and annealed for 1 h at 300 °C + Water Quenched [25]
Mg-1Gd	111 ± 4.8	240 ± 22	29.7	Cast + hot rolled at 400 °C and annealed for 1 h at 350 °C + Water Quenched [25]
Mg-1Gd	71.3 ± 3.4	184.9 ± 2.5	29.6	Cast + hot rolled at 400 °C and annealed for 1 h at 400 °C + Water Quenched [25]
Mg-1Gd	70.4 ± 2.4	220.6 ± 2.9	29.6	Cast + hot rolled at 400 °C and annealed for 1 h at 450 °C + Water Quenched [25]
Mg-1.55Gd	102	214	23.9	Cast + Solution treated at 530 °C for 3 h + 560 °C for 5 h + Extruded at 450 °C [30]
Mg-1.55Gd	130	210	15.8	Cast + Solution Treated at 530 °C for 3 h + 560 °C for 5 h + Extruded at 510 °C [30]

2.4. Lanthanum

Lanthanum has limited solid solubility in Mg and has a very high eutectic temperature of 612 °C. Due to its poor solid solubility in Mg, Mg-La alloys do not undergo age hardening [31]. Chia *et al.* [6] reported that Mg-La eutectic is lamellar and with the increase in La content, strength (Table 6) is seen to increase due to the $Mg_{12}La$ intermetallic. This was observed with the reduction in ductility. The have also reported [6] that La is largely believed to be an effective grain refiner as well as an effective texture modifier in Mg alloys . In a study by Stanford *et al.* [30], it was reported that Mg-La alloys extruded at 450 °C exhibited new texture component that is similar to "Rare Earth Texture" along $\langle 11\bar{2}1 \rangle$ direction parallel to the extrusion direction. However, the development of such texture components was suppressed in those Mg-La alloys extruded at higher temperatures thus resulting in lowering of the ductility (Table 6).

Table 6. Tensile properties of Mg-La binary alloys.

Alloy (wt.%)	Tensile Yield Strength (MPa)	Ultimate Tensile Strength (MPa)	Tensile Ductility (%)	Remarks Processing Condition and Reference
Mg-0.51La	80	135	5	High pressure die-cast [6]
Mg-0.94La	90	-	4.5	High pressure die-cast [6]
Mg-1.71La	110	-	3.72	High pressure die-cast [6]
Mg-3.44La	140	170	1	High pressure die-cast [6]
Mg-5.07La	168	-	0.75	High pressure die-cast [6]
Mg-0.22La	115	232	19.4	Cast + Solution treated at 560 °C for 8 h + Extruded at 450 °C [30]
Mg-0.22La	150	220	13.8	Cast + Solution treated at 560 °C for 8 h + Extruded at 520 °C [30]

2.5. Erbium

Erbium is one of the rare earths that is well soluble in Mg. Wu *et al.* [32] investigated the strengthening of the extrusion texture component <0 0 0 1> parallel to the direction of extrusion occured with the addition of Er. This developed to RE texture after complete recrystallization in Mg-6Er. The tensile ductility (Table 7) of the Mg-Er alloys was reported to be due to the reduction in the c/a ratio and development of texture that led to the activation of different modes of plastic deformation. A high tensile ductility of about 30% was observed in Mg-6Er alloy. In Mg-3.6%Er [33], the authors reported that ageing at 200 °C led to the dynamic strain ageing that gave rise to the serrated flow behaviour.

Table 7. Tensile properties of Mg-Er binary alloys.

Alloy (wt.%)	Tensile Yield Strength (MPa)	Ultimate Tensile Strength (MPa)	Tensile Ductility (%)	Remarks Processing condition and Reference
Mg-3.6Er	90	140	20	Cast + solution treated at 500 °C for 8 h, + aged at 200 °C for 8 h [33]
Mg-2Er	83 **47.0 ± 1.5 ***	251 **205.0 ± 5.0 ***	19.6 **33.0 ± 3.0 ***	Die cast + homogenized at 520 °C for 48 h + Extruded at 400 °C + annealed at 400 °C/60 min [32]
Mg-4Er	80 **47.0 ± 1.3 ***	184 **176.0 ± 3.4 ***	28.4 **35.0 ± 3.8 ***	Die cast + homogenized at 520 °C for 48 h + Extruded at 440 °C + annealed at 450 °C/20 min [32]
Mg-6Er	72 **47.0 ± 1.5 ***	195 **170.0 ± 4.2 ***	29.4 **42.3 ± 3.3 ***	Die cast + homogenized at 520 °C for 48 h + Extruded at 440 °C + annealed at 450 °C/30 min [32]

* indicates the compressive properties of the same alloy.

2.6. Neodymium

Neodymium has the highest solid solubility in Mg and lowest eutectic temperature of about 552 °C and shows best response to age hardening when added to Mg due to its high solid solubility [6]. Chia *et al.* [6] reported that unlike Ce and La, Nd forms Mg_3Nd phase which is a very hard phase and not $Mg_{12}Nd$ phase. They have attributed the same for better strength of Nd containing Mg alloys as compared to that of Ce and La containing Mg alloys. In contrast, Jingli *et al.* [34] reported that the microstructure of Mg-Nd alloys consists of dendritic α-Mg and divorced eutectic $Mg_{12}Nd$. The improvement in strength with increase in Nd content was attributed to both solid solution hardening and precipitation hardening. Seitz *et al.* [35] studied Mg-2 Nd alloys in different extruded and heat treated conditions and reported that the high elongation ratios combined with the low yield strength (Table 8) and low degradation of the Mg-2Nd alloys make them promising for resorbable stent applications and comparable to the conventional WE 43 alloys.

Table 8. Tensile properties of Mg-Nd binary alloys.

Alloy (wt.%)	Tensile Yield Strength (MPa)	Ultimate Tensile Strength (MPa)	Tensile Ductility (%)	Remarks Processing condition and Reference
Mg-0.47Nd	81	189	9.5	High Pressure Die-cast [6]
Mg-0.76Nd	85	-	6.75	High Pressure Die-cast [6]
Mg-1.25Nd	92	-	4.8	High Pressure Die-cast [6]
Mg-2.60Nd	115	-	4.1	High Pressure Die-cast [6]
Mg-3.53Nd	130	140	2.5	High Pressure Die-cast [6]
Mg-1.2Nd	95	123	4.61	As-cast [34]
Mg-1.85Nd	121.2	155.8	2.76	As-cast [34]
Mg-3.59Nd	141.2	153.6	1.08	As-cast [34]
Mg-2Nd	77 **102 ***	193 **327 ***	30 -	Cast +Extruded at 380 °C [35]
Mg-2Nd	123 **106 ***	240 **340 ***	26 -	Cast + Extruded at 380 °C + annealing at 204 °C for 16 h (T5(1)) [35]
Mg-2Nd	102 **110 ***	242 **340 ***	27.5 -	Cast + Extruded at 380 °C + annealing at 204 °C for 48 h (T5(2)) [35]
Mg-2Nd	125 **105 ***	220 **320 ***	15 -	Cast + Extruded at 380 °C + solution treatment at 510 °C for 3 h + annealing at 204 °C for 16 h (T6(1)) [35]
Mg-2Nd	70 **85 ***	230 **335 ***	18.5 -	Cast + Extruded at 380 °C + solution treatment at 510 °C for 3 h + annealing at 204 °C for 48 h (T6(2)) [35]

* indicates the compressive properties of the same alloy.

2.7. Dysprosium

Dysprosium has a high solid solubility in Mg and the melting point of the intermetallic $Mg_{24}Dy_5$ is 560 °C [36]. It is expected to improve the room temperature mechanical properties of Mg by solid solution strengthening and precipitation hardening as the solubility decreases drastically with decrease in temperature. The ductility decreases with increase in Dy content and is very poor at room temperature [37,38]. The authors [37] reported that of all the alloys, Mg-10Dy can be developd further for biomedical applications due to its mechanical and corrosion properties. They also suggested in reference [38] that ageing treatment at 200 °C can be selected for applicability of the Mg-Dy alloys as bone fictures as low ducility and high strength are required. The mechanical properties of various Mg-Dy binary alloys are shown in Table 9.

Table 9. Tensile properties of Mg-Dy binary alloys.

Alloy (wt.%)	Tensile Yield Strength (MPa)	Ultimate Tensile Strength (MPa)	Tensile Ductility (%)	Remarks Processing Condition and Reference
Mg-5Dy	48	77	4.6	As-Cast(F) [37]
Mg-5Dy	40	76	3.3	Cast + Solutionized at 520 °C for 24 h + Water Quenched (T4) [37]
Mg-10Dy	82	130	5.5	As-Cast(F) [37]
Mg-10Dy	63	104	4	Cast + Solutionized at 520 °C for 24 h + Water Quenched (T4) [37]
Mg-10Dy	65	108	3.8	Cast + Solutionized at 520 °C for 24 h + Water Quenched + aged at 250 °C for 16 h (T6-1) [38]
Mg-10Dy	70	95	4.2	Cast + Solutionized at 520 °C for 24 h + Water Quenched + aged at 200 °C for 168 h (T6-2) [38]
Mg-15Dy	105	125	1.9	As-Cast(F) [37]
Mg-15Dy	68	125	3	Cast + Solutionized at 520 °C for 24 h + Water Quenched (T4) [37]
Mg-15Dy	72	113	2	Cast + Solutionized at 520 °C for 24 h + Water Quenched + aged at 250 °C for 16 h (T6-1) [38]
Mg-15Dy	104	137	2.2	Cast + Solutionized at 520 °C for 24 h + Water Quenched + aged at 200 °C for 168 h (T6-2) [38]
Mg-20Dy	120	142	1.5	As-Cast(F) [37]
Mg-20Dy	110	148	1.25	Cast + Solutionized at 520 °C for 24 h + Water Quenched (T4) [37]
Mg-20Dy	120	140	0.6	Cast + Solutionized at 520 °C for 24 h + Water Quenched + aged at 250 °C for 16 h (T6-1) [38]
Mg-20Dy	168	223	0.9	Cast + Solutionized at 520 °C for 24 h + Water Quenched + aged at 200 °C for 168 h (T6-2) [38]
Mg-12.1Dy	83	114	2.8	As-Cast [36]

3. Ternary Systems

Several compositions are tried where rare earths are major as well as minor alloying elements in order to investigate the influence of the rare earths on the properties of the Mg based alloy. In this section, properties of such investigated Mg-RE based ternary alloys containing magnesium, rare earth element and another alloying element *i.e.*, Mg-X-RE (X = RE, Al, Zn, Zr, Sn, Mn, Cu) are discussed.

3.1. Mg-RE Ternary Systems

The Mg-10Y-2.5Sm alloy was investigated by Zhang *et al.* [39] and it was reported that $Mg_{24}Y_5$ phase was distributed in α-Mg matrix uniformly and no phases contained Sm. The tensile properties, in the study, are attributed to the solid solution strengthening effect of Sm and strengthening by $Mg_{24}Y_5$. In the Mg-8.3Gd-1.9Er alloy, the strength increased with ageing due to the β' phase that has formed by precipitation mechanism. Gavras *et al.* [40] investigated Mg-La-(Y,Gd,Nd) systems and reported that the tensile strength of Mg-2.5La-3.6Nd system is highest of all the investigated Mg-La-(Y,Gd,Nd) alloys with a value of 195 MPa. It was reported to be due to the higher amount of intermetallics in the eutectic that settled at the grain boundaries. The tensile properties of the above discussed alloys are shown in Table 10.

Table 10. Tensile properties of Mg-RE ternary alloys.

Alloy (wt.%)	Tensile Yield Strength (MPa)	Ultimate Tensile Strength (MPa)	Tensile Ductility (%)	Remarks Processing Condition and Reference
Mg-10Y-2.5Sm	-	207	2.53	Cast + Solution treated at 540 °C for 6 h + Water Quenched+ Aged at 250 °C for 2 h [39]
Mg-8.3Gd-1.9Er	112	246	6.5	As-Cast [41]
Mg-8.3Gd-1.9Er	190	308	4.9	Cast + solution treated at 570 °C for 6 h + Isothermally aged at 200 °C [41]
Mg-2.5La-3.6Nd	195	-	2.2	High Pressure Die Cast [40]
Mg-2.5La-2.5Y	170		5	
Mg-2.5La-4.1Y	186	-	4	
Mg-2.4La-5.2Gd	184	-	3.9	

3.2. Mg-Al-RE Ternary Systems

Luo *et al.* [24] reported that the addition of 0.2 or 0.5% Ce to Mg-3Al did not show significant improvement in tensile properties due to the affinity of Ce for Al thus forming $Al_{11}Ce_3$ in Mg-Al-Ce ternary alloys resulting in the properties as shown in Table 11.

Table 11. Tensile properties of Mg-Zn-RE ternary alloys.

Alloy (wt.%)	Tensile Yield Strength (MPa)	Ultimate Tensile Strength (MPa)	Tensile Ductility (%)	Remarks Processing condition and Reference
Mg-3Al-0.2Ce	120	235	18	Cast + Homogenized at 400 °C + Extrusion at 350 °C [24]
Mg-3Al-0.5Ce	125	230	20	

3.3. Mg-Zn-RE Ternary Systems

Luo *et al.* [42] investigated the Mg-Zn-Ce alloys and reported that the addition of Zn to Mg-0.2Ce alloys has improved the strength significantly with a slight reduction in ductility and attributed it to the independent effect of solid solution strengthening by Zn. The authors also reported that Zn does not react with Ce thereby producing random texture in extrusion. Le *et al.* [43] investigated Mg-2Zn-0.4RE alloys and reported that the highest strength was seen in Ce containing alloy and highest ductility in Y containing alloy. Wu *et al.* [44] reported that the excellent ductility of the rolled Mg-Gd-Zn alloy sheets is due to the texture weakening effects of Gd. The effects of ultrasonic treatment was studied on Mg-5Zn-2Er alloys [45] and it was reported that the improved mechanical properties are a result of cavitation and acoustic streaming during ultrasonic treatment. The properties of cast Mg-5Zn-0.63Er alloy were studied under heat treated and peak aged conditions [46]. In this study, it was reported that the properties were high in aged condition due to the presence of rod-like $MgZn_2$ particles. Wang *et al.* [47] suggested that the texture of as-extruded Mg-*x*Zn-*x*Er was weakened by the recrystallization via particle stimulated nucleation (PSN). Srinivasan *et al.* [48] investigated Mg-Gd-Zn alloys and reported that the Mg-10Gd-*x*Zn alloys ($x = 2, 6$) exhibited good strength due to the 14H-type LPSO phases present in the matrix while those alloys containing lower Gd exhibited good ductility due to the lower fraction of LPSO. Singh *et al.* [49] attributed the high strength of the Mg-Zn-Y alloy to the nano-quasi-crystalline phase that has formed during extrusion. The properties of all the alloys that are discussed/investigated are listed in Table 12.

Table 12. Tensile properties of Mg-Zn-RE ternary alloys.

Alloy (wt.%)	Tensile Yield Strength (MPa)	Ultimate Tensile Strength (MPa)	Tensile Ductility (%)	Remarks: Processing Condition and Reference
Mg-2Zn-0.2Ce	135	225	27	Cast + Extruded at 400 °C [42]
Mg-5Zn-0.2Ce	135	247	15	
Mg-8Zn-0.2Ce	136	289	16	
Mg-1.11Zn-1.68Gd	129.9	233.4	40.3	Cast + homogenized at 500 °C for 10 h + Rolled at 430 °C + annealed at 400 °C for 1 h (Tested in Rolling Direction) [44]
Mg-1.11Zn-1.68Gd	113.8	221.2	44.5	Cast + homogenized at 500 °C for 10 h + Rolled at 430 °C + annealed at 400 °C for 1 h (Tested in 45° to Rolling Direction) [44]
Mg-1.11Zn-1.68Gd	110.1	218.4	44.6	Cast + homogenized at 500 °C for 10 h + Rolled at 430 °C + annealed at 400 °C for 1 h (Tested in Transverse Direction) [44]
Mg-1.06Zn-2.74Gd	130.6	220.0	40.3	Cast + homogenized at 500 °C for 10 h + Rolled at 430 °C + annealed at 400 °C for 1 h (Tested in Rolling Direction) [44]
Mg-1.06Zn-2.74Gd	121.0	220.3	47.3	Cast + homogenized at 500 °C for 10 h + Rolled at 430 °C + annealed at 400 °C for 1 h (Tested at 45° to Rolling Direction) [44]
Mg-1.06Zn-2.74Gd	118.0	220.9	45.1	Cast + homogenized at 500 °C for 10 h + Rolled at 430 °C + annealed at 400 °C for 1 h (Tested in Transverse Direction) [44]
Mg-2Zn-0.4Ce	190	255	18	Cast + Extruded at 310 °C [43]
Mg-2Zn-0.4Gd	125	220	26	
Mg-2Zn-0.4Y	160	240	30	
Mg-2Zn-0.4Nd	175	245	28	
Mg-2Zn-2Gd	71	135	5.5	Gravity permanent mold cast [48]
Mg-6Zn-2Gd	89	170	4.5	
Mg-2Zn-10Gd	119	146	1.5	
Mg-6Zn-10Gd	116	144	1	
Mg-0.2Zn-12.12Dy	92	125	6.3	As-Cast [36]
Mg-1.2Zn-12Dy	100	145	5.2	
Mg-2.4Zn-11.9Dy	95	128	1.2	

Table 12. *Cont.*

Alloy (wt.%)	Tensile Yield Strength (MPa)	Ultimate Tensile Strength (MPa)	Tensile Ductility (%)	Remarks Processing Condition and Reference
Mg-2Zn-2.3Er	310 ± 6.5	320 ± 7.9	12.8 ± 1.2	Cast + annealed at 400 °C for 10 h + Extruded at 300 °C [47]
Mg-2Zn-2.3Er	247 ± 6.2	279 ± 6.8	12.1 ± 1.6	Cast + annealed at 400 °C for 10 h + Extruded at 400 °C [47]
Mg-3.7Zn-4Er	295 ± 2.8	330 ± 3.0	13.7 ± 2.1	Cast + annealed at 400 °C for 10 h + Extruded at 300 °C [47]
Mg-3.7Zn-4Er	278 ± 2.9	319 ± 3.1	17.6 ± 2.0	Cast + annealed at 400 °C for 10 h + Extruded at 400 °C [47]
Mg-5.5Zn-6.2Er	299 ± 6.3	343 ± 7.0	16.8 ± 1.2	Cast + annealed at 400 °C for 10 h + Extruded at 300 °C [47]
Mg-5.5Zn-6.2Er	283 ± 2.2	328 ± 2.5	19.7 ± 1.2	Cast + annealed at 400 °C for 10 h + Extruded at 400 °C [47]
Mg-4Zn-0.1Ce	109 ± 2.6	234 ± 4.0	17.3 ± 0.94	Cast + homogenized for 3 h at 300 °C + 24 h at 400 °C + hot rolled at 400 °C + annealed at 400 °C for 30 min [50]
Mg-3Zn-0.3Er	70	180	12.5	
Mg-3Zn-0.38Er	75	185	13	
Mg-3Zn-0.5Er	77	186	13.5	
Mg-3Zn-0.75Er	80	155	11.5	
Mg-3Zn-2.5Er	82	164	8	
Mg-3Zn-3Er	104	184	7.5	
Mg-3Zn-3.8Er	103	162	6	
Mg-5Zn-0.5Er	94	205	11.5	
Mg-5Zn-0.63Er	96	210	12.5	
Mg-5Zn-0.83Er	97	209.52	11	
Mg-5Zn-1.25Er	98	187.6	10.5	As-Cast [51]
Mg-5Zn-2.5Er	99	185	9	
Mg-5Zn-5Er	124	213.7	8.5	
Mg-5Zn-6.25Er	117	186.03	7.5	
Mg-7Zn-0.7Er	102.11	195	6.5	
Mg-7Zn-0.88Er	120	197	7.5	
Mg-7Zn-1.17Er	124	196	7	
Mg-7Zn-1.75Er	126	158	6.5	
Mg-7Zn-3.5Er	128	169	6	
Mg-7Zn-7Er	130.24	210	5.5	
Mg-7Zn-8.75Er	128.49	176	3.5	
Mg-5Zn-0.63Er	112.5	223	11.5	As-Cast [46]
Mg-5Zn-0.63Er	106	206.8	13.6	Cast + solution heat-treated at 440, 460, 480 and 500 °C for 10 h (T4) [46]
Mg-5Zn-0.63Er	124	261	10.5	Cast + solution heat-treated at 440, 460, 480 and 500 °C for 10 h + isothermally aged at 175 °C (T6) [46]
Mg-5Zn-2Er	-	151	7	As-Cast [45]

Table 12. *Cont.*

Alloy (wt.%)	Tensile Yield Strength (MPa)	Ultimate Tensile Strength (MPa)	Tensile Ductility (%)	Remarks Processing Condition and Reference
Mg-5Zn-2Er	-	210	11	Ultrasonic Treatment for 100 s and power 600 W + Cast [45]
Mg-6Zn-0.3Er	72	210	12	As-Cast [52]
Mg-6Zn-0.3Er	138	289	25	Cast + homogenized at 400 °C for 10 h + extruded at 300 °C [52]
Mg-6Zn-0.3Er	157	291	18	Cast + homogenized at 400 °C for 10 h + extruded at 300 °C + aged at 200 °C [52]
Mg-6Zn-0.5Er	87	184	12	As-Cast [52]
Mg-6Zn-0.5Er	155	310	17	Cast + homogenized at 400 °C for 10 h + extruded at 300 °C [52]
Mg-6Zn-0.5Er	183	329	12	Cast+ homogenized at 400 °C for 10 h+ extruded at 300 °C + aged at 200 °C [52]
Mg-6Zn-1.0Er	104	224	11	As-Cast [52]
Mg-6Zn-1.0Er	187	295	18	Cast + homogenized at 400 °C for 10 h + extruded at 300 °C [52]
Mg-6Zn-1.0Er	193	302	14	Cast + homogenized at 400 °C for 10 h + extruded at 300 °C + aged at 200 °C [52]
Mg-6Zn-1.5Er	100	203	10	As-Cast [52]
Mg-6Zn-1.5Er	175	296	17	Cast + homogenized at 400 °C for 10 h + extruded at 300 °C [52]
Mg-6Zn-1.5Er	188	300	15	Cast + homogenized at 400 °C for 10 h + extruded at 300 °C + aged at 200 °C [52]
Mg-6Zn-2.0Er	110	198	6	As-Cast [52]
Mg-6Zn-2.0Er	194	304	16	Cast + homogenized at 400 °C for 10 h + extruded at 300 °C [52]
Mg-6Zn-2.0Er	193	301	12	Cast + homogenized at 400 °C for 10 h + extruded at 300 °C + aged at 200 °C [52]
Mg-14.4Zn-3.3Y	171 **213 ***	320 **530 ***	12 **14 ***	Cast + solutionized at 480 °C for 24 h + extruded at 430 °C [49]
Mg-14.4Zn-3.3Y	365.0 ± 3.5 **267.7 ± 0.7 ***	380 **550 ***	8 **12 ***	Cast + solutionized at 480 °C for 24 h + extruded at 430 °C + aged at 150 °C [49]

* indicates the compressive properties of the same alloy.

3.4. Mg-Zr-RE Ternary Systems

Huang *et al.* [4] reported the effect of multi-micro alloying of RE on the ductility of Mg alloys. Different rare earths were studied in the ternary system and the best ductility was observed with Gd addition to Mg-0.5Zr. Investigations on Mg-0.6Zr-8Gd were done in [53] and the properties were attributed to the precipitate β-Mg5RE (Gd/Er) and dispersed β'-$Mg_{15}RE_3$ (Gd/Er) metastable phase. The properties are shown in Table 13.

Table 13. Tensile properties of Mg-Zr-RE ternary alloys.

Alloy (wt.%)	Tensile Yield Strength (MPa)	Ultimate Tensile Strength (MPa)	Tensile Ductility (%)	Remarks: Processing condition and Reference
Mg-0.5Zr-0.4Y	20.9 ± 1.0	119.3 ± 7.2	4.3 ± 2.5	As-Cast [4]
	25.6 ± 2.0 *	**200.8 ± 16.6 ***	**15.6 ± 3.6 ***	
Mg-0.5Zr-0.4Gd	52.2 ± 3.8	144.6 ± 1.7	22.1 ± 1.6	
	40.2 ± 1.1 *	**231.3 ± 8.7 ***	**24.4 ± 1.5 ***	
Mg-0.5Zr-0.4Dy	53.7 ± 3.0	140.7 ± 0.4	17.5 ± 0	
	32.5 ± 2.6 *	**231.6 ± 16 ***	**22.7 ± 2.0 ***	
Mg-0.5Zr-0.4Sm	51.6 ± 3.1	144.2 ± 6.5	17.0 ± 2.5	
	36.2 ± 1.6 *	**237.1 ± 3.0 ***	**22.9 ± 2.6 ***	
Mg-0.6Zr-8Gd	82	141	6.2	Cast + solution treated at 530 °C for 10 h + Aged at 230 °C [53]
Mg-0.6Zr-8Gd	81	155	6.4	

* indicates the compressive properties of the same alloy.

3.5. Mg-Sn-RE Ternary Systems

The properties of Mg-Sn-RE system have been investigated by Zhao *et al.* [54]. In this study, Y was used as the rare earth element and it was reported that when Y is 1.5%, MgSnY phase forms and with increase of Y to 3%, MgSnY+Sn_3Y_5 phases form and at 3.5%Y, Sn_3Y_5 phase forms. The combined effect of intermetallics in Mg-3%Y is responsible for the higher properties as shown in Table 14. Wang *et al.* [55] reported that Mg-8.23Sn-2Nd exhibited the best tensile properties. This was related to the microstructure as α-Mg, Mg_2Sn and Mg-Sn-Nd phases were present in the microstructure of the alloys and the strength was attributed to the change in morphology of the Mg-Sn-Nd phase and size of the Mg_2Sn phase. The properties of the different Mg-Sn-RE alloys that are investigated are shown in Table 14.

Table 14. Tensile properties of Mg-Sn-RE ternary alloys.

Alloy (wt.%)	Tensile Yield Strength (MPa)	Ultimate Tensile Strength (MPa)	Tensile Ducility (%)	Remarks: Processing Condition and Reference
Mg-1Sn-1.5Y	165	199	12.8	Cast + Homogenized at 480 °C for 12 h + Extruded at 350 °C [54]
Mg-1Sn-3Y	295	305	2.4	
Mg-1Sn-3.5Y	136	225	14	
Mg-1.65Sn-2Nd	-	115	8	As-Cast [55]
Mg-4.92Sn-2Nd	-	132.5	8.5	
Mg-8.23Sn-2Nd	-	140	10	
Mg-11.52Sn-2Nd	-	135	8.75	

3.6. Other Ternary Systems

The addition of Er to Mg-1.8Mn resulted in the increase in ductility and this was attributed to the resistance to recrystallize and retard the grain growth with the addition of Er [56]. The highest tensile properties were found in alloy containing 0.7% Er. Du *et al.* [57] reported that 18R LPSO phase is formed in Mg-10Er-2Cu alloy that resulted in the properties as shown in Table 15.

Table 15. Tensile properties of other ternary alloys.

Alloy (wt.%)	Tensile Yield Strength (MPa)	Ultimate Tensile Strength (MPa)	Tensile Ductility (%)	Remarks Processing Condition and Reference
Mg-1.8Mn-0.1Er	173	255	7	Cast + homogenized at 450 °C for 4 h + Extruded at 450 °C + Annealed at 390 °C for 1 h [56]
Mg-1.8Mn-0.4Er	224	276	9	
Mg-1.8Mn-0.7Er	228	275	12.5	
Mg-10Er-2Cu	320	380	15	Cast + homogenized at 450 °C for 24 h + Extruded at 430 °C [57]

4. Higher Alloy Systems

In this section, magnesium rare earth alloys containing more than three elements are discussed. The alloying elements whose effect is studied include REs, Al, Li, Zr, Zn, Sn, Mn, Cu, Ca and V.

4.1. Mg-RE Higher Alloy Systems

Zhang *et al.* [58] reported the properties of WE43 degradable biomaterial in extruded conditions as reported in Table 16. In reference [59], the authors reported the properties of Mg-4Y-3.2RE at room temperature and reported that ageing improved the strength of the alloy significantly. Su *et al.* [60] reported the properties of peak aged WE43 alloy as shown in Table 16. Mukai *et al.* [61] reported the properties of the WE43 alloys in annealed, extruded and aged conditions and reported that the extruded alloy exhibits superior properties in terms of strength and ductility due to the fine grained microstructure and transition from intergranular to transgranular fracture with grain refinement. Panigrahi *et al.* [62] studied the effects of forging and ageing on the mechanical properties of WE43 alloys and reported that the improvement in strength is due to the combined effect of grain refinement, work hardening and precipitation strengthening. The improvement in ductility is also reported to be due to the limited intergranular fracture and activation of non-basal slip prior to twinning.

Table 16. Tensile properties of Mg-RE based higher alloys.

Alloy (wt.%)	Tensile Yield Strength (MPa)	Ultimate Tensile Strength (MPa)	Tensile Ductility (%)	Remarks Processing condition and Reference
Mg-4Y-2.3Nd-0.88Gd	221 ± 1.7	295 ± 3.1	10.7 ± 0.8	Cast + Extruded at 350 °C [58]
Mg-4Y-3.2RE	-	240	-	Cast + Annealed at 525 °C [59]
Mg-4Y-3.2RE	-	320	-	Cast + Annealed at 525 °C + Aged at 200 °C [59]
Mg-4Y-3RE	165	250	2.0	T6 [60]
Mg-4Y-3RE (2.2Nd)	-	230	16	Annealed at 525 °C/5 h [61]
Mg-4Y-3RE (2.2Nd)	-	300	6	Annealed at 525 °C/5 h + Aged at 200 °C [61]
Mg-4Y-3RE (2.2Nd)	-	320	20	Annealed at 525 °C/5 h + Extruded at 300 °C [61]
Mg-4Y-3RE	185 ± 12	261 ± 5	31 ± 1	As-Received [62]
Mg-4Y-3RE	270 ± 15	348 ± 6	16 ± 3	As Received + T5 [62]
Mg-4Y-3RE	263 ± 7	311 ± 11	23 ± 3	Forged [62]
Mg-4Y-3RE	318 ± 9	368 ± 10	17 ± 1	Forged + Aged at 210 °C/32h [62]
Mg-4Y-3RE	344 ± 11	388 ± 12	23 ± 1	Forged + Aged at 180 °C/60 h [62]
Mg-4Y-3RE	286 ± 10	341 ± 4	28 ± 1	Forged + Aged at 150 °C/104 h [62]

4.2. Mg-Al-RE Higher Alloy Systems

Rzychoń *et al.* [63], in their study on Mg-Al-RE alloys, reported that when RE/Al ratio is >0.5, no $Mg_{17}Al_{12}$ phase forms. The $Mg_{17}Al_{12}$ phase has lower melting temperature compared to the other $Al_{11}RE_3$, Al_2RE phases and thus when the ratio between RE and Al is maintained at an optimum level, the thermal stability of Mg alloys can be improved. In reference [64], the authors investigated the microstructure and mechanical properties of Mg-10Gd-3Y-0.8Al alloys and reported that the microstructure of cast Mg-10Gd-3Y alloy was refined with the addition of 0.8%Al and when solution treated at 520 °C for 6 h + 550 °C for 7 h, the ductility improved from 5%–13%. Further, they also discussed the effect of ageing on the properties of the alloy. The strength increased due to the precipitation strengthening and solute dissolution of the intermetallic particles in the solute. Zhang *et al.* [65] investigated the properties of Mg-3.0Al-1.8Ce-0.3Y-0.2Mn alloy and found that it exhibited high structural stability and strength due to the presence of dendrite boundaries with $Al_{11}(Ce,Y)_3$ intermetallics. The strength enhancement, in this study, is also thought to be due to the solid solution strengthening effects of Y. Similar strengthening mechanism was reported in reference [66] by the same authors wherein Ce was mostly present in the $Al_{11}RE_3$ and Y was observed to exist in Al_2RE. Rzychoń *et al.* [67] investigated the properties of AE44 alloys and reported that the high pressure die cast alloys exhibited better properties when compared to the sand cast alloys. They attributed the same to the low solidification rates in the sand cast alloys that led to the unfavorable morphology of the intermetallics $Al_{11}RE_3$ (needle shaped) and Al_2RE (polyhedral) and coarse grained structure. Zhang *et al.* [68] also reported that when La substitutes Ce rich misch metal, α-Mg and $Al_{11}La_3$ phases are observed instead of $Al_{11}RE_3$ and Al_2RE. It was reported that La containing alloy exhibits better properties due to the stability of $Al_{11}La_3$ phase in Mg-4Al-4RE-0.4Mn alloy. The same author also reported in [69] that with an increase in Ce content, the strength was improved and was attributed to the acicular morphology of the main strengthening phase, $Al_{11}Ce_3$. A similar trend was observed with La, Pr and Nd in [70–72]. Wang *et al.* [73] also observed that Mg-5Al-0.3Mn-1.5Ce alloy exhibited the best tensile strength due to the presence of optimized content of $Al_{11}Ce_3$ + β-$Mg_{17}Al_{12}$. Chen *et al.* [74] reported that the addition of Nd led to the formation of higher melting Al-Nd intermetallic and also improved the room temperature strength of the Mg-6Al-2Ca-*x*Nd ($x = 1,2$) alloys. Wu *et al.* [75] reported that the addition of REs and Ca to AZ91 alloys led to the improvement in strength as well as corrosion resistance due to the presence of Al_2Ca phases. The properties of the above discussed alloys are given in Table 17.

Table 17. Tensile properties of Mg-Al-RE based higher alloys.

Alloy (wt.%)	Tensile Yield Strength (MPa)	Ultimate Tensile Strength (MPa)	Tensile Ductiliy (%)	Remarks
				Processing Condition and Reference
Mg-10Gd-3Y-0.8Al	136	215	4.8	As-Cast [64]
Mg-10Gd-3Y-0.8Al	126	226	13	Cast + solution treated at 520 °C for 6 h + 550 °C for 7 h (T4A) [64]
Mg-10Gd-3Y-0.8Al	227	353	3.5	Cast + solution treated at 520 °C for 6 h + 550 °C for 7 h + peak-aged at 200 °C (T6A) [64]

Table 17. *Cont.*

Alloy (wt.%)	Tensile Yield Strength (MPa)	Ultimate Tensile Strength (MPa)	Tensile Ductiliy (%)	Remarks
				Processing Condition and Reference
Mg-10Gd-3Y-0.8Al	213	301	12.1	Cast + solution treated at 520 °C for 6 h + 550 °C for 7 h + peak-aged at 225 °C (T6B) [64]
Mg-3Al-1.8Ce-0.3Y-0.2Mn	158	255	10	Die-cast [65]
Mg-3.4Al-2.4Cemm-0.3Ymm-0.3Mn	166 **158***	267 **372***	11 **17***	Die-cast [66]
Mg-4Al-4RE-0.18Mn	47	146	7.1	Sand Cast [67]
(2.35%Ce, 1.07%La, 0.59%Nd, 0.16%Pr)	178	241	10.8	High Pressure Die Cast [67]
Mg-4Al-4RE-0.4Mn (RE = 52–55Ce, 3–5La, 16–20Nd, 5–6Pr)	140	247	11	Die-cast [68]
Mg-4Al-4La-0.4Mn	146	264	13	Die-cast [68]
Mg-4Al-1Ce-0.3Mn	146	232	9	As-Cast [69]
Mg-4Al-2Ce-0.3Mn	148	247	12	As-Cast [69]
Mg-4Al-4Ce-0.3Mn	157	250	11	As-Cast [69]
Mg-4Al-6Ce-0.3Mn	161	254	10	As-Cast [69]
Mg-4Al-1La-0.3Mn	133	236	12	As-Cast [70]
Mg-4Al-2La-0.3Mn	140	245	13	As-Cast [70]
Mg-4Al-4La-0.3Mn	155	265	12	As-Cast [70]
Mg-4Al-6La-0.3Mn	171	257	7	As-Cast [70]
Mg-4Al-1Pr-0.3Mn	145	241	13	As-Cast [71]
Mg-4Al-2Pr-0.3Mn	148	248	13	As-Cast [71]
Mg-4Al-4Pr-0.3Mn	165	262	16	As-Cast [71]
Mg-4Al-6Pr-0.3Mn	155	251	10	As-Cast [71]
Mg-4Al-1Nd-0.3Mn	150	244	12	As-Cast [72]
Mg-4Al-2Nd-0.3Mn	154	248	13	As-Cast [72]
Mg-4Al-4Nd-0.3Mn	156	258	15	As-Cast [72]
Mg-4Al-6Nd-0.3Mn	165	261	12	As-Cast [72]
Mg-5Al-0.3Mn-0.5Ce	71	173	9	As-Cast [73]
Mg-5Al-0.3Mn-1.0Ce	82	184	15	As-Cast [73]
Mg-5Al-0.3Mn-1.5Ce	88	203	20	As-Cast [73]
Mg-5Al-0.3Mn-1.5Ce	225	318	9	Cast + Hot Rolled at 400 °C [73]
Mg-5Al-0.3Mn-2.0Ce	75	177	13	As-Cast [73]
Mg-5Al-0.3Mn-3.0Ce	68	165	6	As-Cast [73]
Mg-6Al-2Ca-1Nd	180	286	9.5	Cast + Homogenised at 460 °C for 24 h + Extruded at 330 °C [74]
Mg-6Al-2Ca-2Nd	186	306	12.3	
Mg-6Al-2Ca-3Nd	205	310	13	
Mg-6Al-2Ca-4Nd	210	319	12.8	
Mg-9Al-0.5Zn-0.5RE	91	158	1.65	As-Cast [75]
Mg-9Al-0.5Zn-1.0RE	90	165	1.62	
Mg-9Al-0.5Zn-1.2RE	88	170	1.6	
Mg-9Al-0.5Zn-1.5RE	93	174	1.5	
Mg-9Al-0.5Zn-1Ca-1RE	90	169	1.6	
Mg-9Al-0.5Zn-2Ca-1RE	78	150	1.4	
Mg-9Al-0.5Zn-3Ca-1RE	75	129	1.3	
Mg-9Al-0.5Zn-4Ca-1RE	70	115	0.9	

* indicates the compressive properties of the same alloy.

4.3. Mg-Li-RE Higher Alloy System

Krausse *et al.* [76] investigated the biodegradation behavior of LAE442 alloys and reported that the alloys have sufficient initial strength to be used in weight bearing applications in bones. In reference [77], the authors characterized Mg-1.21Li-1.12Ca-1Y alloy and reported that the alloy exhibited better tensile properties in extruded state due to the refinement of microstructure. The authors also indicated that the corrosion resistance of extruded alloy is better than the as-cast alloy due to the delay of the initiation of the corrosion pits. Tao *et al.* [78] investigated the structural and mechanical properties of Mg-Li-Al-Zn-*x*RE alloys containing varying RE content between 0.2% and 1%. In this study, it was reported that the microstructure of the alloys mainly comprised of α-phase, β-phase, $Mg_{17}Al_{12}$ phase, $Mg_{64.5}Li_{34.3}Al_{0.9}Zn_{0.3}$ and Al_2Zn_2La intermetallic compounds. They also reported that besides reducing the laminar spacing of the matrix, RE also acted as an effective grain refiner. The improvement of strength at both room temperature and high temperature was attributed to the formation of Al_2Zn_2La strengthening phase. Zhou *et al.* [79] studied the Mg-Li-Al-RE alloys and reported that the high properties of the alloys are due to the addition of Al and rare earths that result in grain refinement, solid solution strengthening and dispersion strengthening. They also reported the applicability of the alloys for cardiovascular stent materials due to the good corrosion resistance and good cytotoxicity test results. Wang *et al.* [80] studied the Mg-8Li-1Al-1Ce alloy in as-cast and extruded condition and reported that the α(Mg) phase and β(Li) phase are refined after extrusion and the long rod-like Al_2Ce present in the as-cast state becomes short and rod-like after extrusion which is responsible for the properties as shown in Table 18.

Table 18. Tensile properties of Mg-Li-RE based higher alloys.

Alloy (RE = 85% La, 10% Pr, 5% Ce) (wt.%)	Tensile Yield Strength (MPa)	Ultimate Tensile Strength (MPa)	Tensile Ductility (%)	Remarks Processing condition and Reference
Mg-1.21Li-1.12Ca-1Y	44.00	51.71	1.47	As-Cast [77]
Mg-1.21Li-1.12Ca-1Y	115.02	183.72	14.45	Cast + Extruded [77]
Mg-7Li-6A1-6Zn-0.2 RE	-	194	2.8	
Mg-7Li-6A1-6Zn-0.4 RE	-	200	2.4	
Mg-7Li-6A1-6Zn-0.6 RE	-	204	2.6	As-cast [78]
Mg-7Li-6A1-6Zn-0.8 RE	-	205	2.5	
Ma-7Li-6A1-6Zn-1RE	-	209	2.1	
Mg-3.5Li-2Al-2RE	95	190	22	
Mg-5.5Li-2Al-2RE	140	235	23	Cast + Extruded [79]
Mg-8.5Li-2Al-2RE	100	150	32	
Mg-8Li-1Al-1Ce	141	160	16	As-Cast [80]
Mg-8Li-1Al-1Ce	175	187	33	Cast + Extruded at 220 °C [80]
Mg-8Li-7Al-Si-4.5RE	200	260	14	Wrought [81]

4.4. Mg-Zr-RE Higher Alloy Systems

For Mg-10Gd-3Y-0.4Zr, the mechanism of strengthening is similar to as discussed in the case of Mg-10Gd-3Y-0.8Al [64]. Mg-*x*Y-1.5LPC-0.4Zr (x = 0, 2, 4, 6) (LPC = 85% La, 8% Ce, 7% Pr) alloys were investigated in [82]. In this study, the authors showed that the tensile properties of the Mg-Zr-RE alloys improved with an increase in Y content and it was attributed to the distribution of the cubic β-$Mg_{24}Y_5$ precipitate phases and prismatic β′ phases in Mg matrix. Mohri *et al.* [83] investigated the Mg-4Y-3Re-0.5Zr alloy in different conditions and reported that the material extruded at 400 °C exhibited the best properties due to the presence of fine spherical precipitates in the grains. Zhang *et al.* [84] reported that the solution treatment and ageing treatment can enhance the strength of the alloy. The authors also reported that the *in vitro* degradation rate of the alloy increases by solution treatment and decreases by aging due to coarse microstructure and relief of internal stresses in the precipitation phase, respectively. Zhang *et al.* [58] reported that Mg-0.44Zr-3.09Nd, 0.22Zn (JDBM) alloy exhibited superior mechanical properties (due to finer grain size) as well as biocorrosion properties as compared to WE43 and AZ31 alloys and thus is a promising degradable biomaterial. Su *et al.* [60] reported the age hardening behavior and mechanical properties of the Mg-4Y-2.4Nd-0.2Zn-0.4Zr alloy and suggested that the presence of fine β" precipitates in the matrix result in the high mechanical properties in the peak aged condition. Zhang *et al.* [85] reported that the cyclic extrusion resulted in better mechanical properties and bio corrosion of the Mg-2.73Nd-0.16Zn-0.45Zr alloy. In reference [86], the authors studied the mechanical properties and biocorrosion behavior of the Mg-Nd-Zr-Zn alloys at different extrusion ratios. With extrusion ratio 8, the alloys exhibited high strength and moderate elongation while with extrusion ratio 25, the alloys exhibited high elongation and moderate strength. Thus, the authors suggested the optimal properties of the alloy when the alloy undergoes dynamic recrystallization and the growth is suppressed. The authors also indicated that the corrosion properties and cytotoxicity of the alloy meet the requirement of the cell toxicity and hence this makes it a potential biomaterial. Zhang *et al.* [87] reported the effect of double extrusion on the improved mechanical properties and improved corrosion resistance of the biodegradable Mg-Nd-Zn-Zr alloy due to the grain refinement. Kielbus *et al.* [88] investigated Mg-4Y-3RE and Mg-3Nd-1Gd alloys and reported that Mg-4Y-3RE alloys exhibit higher tensile and creep properties due to the presence of higher stable Y containing phases in Mg-4Y-3RE alloys. Zheng *et al.* [89] investigated the effect of thermomechanical treatment on Mg-6Gd-2Nd-0.5Zr and reported that cold deformation increase from 5%–10% accelerated the age hardening response of the alloy and improved the strength. They also reported that hot extrusion and ageing lead to the very high tensile properties of the alloy as reported in Table 19. Xiao *et al.* [90] investigated the Mg-10Gd-3Y-0.5Zr and reported that the Friction Stir processing (FSP) led to grain refinement and dissolution of the eutectic Mg5(Gd,Y) thereby improving the ductility. Ageing treatments done after the FSP led to the improvement in the strengths. Similarly, Li *et al.* [91] studied the effects of FSP on WE43 alloy and reported that the improvement in mechanical properties is due to the refinement of grains in the alloy's microstructure. He *et al.* [92] studied WE 93 alloy and it was observed that the extruded and aged alloy exhibited the best combination of tensile properties. This was reported to be due to the $Mg_{24}Y_5$ phase that is completely dissolved in Mg_{12}(MM) (MM = Misch Metal) which is present around the grain boundaries. In [93], the authors reported that the rolling process has effectively reduced the grain size and improved the mechanical properties. The properties of the different alloys discussed above are reported in Table 19.

Table 19. Tensile properties of Mg-Zr-RE based higher alloys.

Alloy (wt.%)	Tensile Yield Strength (MPa)	Ultimate Tensile Strength (MPa)	Tensile Ductility (%)	Remarks Processing condition and Reference
Mg-10Gd-3Y-0.4Zr	151	230	4.4	As-Cast [64]
Mg-10Gd-3Y-0.4Zr	131	247	14.4	Cast + Solution Treated at 500 °C for 6 h (T4) [64]
Mg-10Gd-3Y-0.4Zr	231	349	2.2	Cast + Solution Treated at 500 °C for 6 h + peak-aged at 225 °C [64]
Mg-2Y-1.5LPC-0.4Zr (LPC = 85% La, 8% Ce, 7% Pr)	90	180	25	
Mg-4Y-1.5LPC-0.4Zr (LPC = 85% La, 8% Ce, 7% Pr)	110	195	20	Cast + Solution treated for 10 h at 525 °C [82]
Mg-6Y-1.5LPC-0.4Zr (LPC = 85% La, 8% Ce, 7% Pr)	135	250	10	
Mg-4Y-3RE-0.4Zr	-	235	17	Cast + Annealed at 525 °C/5 h [83]
Mg-4Y-3RE-0.4Zr	-	300	6	Cast + Annealed at 525 °C/5h + Aged at 200 °C [83]
Mg-4Y-3RE-0.4Zr	-	325	20	Cast + Annealed at 525 °C/5 h + Extruded at 100:1 at 400 °C [83]
Mg-4Y-3RE-0.4Zr	-	330	20	Cast + Annealed at 525 °C/5 h + Extruded at 100:1 at 400 °C + Aged at 200 °C [83]
Mg-4Y-3RE-0.4Zr	-	350	13	Cast + Annealed at 525 °C/5 h + Extruded at 2.8:1 at 400 °C [83]
Mg-4Y-3RE-0.4Zr	-	370	5	Cast + Annealed at 525 °C/5 h + Extruded at 2.8:1 at 400 °C + Aged at 200 °C [83]
Mg-4Y-2.4Nd-0.2Zn-0.4Zr	150	197	7.5	As-cast (F) [60]
Mg-4Y-2.4Nd-0.2Zn-0.4Zr	162	240	15	Cast + solution treated at 490 °C+ water quenched (T4) [60]
Mg-4Y-2.4Nd-0.2Zn-0.4Zr	268	339	4.0	Cast + solution treated at 490 °C + water quenched + Aged at 200 °C (T60) [60]
Mg-4Y-2.4Nd-0.2Zn-0.4Zr	265	330	6.5	Cast + solution treated at 500 °C + water quenched + Aged at 225 °C (T61) [60]
Mg-4Y-2.4Nd-0.2Zn-0.4Zr	195	260	3.0	Cast + solution treated at 510 °C + water quenched + Aged at 250 °C (T62) [60]
Mg-2.7Nd-0.2Zn-0.4Zr	363 ± 6.3	376 ± 4.3	8.4 ± 2.2	Cast + solution-treated at 540 °C for 10 h + Water quenched + Extruded at 250 °C [84]
Mg-2.7Nd-0.2Zn-0.4Zr	394 ± 5.2	417 ± 7.6	2.6 ± 0.2	Cast + solution-treated at 540 °C for 10 h + Water quenched + Extruded at 250 °C + Aged at 200 °C for 8 h [84]
Mg-2.7Nd-0.2Zn-0.4Zr	121 ± 4.8	217 ± 3.3	22.2 ± 2.4	Cast + solution-treated at 540 °C for 10 h + Water quenched + Extruded at 250 °C + solution-treated at 530 °C for 30 min [84]

Table 19. *Cont.*

Alloy (wt.%)	Tensile Yield Strength (MPa)	Ultimate Tensile Strength (MPa)	Tensile Ductility (%)	Remarks Processing condition and Reference
Mg-2.7Nd-0.2Zn-0.4Zr	191 ± 2.6	326 ± 2.8	12.2 ± 1.2	Cast + solution-treated at 540 °C for 10 h + Water quenched + Extruded at 250 °C + solution-treated at 530 °C for 30 min + Aged at 200 °C for 8 h [84]
Mg-0.44Zr-3.09Nd, 0.22Zn	293 ± 5.1	307 ± 1.9	15.9 ± 3.1	Cast + Extruded at 350 °C [58]
Mg-0.45 Zr-2.73Nd-0.16Zn	175	240	11	Cast + solution-treated at 540 °C for 10 h + Water quenched + Extruded at 350 °C [85]
Mg-0.45 Zr-2.73Nd-0.16Zn	260	300	29	Cast + solution-treated at 540 °C for 10 h + Water quenched + Extruded at 350 °C (Cyclic Extrusion and Compression) [85]
Mg-0.4Zr-3Nd-1.6Zn	90 ± 7	194 ± 3	12.0 ± 0.8	Cast + Solution treated at 540 °C for 10 h + Water quenched (T4) [86]
Mg-0.4Zr-3Nd-1.6Zn	308 ± 6	312 ± 2	12.2 ± 0.6	T4 + Extruded at 320 °C with ratio 8 (R8) [86]
Mg-0.4Zr-3Nd-1.6Zn	333 ± 4	334 ± 4	7.9 ± 0.2	R8 + Aging [86]
Mg-0.4Zr-3Nd-1.6Zn	156 ± 1	233 ± 4	25.9 ± 0.8	T4 + Extruded at 320 °C with ratio 25 (R25) [86]
Mg-0.4Zr-3Nd-1.6Zn	177 ± 2	238 ± 3	20.4 ± 0.3	R25 + Aging [86]
Mg-2.25Nd-0.11Zn-0.43Zr	204 ± 5.3	247 ± 4.4	20.6 ± 1.6	Cast + solution-treated at 540 °C for 10 h + Water quenched + Single Extruded at 290 °C [87]
Mg-2.25Nd-0.11Zn-0.43Zr	276 ± 6.0	309 ± 6.4	34.3 ± 3.4	Cast + solution-treated at 540 °C for 10 h + Water quenched + Double Extruded at 320 °C [87]
Mg-2.70Nd-0.20Zn-0.41Zr	163 ± 1.9	245 ± 2.2	14 ± 1.5	Cast + solution-treated at 540 °C for 10 h + Water quenched + Single Extruded at 350 °C [87]
Mg-2.70Nd-0.20Zn-0.41Zr	275 ± 4.7	308 ± 2.3	32.8 ± 1.4	Cast + solution-treated at 540 °C for 10 h + Water quenched + Single Extruded at 320 °C [87]
Mg-4Y-3RE-0.5Zr	225	331	6	Cast + solution-treated at 520 °C for 8 h + Water quenched + Aged at 250 °C for 16 h [88]
Mg-3Nd-1Gd-0.5Zr-0.4Zn	163	293	7	Cast + solution-treated at 520 °C for 8 h + Water quenched + Aged at 200 °C for 16 h [88]
Mg-0.5Zr-0.4Y-0.4Gd	51.7 ± 2.8 **43.7 ± 2.7 ***	140.2 ± 0.6 **242.4 ± 16.2 ***	27.7 ± 1.5 **24.9 ± 0.2 ***	As-Cast [4]
Mg-0.5Zr-0.4Y-0.4Dy	48.9 ± 2.5 **43.8 ± 1.6 ***	132.2 ± 1.5 **247.7 ± 7.1 ***	29.3 ± 1.8 **25.0 ± 0.3 ***	
Mg-0.5Zr-0.4Y-0.4Sm	55.7 ± 2.7 **47.0 ± 5.7 ***	148.6 ± 2.9 **260.8 ± 10.2 ***	27.0 ± 2.3 **25.1 ± 0.2 ***	
Mg-0.5Zr-0.4Gd-0.4Dy	47.6 ± 2.7 **38.1 ± 2.3 ***	143.7 ± 2.4 **243.5 ± 4.7 ***	22.2 ± 2.2 **25.6 ± 0.3 ***	
Mg-0.5Zr-0.4Gd-0.4Sm	51.7 ± 0.3 **44.1 ± 0.8 ***	145.1 ± 3.6 **247.5 ± 1.0 ***	26.4 ± 0.7 **24.8 ± 1.0 ***	

Table 19. *Cont.*

Alloy (wt.%)	Tensile Yield Strength (MPa)	Ultimate Tensile Strength (MPa)	Tensile Ductility (%)	Remarks Processing condition and Reference
Mg-0.5Zr-0.4Dy-04Sm	49.2 ± 2.1 **38.2 ± 1.0 ***	148.4 ± 2.0 **250.0 ± 9.3 ***	19.6 ± 2.5 **24.7 ± 1.5 ***	
Mg-0.5Zr-0.4Gd-0.4Dy-0.4Sm	57.8 ± 1.9 **50.9 ± 0.4 ***	140.8 ± 4.4 **264.7 ± 2.0 ***	30.8 ± 0.6 **26.5 ± 0.4 ***	
Mg-0.5Zr-0.4Y-0.4Gd-0.4Dy-0.4Sm	49.6 ± 1.1 **45.6 ± 1.6 ***	146.0 ± 1.3 **249.4 ± 7.6 ***	17.4 ± 2.0 **24.2 ± 2.0 ***	
Mg-6Gd-2Nd-0.5Zr	118	220	17	Cast + Solution Treated at 500 °C + Quenched [89]
Mg-6Gd-2Nd-0.5Zr	175	345	7.5	Cast + Solution Treated at 500 °C + Quenched + peak-aged (200 °C for 24 h) [89]
Mg-6Gd-2Nd-0.5Zr	245	340	7	Cast + Solution Treated at 500 °C + Quenched + deformed (5%) and peak-aged at 200 °C for 12 h [89]
Mg-6Gd-2Nd-0.5Zr	270	350	4	Cast + Solution Treated at 500 °C + Quenched + deformed (10%) and peak-aged at 200 °C for 8 h [89]
Mg-6Gd-2Nd-0.5Zr	200	275	21	as-extruded at 450 °C [89]
Mg-6Gd-2Nd-0.5Zr	250	350	8	as-extruded at 350 °C [89]
Mg-6Gd-2Nd-0.5Zr	245	290	29	extruded at 450 °C and peak-aged at 200 °C for 24 h [89]
Mg-6Gd-2Nd-0.5Zr	275	375	17.5	extruded at 350 °C and peak-aged 200 °C for 24 h [89]
Mg-10Gd-3Y-0.5Zr	178	187	3.2	As-Cast [90]
Mg-10Gd-3Y-0.5Zr	210	312	19	Cast + Friction Stir Processed [90]
Mg-10Gd-3Y-0.5Zr	330	439	3.4	Cast + Friction Stir Processed+ Aged at 225 °C for 13 h [90]
Mg-3.99Y-3.81Nd-0.53Zr	-	167	7.4	As-cast [91]
Mg-3.99Y-3.81Nd-0.53Zr	-	260	8	As-Cast + Frictrion stir processed at 60 mm·min^{-1} and tool rotation rates of 400 r·min^{-1} [91]
Mg-3.99Y-3.81Nd-0.53Zr	-	290	17.2	As-Cast + Frictrion stir processed at 60 mm·min^{-1} and tool rotation rates of 800 r·min^{-1} [91]
Mg-3.99Y-3.81Nd-0.53Zr	-	280	11.4	As-Cast + Frictrion stir processed at 60 mm·min^{-1} and tool rotation rates of 1200 r·min^{-1} [91]
Mg-3.99Y-3.81Nd-0.53Zr	-	265	9.3	As-Cast + Frictrion stir processed at 60 mm·min^{-1} and tool rotation rates of 1500 r·min^{-1} [91]
Mg-0.56Zr-9Y-3.24MM	230	240	1.0	As-Cast [92]
Mg-0.56Zr-9Y-3.24MM	215	245	2.5	Cast + Homogenized at 535 °C for 18 h [92]
Mg-0.56Zr-9Y-3.24MM	245	305	12.5	Cast + Homogenized at 535 °C for 18 h + Extruded at 420 °C [92]

Table 19. *Cont.*

Alloy (wt.%)	Tensile Yield Strength (MPa)	Ultimate Tensile Strength (MPa)	Tensile Ductility (%)	Remarks Processing condition and Reference
Mg-0.56Zr-9Y-3.24MM	315	385	6.5	Cast + Homogenized at 535 °C for 18 h + Extruded at 420 °C + aged at 225 °C for 10 h in air [92]
Mg-12Gd-3Y-0.4Zr	187.3	282.5	6.3	Cast + Extruded at 400 °C [93]
Mg-12Gd-3Y-0.4Zr	309.6	381.8	4.4	Cast + Extruded at 400 °C and Hot rolled at 200 °C [93]
Mg-12Gd-3Y-0.4Zr	162.0	285.4	10.9	Cast + Extruded at 400 °C and Hot rolled at 200 °C + Annealed at 450 °C for 2 h [93]
Mg-12Gd-3Y-0.4Zr	141.8	252.8	8.1	Cast + Extruded at 400 °C and Hot rolled at 200 °C + Annealed at 500 °C for 2 h [93]
Mg-12Gd-3Y-0.4Zr	342.8	457.6	3.8	Cast + Extruded at 400 °C and Hot rolled at 200 °C + aged at 225 °C for 17 h (T5) [93]
Mg-8Gd-0.6Zr-1Er	96	190	5.6	As-Cast [53]
Mg-8Gd-0.6Zr-1Er	156	234	5.8	Cast + solution treated at 530 °C for 10 h + Aged at 230 °C [53]
Mg-8Gd-0.6Zr-3Er	101	210	5.3	As-Cast [53]
Mg-8Gd-0.6Zr-3Er	173	261	5.1	Cast + solution treated at 530 °C for 10 h + Aged at 230 °C [53]
Mg-8Gd-0.6Zr-5Er	99	205	4.9	As-Cast [53]
Mg-8Gd-0.6Zr-5Er	160	232	3.7	Cast + solution treated at 530 °C for 10 h + Aged at 230 °C [53]

* indicates the compressive properties of the same alloy.

4.5. Mg-Zn-RE Higher Alloy Systems

Xu *et al.* [94] reported that the as cast Mg-8.2Gd-3.8Y-1.0Zn-0.4Zr alloy consisted of α-Mg grains surrounded by Mg_3(Gd,Y) eutectic compounds while the as-homogenized alloy consisted of 18R and 14 H type LPSO phases which was attributed to the higher strength of as-homogenized alloy. The same authors reported in [95] that the Mg-8.2Gd-3.8Y-1.0Zn-0.4Zr alloy exhibited better properties when it is rolled with 96% reduction. Xu *et al.* [96] reported that upon ageing, the microstructure consisted of β' phase in the α-Mg grains and LPSO and Mg-Gd-Y containing phases at grain boundaries. Freeney *et al.* [97] reported the effects of Friction stir processing and ageing on the grain refinement and breakage and dissolution of second phase particles that resulted in the increase in the strength of the aged alloy. Yang *et al.* [98] attributed the high strength of the GWZ930 (Mg-9Gd-3Y-0.5Zn-0.5Zr) alloy mainly to precipitation strengthening and slightly to grain boundary strengthening. Al-Samman *et al.* [99] investigated sheet texture modification in ZEK100 alloys containing rare earths Ce, Nd, La, Gd and Ce rich Misch metal and reported that the Gd containing alloy has the highest tensile ductility of 30% and a very low tensile yield strength. This was because the Ce, La, Nd and Misch metal containing alloys were said to depict a common rare earth texture while Gd containing alloy revealed a different type of rare earth texture. In Mg-Zn-Mn based alloys,

Stulikova *et al.* [100] reported that the as-cast MgY4Zn1Mn1 alloy contained 18R LPSO which were responsible for the high strength. For MgCe4Zn1Mn1 alloy, it was reported that $Mg_{12}Ce$ phase is present along with small particles of Zn and Mg which pins the dislocations thereby increasing the thermal stability of the alloy. Dobron *et al.* [101] investigated the effect of variation of extrusion speeds on the mechanical properties of ZEK100 alloys and reported that the increase in extrusion speed leads to the recrystallized and homogenized microstructure without much effect on the texture. Garcia *et al.* [102] investigated ZEK 100 alloys (with RE = Ce rich Misch metal) which are considered important biomaterials and reported the tensile and compressive properties with respect to the extrusion speeds. The addition of Ce rich misch metal led to the grain refinement leading to very high tensile and compressive properties and also inhibited grain growth due to the presence of intermetallic particles distributed in the matrix. With the addition of Ca to Mg-Zn-RE based alloys, Kamrani *et al.* [103] studied that the microstructure consists of Mg-Ca besides the Mg-RE precipitates that are responsible for the strength. The tension-compression asymmetry was studied and analyzed to be due to the grain size, texture, precipitates that are present at the grain boundaries and inside the grains. Zhang *et al.* [104] studied the effects of Er and reported that Er played a major role in enhancing the ductility as well as solid solution strengthening effects. Zhang *et al.* [105] also reported that the addition of Er has caused the interactions between dislocations and solute and thereby caused yield point phenomenon. Yu *et al.* [106] investigated Mg-Zn-Zr alloys and reported that with the addition of Gd to the alloys, quasicrystal I-phases (Mg-Zn-Gd ternary phase) are formed along the grain boundaries and increased with increasing the Gd content. In a similar study, Xiao *et al.* [107] reported that the Al_2REZn_2 phases are formed with increase in RE content. They also reported in [108] that after ageing, precipitates such as $Mg_{24}Y_5$, W-phase and I-phase were formed and affected the tensile properties of Mg-alloys as shown in Table 20. Langelier *et al.* [50] reported that the combined micro alloying of Ce and Ca results in the formation of $Mg_6Ca_2Zn_3$ particles and MgZnCe T-phase in annealed alloys that limit the texture effects due to their large size and coarse distribution. The improved properties of the alloys are due to the grain boundary pinning by $Mg_6Ca_2Zn_3$ precipitates. Li *et al.* [109] reported that the addition of Nd to Mg-5Zn-0.6Zr led to the change in the morphologies of the phases. The continuous intergranular phases in Mg-Zn-Nd-Zr led to the significant deterioration in the strength and ductility in the as-cast alloys. Yu *et al.* [110] studied the effect of extrusion speed on the mechanical properties and reported that the texture intensity decreased with the increase in extrusion speed thereby improving the tensile properties due to the increased fraction of unDRXed grains. Xu *et al.* [111] reported the properties of forged ZK60-Y alloys as shown in Table 20. Wang *et al.* [112] studied the effects of addition of RE to Mg-8Zn-4Al alloy and reported that a new quaternary Mg3Al4Zn2RE phase is formed and the microstructure is refined with increase in the RE due to the crystal multiplication mechanism and prevention of the grain growth by the quarternary phase that result in the tensile properties as shown in Table 20.

Table 20. Tensile properties of Mg-Zn-RE based higher alloys.

Alloy (wt.%)	Tensile Yield Strength (MPa)	Ultimate Tensile Strength (MPa)	Tensile Ductiilty (%)	Remarks Processing Condition and Reference
Mg-8.2Gd-3.8Y-1.0Zn-0.4Zr	119	187	2.1	As-Cast [94]
Mg-8.2Gd-3.8Y-1.0Zn-0.4Zr	130	206	5.5	Cast + homogenized at 520 °C for 12 h [94]
Mg-8.2Gd-3.8Y-1.0Zn-0.4Zr	186	297	7.3	Cast + hot rolled at 400 °C [94]
Mg-8.2Gd-3.8Y-1.0Zn-0.4Zr	313	373	6.4	Cast + homogenized at 520 °C for 12 h + hot rolled at 400 °C (Reduction 68%) [94]
Mg-8.2Gd-3.8Y-1Zn-0.4Zr	318	403	13.7	Cast + solution treated at 510 °C for 12 h + hot rolled at 400 °C (Reduction 96%) [95]
Mg-8.2Gd-3.8Y-1.0Zn-0.4Zr	455	469	1.3	Cast + Solution treated at 510 °C for 12 h + Rolled at 300 °C + Aged at 200 °C [96]
Mg-8.2Gd-3.8Y-1.0Zn-0.4Zr	393	423	1.5	Cast + Solution treated at 510 °C for 12 h + Rolled at 300 °C + Aged at 225 °C [96]
Mg-8.2Gd-3.8Y-1.0Zn-0.4Zr	372	473	10.2	Cast + Solution treated at 510 °C for 12 h + Rolled at 400°C + Aged at 200 °C [96]
Mg-8.2Gd-3.8Y-1.0Zn-0.4Zr	331	436	17.8	Cast + Solution treated at 510 °C for 12 h + Rolled at 400 °C + Aged at 225 °C [96]
Mg-0.5Zn-3,1Nd-1.7Gd-0.3Zr	188	290	7	Cast + Solutionized at 520 °C/8 h [97]
Mg-0.5Zn-3,1Nd-1.7Gd-0.3Zr	220	275	27	Cast + Friction Stir processed [97]
Mg-0.5Zn-3,1Nd-1.7Gd-0.3Zr	200	330	11.5	Cast + Friction stir processed + Solutionized at 520 °C /8 h (T6) [97]
Mg-0.5Zn-3,1Nd-1.7Gd-0.3Zr	180	320	14	Cast + Friction stir processed + SS + Aged at 200 °C/16 h [97]
Mg-0.5Zn-3,1Nd-1.7Gd-0.3Zr	270	305	14	Cast + Friction stir processed + Aged at 200 °C/16 h [97]
Mg-0.5Zn-0.4Zr-2.5Ce	-	139	-	As-Cast [113]
Mg-0.5Zn-0.4Zr-2.5Ce	-	251.3	-	Cast + homogenized at 320 °C for 18 h + extruded at 350 °C [113]
Mg-0.5Zn-0.4Zr-2.5Nd	-	212.9	-	As-Cast [113]
Mg-0.5Zn-0.4Zr-2.5Ce	-	276	-	Cast + homogenized at 320 °C for 18 h + extruded at 350 °C [113]
Mg-0.5Zn-0.4Zr-2.5Nd-2.5Y	-	244.8	-	As-Cast [113]
Mg-0.5Zn-0.4Zr-2.5Ce	-	258	-	Cast + homogenized at 320 °C for 18 h + extruded at 350 °C [113]
Mg-8.8Gd-3.1Y-0.6Zn-0.5Zr	170	230	7.0	As-Cast [98]
Mg-8.8Gd-3.1Y-0.6Zn-0.5Zr	208	297	17.6	Cast + Extruded at 250 °C [98]
Mg-8.8Gd-3.1Y-0.6Zn-0.5Zr	310	395	13.7	Cast + Extruded at 250 °C + Aged at 200 °C/40 h [98]

Table 20. *Cont.*

Alloy (wt.%)	Tensile Yield Strength (MPa)	Ultimate Tensile Strength (MPa)	Tensile Ductiilty (%)	Remarks
				Processing Condition and Reference
Mg-8.8Gd-3.1Y-0.6Zn-0.5Zr	375	430	9.5	Cast + Extruded at 250 °C + Aged at 200 °C/63 h [98]
Mg-8.8Gd-3.1Y-0.6Zn-0.5Zr	340	422	12.9	Cast + Extruded at 250 °C + Aged at 200 °C/100 h [98]
Mg-8.8Gd-3.1Y-0.6Zn-0.5Zr	320	407	14.3	Cast + Extruded at 250 °C + Aged at 200 °C/126 h [98]
Mg-0.7Zn-0.2Zr-0.8Ce	117	232	18.84	Cast + homogenized at 450 °C for 12 h + water quenched+ Rolled at 400 °C + Annealed at 400 °C for 1 h [99]
Mg-0.9Zn-0.2Zr-0.7La	109	241	24.41	
Mg-0.6Zn-0.3Zr-0.6Nd	99	237	28.02	
Mg-0.7Zn-0.2Zr-0.7Gd	78	229	30.34	
Mg-0.8Zn-0.3Zr-MM (MM = 0.6Ce + 0.2La + 0.06Nd)	112	242	23.58	
Mg-1Zn-1Mn-4Y	135	175	6	As-Cast [100]
Mg-1Zn-1Mn-4Y	123	165	10	Cast + Heat Treated at 250 °C/42 h (T5) [100]
Mg-1Zn-1Mn-4Ce	90	120	5	As-Cast [100]
Mg-1Zn-1Mn-4Ce	112	170	9	Cast + Heat Treated at 275 °C/36 h (T5) [100]
Mg-1Zn-0.6Zr-1Ce	199 ± 3.6	283 ± 1.6	6.3	Cast + hot rolled at 400 °C [25]
Mg-1Zn-0.6Zr-1Ce	192.5 ± 2.5	274 ± 2.5	7.9	Cast + hot rolled at 400 °C and annealed for 1 h at 250 °C + Water Quenched [25]
Mg-1Zn-0.6Zr-1Ce	176 ± 2.7	276.4 ± 3.3	12.8	Cast + hot rolled at 400 °C and annealed for 1 h at 300 °C + Water Quenched [25]
Mg-1Zn-0.6Zr-1Ce	169.6 ± 1.5	268 ± 2.4	29	Cast + hot rolled at 400 °C and annealed for 1 h at 350 °C + Water Quenched [25]
Mg-1Zn-0.6Zr-1Ce	155 ± 1.7	263.6 ± 2.5	32.6	Cast + hot rolled at 400 °C and annealed for 1 h at 400 °C + Water Quenched [25]
Mg-1Zn-0.6Zr-1Ce	135.7 ± 0.4	264 ± 2.9	31.8	Cast + hot rolled at 400 °C and annealed for 1 h at 450 °C + Water Quenched [25]
Mg-1Zn-0.6Zr-1Gd	194.7 ± 3.8	236.4 ± 4.1	10	Cast + hot rolled at 400 °C
Mg-1Zn-0.6Zr-1Gd	193.8 ± 4.5	238.6 ± 5.9	4.3	Cast + hot rolled at 400 °C and annealed for 1 h at 250 °C+ Water Quenched [25]
Mg-1Zn-0.6Zr-1Gd	173.4 ± 2.7	236.4 ± 3.6	15.7	Cast + hot rolled at 400 °C and annealed for 1 h at 300 °C + Water Quenched [25]

Table 20. *Cont.*

Alloy (wt.%)	Tensile Yield Strength (MPa)	Ultimate Tensile Strength (MPa)	Tensile Ductiilty (%)	Remarks Processing Condition and Reference
Mg-1Zn-0.6Zr-1Gd	153.7 ± 4.4	277.8 ± 3.2	37.2	Cast + hot rolled at 400 °C and annealed for 1 h at 350 °C + Water Quenched [25]
Mg-1Zn-0.6Zr-1Gd	124.7 ± 7	276.5 ± 5	38.8	Cast + hot rolled at 400 °C and annealed for 1 h at 400 °C + Water Quenched [25]
Mg-1Zn-0.6Zr-1Gd	101.9 ± 1.7	249.5 ± 2.6	30	Cast + hot rolled at 400 °C and annealed for 1 h at 450 °C + Water Quenched [25]
Mg-1.3Zn-0.2Ce-0.5Zr	305 ± 3	313 ± 3	7.5 ± 0.1	Cast + Extruded at 300 °C at a speed of 1 m/min[101]
Mg-1.3Zn-0.2Ce-0.5Zr	204 ± 1	257 ± 1	9.3 ± 0.1	Cast + Extruded at 300 °C at a speed of 10 m/min [101]
Mg-1.3Zn-0.2Ce-0.5Zr	209 ± 2	259 ± 2	9.2 ± 0.1	Cast + Extruded at 300 °C at a speed of 20 m/min [101]
Mg-1Zn-0.8RE-0.4Zr	296 184 *	299 434 *	18 9 *	Cast + Extruded at 300 °C at 1 m/min + Annealed at 400 °C/1 h [102]
Mg-1Zn-0.8RE-0.4Zr	221 156 *	260 381 *	21 10 *	Cast + Extruded at 300 °C at 5 m/min + Annealed at 400 °C/1 h [102]
Mg-1Zn-0.8RE-0.4Zr	201 142 *	251 369 *	19 11*	Cast + Extruded at 300 °C at 10 m/min + Annealed at 400 °C/1 h [102]
Mg-2Zn-0.8RE-0.6Zr	308 201 *	311 462 *	19 9 *	Cast + Extruded at 300 °C at 1 m/min + Annealed at 400 °C/1 h [102]
Mg-2Zn-0.8RE-0.6Zr	246 162 *	275 435 *	20 10 *	Cast + Extruded at 300 °C at 5 m/min + Annealed at 400 °C/1 h [102]
Mg-2Zn-0.8RE-0.6Zr	225 154 *	264 412 *	19 10 *	Cast + Extruded at 300 °C at 10 m/min + Annealed at 400 °C/1 h [102]
Mg-2.8Zn-0.8RE-0.6Zr	260 185 *	290 450 *	19 9.5 *	Cast + Extruded at 300 °C at 1m/min + Annealed at 400 °C/1 h [102]
Mg-2.8Zn-0.8RE-0.6Zr	243 169 *	279 447 *	21 10 *	Cast + Extruded at 300 °C at 5 m/min + Annealed at 400 °C/1 h [102]
Mg-2.8Zn-0.8RE-0.6Zr	218 156 *	267 420 *	20 10 *	Cast + Extruded at 300 °C at 10 m/min + Annealed at 400 °C/1 h [102]
Mg-1.4Zn-0.1Zr-0.1RE (RE: 49.1 Ce, 35.9 La, 11.0 Nd, 4.0 Pr)	200 ± 7 **150 ± 6 ***	250 ± 5 **441 ± 3 ***	15.3 ± 0.3 **12.0 ± 0.2 ***	Die Cast + Extruded at 300 °C + Annealed at 300 °C for 30 min [103]
Mg-1.4Zn-0.1Zr-0.1RE-0.4Ca (RE: 49.1 Ce, 35.9 La, 11.0 Nd, 4.0 Pr)	171 ± 2 **148 ± 1 ***	243 ± 1 **432 ± 6 ***	14.6 ± 0.0 **11.7 ± 0.2 ***	
Mg-1.4Zn-0.1Zr-0.1RE-0.8Ca (RE: 49.1 Ce, 35.9 La, 11.0 Nd, 4.0 Pr)	174 ± 1 **149 ± 2 ***	243 ± 1 **410 ± 4 ***	15.1 ± 1.1 **22.0 ± 5.0 ***	

Table 20. *Cont.*

Alloy (wt.%)	Tensile Yield Strength (MPa)	Ultimate Tensile Strength (MPa)	Tensile Ductiilty (%)	Remarks Processing Condition and Reference
Mg-1.5Zn-0.6Zr-0.5Er	261	300	27	Cast + homogenized for 12 h at 410 °C + Extruded at 350 °C [104]
Mg-1.5Zn-0.6Zr-0.5Er	261	290	27	Cast + homogenized for 12 h at 410 °C + Extruded at 420 °C [105]
Mg-1.5Zn-0.6Zr-1Er	205	385	25	Cast + homogenized for 12 h at 410 °C + Extruded at 350 °C [104]
Mg-1.5Zn-0.6Zr-1Er	285	305	24	Cast + homogenized for 12 h at 410 °C + Extruded at 420 °C [105]
Mg-1.5Zn-0.6Zr-2Er	195	340	31	Cast + homogenized for 12 h at 410 °C + Extruded at 350 °C [104]
Mg-1.5Zn-0.6Zr-2Er	255	275	30	Cast + homogenized for 12 h at 410 °C + Extruded at 420 °C [105]
Mg-1.5Zn-0.6Zr-4Er	230	270	37	Cast + homogenized for 12 h at 410 °C + Extruded at 350 °C [104]
Mg-1.5Zn-0.6Zr-4Er	230	260	37	Cast + homogenized for 12 h at 410 °C + Extruded at 420 °C [105]
Mg-9Gd-1Er-1.6Zn-0.6Zr	220	302	19	Cast + annealed at 400 °C for 24 h + Extruded at 400 °C [114]
Mg-9Gd-1Er-1.6Zn-0.6Zr	269	344	10	Cast + annealed at 525 °C for 4 h + Extruded at 400 °C [114]
Mg-9Gd-2Er-1.6Zn-0.6Zr	221	306	17.8	Cast + annealed at 400 °C for 24 h + Extruded at 400 °C [114]
Mg-9Gd-2Er-1.6Zn-0.6Zr	262	342	11.7	Cast + annealed at 525 °C for 4 h + Extruded at 400 °C [114]
Mg-9Gd-3Er-1.6Zn-0.6Zr	223	308	14.6	Cast + annealed at 400 °C for 24 h + Extruded at 400 °C [114]
Mg-9Gd-3Er-1.6Zn-0.6Zr	263	339	10.4	Cast + annealed at 525 °C for 4 h + Extruded at 400 °C [114]
Mg-9Gd-4Er-1.6Zn-0.6Zr	235	321	14.2	Cast + annealed at 400 °C for 24 h + Extruded at 400 °C [114]
Mg-9Gd-4Er-1.6Zn-0.6Zr	261	333	8.4	Cast + annealed at 525 °C for 4 h + Extruded at 400 °C [114]
Mg-2Zn-0.3Ca-0.1Ce	131 ± 12.3	222 ± 7.0	23.9 ± 0.27	Cast + homogenized for 3 h at 300 °C + 24 h at 400 °C + hot rolled at 400 °C + annealed at 400 °C for 30 min [50]
Mg-4Zn-0.3Ca-0.1Ce	119 ± 2.1	240 ± 1.5	18.3 ± 1.30	
Mg-5Zn-1Nd-0.6Zr	100	200	7.5	As-Cast [109]
Mg-5Zn-2Nd-0.6Zr	90	135	3	
Mg-5Zn-2Nd-0.5Y-0.6Zr	95	205	9.5	
Mg-5Zn-2Nd-1Y-0.6Zr	105	220	12	
Mg-5.5Zn-0.6Zr-0.2Gd	227	307	25.3	Cast + homogenized at 300 °C for 20 h and 400 °C for 12 h + high strain rate rolled at 400 °C [106]
Mg-5.5Zn-0.6Zr-0.5Gd	235	318	23.2	
Mg-5.5Zn-0.6Zr-0.8Gd	242	327	22	

Table 20. *Cont.*

Alloy (wt.%)	Tensile Yield Strength (MPa)	Ultimate Tensile Strength (MPa)	Tensile Ductiilty (%)	Remarks Processing Condition and Reference
Mg-6Zn-0.5Zr-1Ce	293	337	26.9	Cast + homogenized at 440 °C for 8 h + water quenched+ extruded at 250 °C at 0.3 mm/s [110]
Mg-6Zn-0.5Zr-1Ce	286	333	25.4	Cast + homogenized at 440 °C for 8 h + water quenched+ extruded at 250 °C at 1.0 mm/s [110]
Mg-6Zn-0.5Zr-1Ce	247	311	22.6	Cast + homogenized at 440 °C for 8 h + water quenched+ extruded at 250 °C at 3 mm/s [110]
Mg-6.3Zn-2Zr-1Y	127	267	12.1	Cast + Forged [111]
Mg-6.3Zn-2Zr-1Y	84	244	13.2	Cast + Forged +Solid solution for 2.5 h at 500 °C (T_4) [111]
Mg-6.3Zn-2Zr-1Y	124	259	10.8	Cast + Forged + Solid solution for 2.5 h at 500 °C + aged for 15 h at 180 °C (T6) [111]
Mg-6.3Zn-2Zr-1Y	129	309	18.7	Cast + Forged + aged for15 h at 180 °C (T5) [111]
Mg-7.5Zn-5Al-0.123RE	100	175	2.2	As-Cast [107]
Mg-7.5Zn-5Al-0.123RE	100	160	1.5	Cast + heat treated at 350 °C for 96 h + water quenched + aged at 175 °C for 16 h [107]
Mg-7.6Zn-5Al-0.763RE	104	198	2.8	As-Cast [107]
Mg-7.6Zn-5Al-0.763RE	120	232	3.4	Cast + heat treated at 350 °C for 96 h + water quenched + aged at 175 °C for 16 h [107]
Mg-7Zn-5Al-1.753RE	100	186	1.9	As-Cast [107]
Mg-7Zn-5Al-1.753RE	120	240	5.2	Cast + heat treated at 350 °C for 96 h + water quenched + aged at 175 °C for 16 h [107]
Mg-8Zn-4Al-0.5RE	110	145	4.5	As-Cast [112]
Mg-8Zn-4Al-1.0RE	118	158	4.2	
Mg-8Zn-4Al-1.5RE	123	165	4	
Mg-12.3Zn-5.8Y-1.4Al	191	100	6.9	As-Cast [108]
Mg-12.3Zn-5.8Y-1.4Al	203	106	4.9	Cast + Solution treated at 335 °C for 12 h + quenched in water + Aged at 200 °C [108]

* indicates the compressive properties of the same alloy.

4.6. Mg-Sn-RE Higher Alloy System

Cheng *et al.* [115] reported that peak ageing occurred faster with the addition of Ce to the Mg-5Sn-4Zn alloy. They showed that the microstructure consisted of Sn-Ce precipitates in the aged alloy in addition to the α-Mg phase and Mg_2Sn phase. It was also reported that with the addition of Ce, the Sn-Ce phase predominates reducing the Mg_2Sn phase thereby affecting the properties as shown in Table 21.

Table 21. Tensile properties of Mg-Sn-RE based higher alloys.

Alloy (wt.%)	TensileYield Strength (MPa)	Ultimate Tensile Strength (MPa)	Tensile Ductility (%)	Remarks Processing condition and Reference
Mg-4Zn-5Sn-1Ce	155	275	27.5	Cast + homogenized at 420 °C for 24 h + Extruded at 250 °C [115]
Mg-4Zn-5Sn-1Ce	190	200	30	Cast + homogenized at 420 °C for 24 h+ Extruded at 250 °C + solutionized at 450 °C for 1 h + aged at 200 °C (T6) [115]

4.7. Other Higher Alloy Systems

The combined addition of Er/Al by microalloying has led to improved mechanical properties due to the reduced grain size and homogenous microstructure after particle-stimulated nucleation assisted dynamic recrystallization [56]. Du *et al.* [57] reported that the addition of V resulted in improvement in morphology of 18R LPSO thereby resulting in improved properties as compared to Mg-10Er-2Cu stated in Table 15. The properties are as shown in Table 22.

Table 22. Tensile properties of other higher alloys.

Alloy (wt.%)	Tensile Yield Strength (MPa)	Ultimate Tensile Strength (MPa)	Tensile Ductility (%)	Remarks Processing condition and Reference
Mg-1.8Mn-0.1Er-0.05Al	224	244	4	Cast + homogenized at 450 °C for 4 h + Extruded at 450 °C + Annealed at 390 °C for 1 h [56]
Mg-1.8Mn-0.4Er-0.2Al	224	245	8	
Mg-1.8Mn-0.7Er-0.34Al	226	250	19	
Mg-10Er-2Cu-V	370	430	11	Cast + homogenized at 450 °C for 24 h + Extruded at 430 °C [57]

Hence, the rare earths have an immense effect on the properties of the Mg alloys but the cost of most of the rare earths is higher compared to the conventional alloying elements due to the scarcity in their availability.

5. Conclusions

In this review, the mechanical properties of various magnesium-rare earth alloys processed under different conditions, investigated by various researchers, are reviewed, and the reasons for their mechanical behavior are studied. The tensile properties of the investigated binary alloys which include Mg-Y, Mg-Ce, Mg-Gd, Mg-La, Mg-Er, Mg-Nd, and Mg-Dy systems are reported. The report also includes the tensile and compressive properties of Mg-RE ternary system containing two different rare earths as alloying elements and Mg-Zn-RE, Mg-Zr-RE, Mg-Sn-RE, and other ternary systems are reported. Finally, the properties of Mg-Al-RE, Mg-Li-RE, Mg-Zr-RE, Mg-Zn-RE, Mg-Sn-RE and other higher alloy systems containing three or more alloying elements are also briefly studied.

Rare earths are widely added to Mg in the form of Misch Metals which are unspecified blends of RE due to the low cost of Misch metals. However, considering the microstructure and mechanical properties of different Mg-RE systems, each one behaves differently from the others. So, it would be necessary to indicate the actual type and composition of RE addition in order to attribute the effect of RE addition to the properties. Further, in biodegradable Mg-RE alloys, each RE element has unique toxicity level and self-degradation period and therefore the use of Misch metals would make the alloy design complex.

Of all the rare earth elements, Y and Ce are being researched widely in combination with other alloying elements due to their significant influence on mechanical properties and texture effects. Besides having good tensile properties, some of the alloys like Mg-Y, Mg-Gd, Mg-Dy, Mg-Nd, WE43, LAE442, ZEK100, JDBM, *etc*., have good biodegradability and the properties of these biodegradable materials are also presented.

The research on Mg-RE systems conducted so far revealed that with regard to specific Mg-RE binary alloy systems, Mg-Y alloys exhibited the best strength while the Mg-Er alloys exhibited the best ductility. In ternary alloys, Mg-Zn-RE system exhibited the highest strength and ductility. Similarly among the higher alloy systems, Mg-Zn-RE based higher alloy system containing three or more alloying elements exhibited best tensile strength and ductility levels.

Overall, the best combination of both strength and ductility was observed in Mg-Y alloys in binary systems, Mg-Zn-RE alloys in ternary systems and Mg-Zn-RE based alloys in higher alloy systems.

Owing to the high cost of rare earths, it is not economical to use rare earths in high concentrations. Hence, it is suggested that further research be done by micro alloying ($<1\%$) to offset the cost of the rare earths without compromising the properties.

Acknowledgements

One of the authors, Sravya Tekumalla, sincerely thanks NUS research scholarship for supporting her graduate program. The authors also gratefully acknowledge Ministry of Education Academic Research Funding (WBS# R-265-000-498-112) for the financial support.

Author Contributions

All authors contributed to the paper. Sravya Tekumalla acquired and interpreted the data. Sravya Tekumalla and Sankaranarayanan Seetharaman provided the analysis in the paper and prepared the manuscript. Manoj Gupta designed the scope of the paper and Manoj Gupta and Abdulhakim Almajid revised the paper. All authors discussed the conclusions and reviewed the manuscript.

Conflicts of Interest

The authors declare no conflict of interest.

References

1. Mordike, B.L.; Ebert, T. Magnesium: Properties—Applications—Potential. *Mater. Sci. Eng. A* **2001**, *302*, 37–45.
2. Somekawa, H.; Mukai, T. Effect of texture on fracture toughness in extruded AZ31 magnesium alloy. *Scripta Mater.* **2005**, *53*, 541–545.
3. Yang, Z.; Li, J.P.; Zhang, J.X.; Lorimer, G.W.; Robson, J. Review on research and development of magnesium alloys. *Acta Metall. Sinica (Engl. Lett.)* **2008**, *21*, 313–328.
4. Huang, Y.; Gan, W.; Kainer, K.U.; Hort, N. Role of multi-microalloying by rare earth elements in ductilization of magnesium alloys. *J. Magnes. Alloys* **2014**, *2*, 1–7.
5. Stanford, N.; Atwell, D.; Beer, A.; Davies, C.; Barnett, M.R. Effect of microalloying with rare-earth elements on the texture of extruded magnesium-based alloys. *Scripta Mater.* **2008**, *59*, 772–775.
6. Chia, T.L.; Easton, M.A.; Zhu, S.M.; Gibson, M.A.; Birbilis, N.; Nie, J.F. The effect of alloy composition on the microstructure and tensile properties of binary Mg-rare earth alloys. *Intermetallics* **2009**, *17*, 481–490.
7. Gao, L.; Chen, R.S.; Han, E.H. Solid solution strengthening behaviors in binary Mg–Y single phase alloys. *J. Alloy. Compd.* **2009**, *472*, 234–240.
8. Hirsch, J.; Al-Samman, T. Superior light metals by texture engineering: Optimized aluminum and magnesium alloys for automotive applications. *Acta Mater.* **2013**, *61*, 818–843.
9. Luo, A.A.; Sachdev, A.K. 12—Applications of magnesium alloys in automotive engineering. In *Advances in Wrought Magnesium Alloys*; Bettles, C., Barnett, M., Eds.; Woodhead Publishing: Cambridge, UK, 2012; pp. 393–426.
10. Witte, F. The history of biodegradable magnesium implants: A review. *Acta Biomater.* **2010**, *6*, 1680–1692.
11. Zheng, Y.F.; Gu, X.N.; Witte, F. Biodegradable metals. *Mater. Sci. Eng. R. Rep.* **2014**, *77*, 1–34.
12. Hort, N.; Huang, Y.; Fechner, D.; Störmer, M.; Blawert, C.; Witte, F.; Vogt, C.; Drücker, H.; Willumeit, R.; Kainer, K.U.; *et al.* Magnesium alloys as implant materials—Principles of property design for Mg–RE alloys. *Acta Biomater.* **2010**, *6*, 1714–1725.
13. Rokhlin, L.L. Structure and properties of alloys of the Mg-RE system. *Met. Sci. Heat Treat.* **2006**, *48*, 487–490.
14. Zhao, H.D.; Qin, G.W.; Ren, Y.P.; Pei, W.L.; Chen, D.; Guo, Y. The maximum solubility of Y in α-Mg and composition ranges of $Mg_{24}Y_{5-x}$ and Mg_2Y_{1-x} intermetallic phases in Mg–Y binary system. *J. Alloy. Compd.* **2011**, *509*, 627–631.
15. Gu, X.; Zheng, Y.; Cheng, Y.; Zhong, S.; Xi, T. *In vitro* corrosion and biocompatibility of binary magnesium alloys. *Biomaterials* **2009**, *30*, 484–498.
16. Zhou, N.; Zhang, Z.; Jin, L.; Dong, J.; Chen, B.; Ding, W. Ductility improvement by twinning and twin–slip interaction in a Mg-Y alloy. *Mater. Des.* **2014**, *56*, 966–974.
17. Essadiqi, E.; Shehata, M.T.; Javaid, A.; Shen, G.; Aljarrah, M.; Verma, R.; Mishra, R. In *Alloying and Process Design of Mg Sheet*; CANMET Materials: Ottawa, ON, Canada, 2011.
18. Sandlöbes, S.; Zaefferer, S.; Schestakow, I.; Yi, S.; Gonzalez-Martinez, R. On the role of non-basal deformation mechanisms for the ductility of mg and Mg–Y alloys. *Acta Mater.* **2011**, *59*, 429–439.

19. Wu, B.L.; Zhao, Y.H.; Du, X.H.; Zhang, Y.D.; Wagner, F.; Esling, C. Ductility enhancement of extruded magnesium via yttrium addition. *Mater. Sci. Eng. A* **2010**, *527*, 4334–4340.
20. Sugamata, M.; Hanawa, S.; Kaneko, J. Structures and mechanical properties of rapidly solidified Mg-Y based alloys. *Mater. Sci. Eng. A* **1997**, *226–228*, 861–866.
21. Zhang, B.; Nagasekhar, A.V.; Tao, X.; Ouyang, Y.; Cáceres, C.H.; Easton, M. Strengthening by the percolating intergranular eutectic in an hpdc Mg–Ce alloy. *Mater. Sci. Eng. A* **2014**, *599*, 204–211.
22. Mishra, R.K.; Gupta, A.K.; Rao, P.R.; Sachdev, A.K.; Kumar, A.M.; Luo, A.A. Influence of cerium on the texture and ductility of magnesium extrusions. *Scripta Mater.* **2008**, *59*, 562–565.
23. Chino, Y.; Kado, M.; Mabuchi, M. Enhancement of tensile ductility and stretch formability of magnesium by addition of 0.2 wt%(0.035 at%)Ce. *Mater. Sci. Eng. A* **2008**, *494*, 343–349.
24. Luo, A.A.; Wu, W.; Mishra, R.K.; Jin, L.; Sachdev, A.K.; Ding, W. Microstructure and mechanical properties of extruded magnesium-aluminum-cerium alloy tubes. *Metall. Mater. Trans. A* **2010**, *41*, 2662–2674.
25. Basu, I.; Al-Samman, T. Triggering rare earth texture modification in magnesium alloys by addition of zinc and zirconium. *Acta Mater.* **2014**, *67*, 116–133.
26. Nayeb-Hashemi, A.A. *Phase Diagrams of Binary Magnesium Alloys*; ASM International: Metals Park, OH, USA, 1998.
27. Gao, L.; Chen, R.S.; Han, E.H. Effects of rare-earth elements Gd and Y on the solid solution strengthening of mg alloys. *J. Alloy. Compd.* **2009**, *481*, 379–384.
28. Peng, Q.; Wu, Y.; Fang, D.; Meng, J.; Wang, L. Microstructures and properties of melt-spun and as-cast Mg-20Gd binary alloy. *J. Rare Earths* **2006**, *24*, 466–470.
29. Stanford, N.; Atwell, D.; Barnett, M.R. The effect of gd on the recrystallisation, texture and deformation behaviour of magnesium-based alloys. *Acta Mater.* **2010**, *58*, 6773–6783.
30. Stanford, N.; Barnett, M.R. The origin of "rare earth" texture development in extruded Mg-based alloys and its effect on tensile ductility. *Mater. Sci. Eng. A* **2008**, *496*, 399–408.
31. Rokhlin, L.L. *Magnesium Alloys Containing Rare Earth Metals*; Taylor & Francis: New York, NY, USA, 2003.
32. Wu, B.L.; Wan, G.; Du, X.H.; Zhang, Y.D.; Wagner, F.; Esling, C. The quasi-static mechanical properties of extruded binary Mg–Er alloys. *Mater. Sci. Eng. A* **2013**, *573*, 205–214.
33. Wang, Z.; Jia, W.; Cui, J. Study on the deformation behavior of Mg-3.6%Er magnesium alloy. *J. Rare Earths* **2007**, *25*, 744–748.
34. Yan, J.; Sun, Y.; Xue, F.; Xue, S.; Tao, W. Microstructure and mechanical properties in cast magnesium–neodymium binary alloys. *Mater. Sci. Eng. A* **2008**, *476*, 366–371.
35. Seitz, J.M.; Eifler, R.; Stahl, J.; Kietzmann, M.; Bach, F.W. Characterization of $MgNd_2$ alloy for potential applications in bioresorbable implantable devices. *Acta Biomater.* **2012**, *8*, 3852–3864.
36. Bi, G.; Li, Y.; Zang, S.; Zhang, J.; Ma, Y.; Hao, Y. Microstructure, mechanical and corrosion properties of Mg–2Dy–xZn (x = 0, 0.1, 0.5 and 1 at.%) alloys. *J. Magnes. Alloy.* **2014**, *2*, 64–71.
37. Yang, L.; Huang, Y.; Peng, Q.; Feyerabend, F.; Kainer, K.U.; Willumeit, R.; Hort, N. Mechanical and corrosion properties of binary Mg–Dy alloys for medical applications. *Mater. Sci. Eng. B* **2011**, *176*, 1827–1834.

38. Yang, L.; Huang, Y.; Feyerabend, F.; Willumeit, R.; Kainer, K.U.; Hort, N. Influence of ageing treatment on microstructure, mechanical and bio-corrosion properties of Mg–Dy alloys. *J. Mech. Behav. Biomed. Mater.* **2012**, *13*, 36–44.
39. Zhang, Q.; Li, Q.; Jing, X.; Zhang, X. Microstructure and mechanical properties of Mg-10Y-2.5Sm alloy. *J. Rare Earths* **2010**, *28* (Suppl. 1), 375–377.
40. Gavras, S.; Easton, M.A.; Gibson, M.A.; Zhu, S.; Nie, J.-F. Microstructure and property evaluation of high-pressure die-cast Mg–La–rare earth (Nd, Y or Gd) alloys. *J. Alloy. Compd.* **2014**, *597*, 21–29.
41. Wang, X.; Wang, Z.; Du, W.; Liu, K.; Li, S. Microstructure evolutions of Mg-8Gd-2Er (wt.%) alloy during isothermal ageing at 200 °C. *J. Rare Earths* **2012**, *30*, 1168–1171.
42. Luo, A.A.; Mishra, R.K.; Sachdev, A.K. Development of high ductility magnesium-zinc-cerium extrusion alloys. In Proceedings of the 2010 TMS Annual Meeting & Exhibition, Seattle, WA, USA, 14–18 February 2010; pp. 313–318.
43. Le, Q.-C.; Zhang, Z.-Q.; Shao, Z.-W.; Cui, J.-Z.; Xie, Y. Microstructures and mechanical properties of Mg-2%Zn-0.4%Re alloys. *Trans. Nonferrous Metals Soc. China* **2010**, *20*, s352–s356.
44. Wu, D.; Chen, R.S.; Han, E.H. Excellent room-temperature ductility and formability of rolled Mg–Gd–Zn alloy sheets. *J. Alloy. Compd.* **2011**, *509*, 2856–2863.
45. Wang, Z.-H.; Wang, X.-D.; Wang, Q.-F.; Du, W.-B. Effects of ultrasonic treatment on microstructure and mechanical properties of Mg-5Zn-2Er alloy. *Trans. Nonferrous Metals Soc. China* **2011**, *21*, 773–777.
46. Zhao, X.-F.; Li, S.-B.; Wang, Q.-F.; Du, W.-B.; Liu, K. Effects of heat treatment on microstructure and mechanical properties of Mg–5Zn–0.63Er alloy. *Trans. Nonferrous Metals Soc. China* **2013**, *23*, 59–65.
47. Wang, Q.; Liu, K.; Wang, Z.; Li, S.; Du, W. Microstructure, texture and mechanical properties of as-extruded Mg–Zn–Er alloys containing W-phase. *J. Alloy. Compd.* **2014**, *602*, 32–39.
48. Srinivasan, A.; Huang, Y.; Mendis, C.L.; Blawert, C.; Kainer, K.U.; Hort, N. Investigations on microstructures, mechanical and corrosion properties of Mg–Gd–Zn alloys. *Mater. Sci. Eng. A* **2014**, *595*, 224–234.
49. Singh, A.; Somekawa, H.; Mukai, T. High temperature processing of Mg–Zn–Y alloys containing quasicrystal phase for high strength. *Mater. Sci. Eng. A* **2011**, *528*, 6647–6651.
50. Langelier, B.; Nasiri, A.M.; Lee, S.Y.; Gharghouri, M.A.; Esmaeili, S. Improving microstructure and ductility in the Mg–Zn alloy system by combinational Ce–Ca microalloying. *Mater. Sci. Eng. A* **2015**, *620*, 76–84.
51. Li, H.; Du, W.; Li, S.; Wang, Z. Effect of Zn/Er weight ratio on phase formation and mechanical properties of as-cast Mg–Zn–Er alloys. *Mater. Des.* **2012**, *35*, 259–265.
52. Liu, K.; Wang, Q.-F.; Du, W.-B.; Wang, Z.-H.; Li, S.-B. Microstructure and mechanical properties of extruded Mg–6Zn–*x*Er alloys. *Trans. Nonferrous Metals Soc. China* **2013**, *23*, 2863–2873.
53. Peng, Q.; Dong, H.; Wu, Y.; Wang, L. Age hardening and mechanical properties of Mg–Gd–Er alloy. *J. Alloy. Compd.* **2008**, *456*, 395–399.
54. Zhao, H.-D.; Qin, G.-W.; Ren, Y.-P.; Pei, W.-L.; Chen, D.; Guo, Y. Microstructure and tensile properties of as-extruded Mg-Sn-Y alloys. *Trans. Nonferrous Metals Soc. China* **2010**, *20*, s493–s497.

55. Wang, Q.; Chen, Y.; Xiao, S.; Zhang, X.; Tang, Y.; Wei, S.; Zhao, Y. Study on microstructure and mechanical properties of as-cast Mg-Sn-Nd alloys. *J. Rare Earths* **2010**, *28*, 790–793.
56. Zhang, J.; Yuan, F.; Liu, M.; Pan, F. Microstructure and mechanical properties of Mg–1.8%Mn alloy modified by single Er and composite Er/Al microalloying. *Mater. Sci. Eng. A* **2013**, *576*, 185–191.
57. Du, X.H.; Duan, G.S.; Hong, M.; Wang, D.P.; Wu, B.L.; Zhang, Y.D.; Esling, C. Effect of V on the microstructure and mechanical properties of Mg–10Er–2Cu alloy with a long period stacking ordered structure. *Mater. Lett.* **2014**, *122*, 312–314.
58. Zhang, X.; Yuan, G.; Mao, L.; Niu, J.; Ding, W. Biocorrosion properties of as-extruded Mg–Nd–Zn–Zr alloy compared with commercial AZ31 and WE43 alloys. *Mater. Lett.* **2012**, *66*, 209–211.
59. Mabuchi, M.; Chino, Y.; Iwasaki, H. Tensile properties at room temperature to 823 K of Mg-4Y-3RE alloy. *Mater. Trans.* **2002**, *43*, 2063–2068.
60. Su, Z.; Liu, C.; Wan, Y. Microstructures and mechanical properties of high performance Mg–4Y–2.4Nd–0.2Zn–0.4Zr alloy. *Mater. Des.* **2013**, *45*, 466–472.
61. Mukai, T.; Mohri, T.; Mabuchi, M.; Nakamura, M.; Ishikawa, K.; Higashi, K. Experimental study of a structural magnesium alloy with high absorption energy under dynamic loading. *Scripta Mater.* **1998**, *39*, 1249–1253.
62. Panigrahi, S.K.; Yuan, W.; Mishra, R.S.; DeLorme, R.; Davis, B.; Howell, R.A.; Cho, K. A study on the combined effect of forging and aging in Mg–Y–RE alloy. *Mater. Sci. Eng. A* **2011**, *530*, 28–35.
63. Rzychoń, T.; Kiełbus, A.; Cwajna, J.; Mizera, J. Microstructural stability and creep properties of die casting Mg–4Al–4RE magnesium alloy. *Mater. Charact.* **2009**, *60*, 1107–1113.
64. Dai, J.; Zhu, S.; Easton, M.A.; Zhang, M.; Qiu, D.; Wu, G.; Liu, W.; Ding, W. Heat treatment, microstructure and mechanical properties of a Mg–Gd–Y alloy grain-refined by Al additions. *Mater. Sci. Eng. A* **2013**, *576*, 298–305.
65. Zhang, J.-J.; Liu, S.-J.; Leng, Z.; Zhang, M.-L.; Meng, J.; Wu, R.-Z. Structure stability and mechanical properties of high-pressure die-cast Mg–Al–Ce–Y-based alloy. *Trans. Nonferrous Metals Soc. China* **2012**, *22*, 262–267.
66. Zhang, J.; Leng, Z.; Liu, S.; Zhang, M.; Meng, J.; Wu, R. Structure stability and mechanical properties of Mg–Al-based alloy modified with Y-rich and Ce-rich misch metals. *J. Alloys Compd.* **2011**, *509*, L187-L193.
67. Rzychoń, T.; Kiełbus, A.; Lityńska-Dobrzyńska, L. Microstructure, microstructural stability and mechanical properties of sand-cast Mg–4Al–4RE alloy. *Mater. Charact.* **2013**, *83*, 21–34.
68. Zhang, J.; Yu, P.; Liu, K.; Fang, D.; Tang, D.; Meng, J. Effect of substituting cerium-rich mischmetal with lanthanum on microstructure and mechanical properties of die-cast Mg–Al–RE alloys. *Mater. Des.* **2009**, *30*, 2372–2378.
69. Zhang, J.; Leng, Z.; Zhang, M.; Meng, J.; Wu, R. Effect of ce on microstructure, mechanical properties and corrosion behavior of high-pressure die-cast Mg–4Al-based alloy. *J. Alloy. Compd.* **2011**, *509*, 1069–1078.
70. Zhang, J.; Zhang, M.; Meng, J.; Wu, R.; Tang, D. Microstructures and mechanical properties of heat-resistant high-pressure die-cast Mg–4Al–xLa–0.3Mn (x = 1, 2, 4, 6) alloys. *Mater. Sci. Eng. A* **2010**, *527*, 2527–2537.

71. Zhang, J.; Liu, K.; Fang, D.; Qiu, X.; Yu, P.; Tang, D.; Meng, J. Microstructures, mechanical properties and corrosion behavior of high-pressure die-cast Mg–4Al–0.4Mn–*x*Pr (*x* = 1, 2, 4, 6) alloys. *J. Alloy. Compd.* **2009**, *480*, 810–819.
72. Zhang, J.; Wang, J.; Qiu, X.; Zhang, D.; Tian, Z.; Niu, X.; Tang, D.; Meng, J. Effect of Nd on the microstructure, mechanical properties and corrosion behavior of die-cast Mg–4Al-based alloy. *J. Alloy. Compd.* **2008**, *464*, 556–564.
73. Wang, J.; Liao, R.; Wang, L.; Wu, Y.; Cao, Z.; Wang, L. Investigations of the properties of Mg–5Al–0.3Mn–*x*Ce (*x* = 0–3 wt.%) alloys. *J. Alloy. Compd.* **2009**, *477*, 341–345.
74. Chen, Y.; Hao, L.; Ruiyu, Y.; Liu, G.; Xia, T. Effects of Nd on microstructure and mechanical properties of Mg–Al–Ca alloy. *Mater. Sci. Technol.* **2013**, *30*, 495–500.
75. Wu, G.; Fan, Y.; Gao, H.; Zhai, C.; Zhu, Y.P. The effect of Ca and rare earth elements on the microstructure, mechanical properties and corrosion behavior of AZ91D. *Mater. Sci. Eng. A* **2005**, *408*, 255–263.
76. Krause, A.; von der Höh, N.; Bormann, D.; Krause, C.; Bach, F.-W.; Windhagen, H.; Meyer-Lindenberg, A. Degradation behaviour and mechanical properties of magnesium implants in rabbit tibiae. *J. Mater. Sci.* **2010**, *45*, 624–632.
77. Zeng, R.; Qi, W.; Zhang, F.; Cui, H.; Zheng, Y. *In vitro* corrosion of Mg–1.21Li–1.12Ca–1Y alloy. *Prog. Nat. Sci. Mater. Int.* **2014**, *24*, 492–499.
78. Tao, W.; Zhang, M.; Niu, Z.; Liu, B. Influence of rare earth elements on microstructure and mechanical properties of Mg-Li alloys. *J. Rare Earths* **2006**, *24*, 797–800.
79. Zhou, W.R.; Zheng, Y.F.; Leeflang, M.A.; Zhou, J. Mechanical property, biocorrosion and *in vitro* biocompatibility evaluations of Mg–Li–(Al)–(RE) alloys for future cardiovascular stent application. *Acta Biomater.* **2013**, *9*, 8488–8498.
80. Wang, T.; Zhang, M.; Wu, R. Microstructure and properties of Mg–8Li–1Al–1Ce alloy. *Mater. Lett.* **2008**, *62*, 1846–1848.
81. Wu, R.Z.; Qu, Z.K.; Zhang, M.L. Reviews on the influences of alloying elements on the microstructure and mechanical properties of Mg-Li based alloys. *Rev. Adv. Mater. Sci.* **2010**, *24*, 14–34.
82. Wang, J.; Nie, J.-J.; Wang, R.; Xu, Y.-D.; Zhu, X.-R.; Ling, G.-P. Effect of Y on age hardening response and mechanical properties of Mg-*x*Y–1.5LPC−0.4Zr alloys. *Trans. Nonferrous Metals Soc. China* **2012**, *22*, 1549–1555.
83. Mohri, T.; Mabuchi, M.; Saito, N.; Nakamura, M. Microstructure and mechanical properties of a Mg-4Y-3RE alloy processed by thermo-mechanical treatment. *Mater. Sci. Eng. A* **1998**, *257*, 287–294.
84. Zhang, X.-B.; Xue, Y.-J.; Wang, Z.-Z. Effect of heat treatment on microstructure, mechanical properties and *in vitro* degradation behavior of as-extruded Mg-2.7Nd-0.2Zn-0.4Zr alloy. *Trans. Nonferrous Metals Soc. China* **2012**, *22*, 2343–2350.
85. Zhang, X.; Yuan, G.; Wang, Z. Mechanical properties and biocorrosion resistance of Mg-Nd-Zn-Zr alloy improved by cyclic extrusion and compression. *Mater. Lett.* **2012**, *74*, 128–131.
86. Zhang, X.; Yuan, G.; Niu, J.; Fu, P.; Ding, W. Microstructure, mechanical properties, biocorrosion behavior, and cytotoxicity of as-extruded Mg–Nd–Zn–Zr alloy with different extrusion ratios. *J. Mech. Behav. Biomed. Mater.* **2012**, *9*, 153–162.

87. Zhang, X.; Wang, Z.; Yuan, G.; Xue, Y. Improvement of mechanical properties and corrosion resistance of biodegradable Mg–Nd–Zn–Zr alloys by double extrusion. *Mater. Sci. Eng. B* **2012**, *177*, 1113–1119.
88. Kielbus, A.; Rzychon, T. Mechanical and creep properties of Mg-4Y-3RE and Mg-3Nd-1Gd magnesium alloy. *Procedia Eng.* **2011**, *10*, 1835–1840.
89. Zheng, K.Y.; Dong, J.; Zeng, X.Q.; Ding, W.J. Effect of thermo-mechanical treatment on the microstructure and mechanical properties of a Mg–6Gd–2Nd–0.5Zr alloy. *Mater. Sci. Eng. A* **2007**, *454–455*, 314–321.
90. Xiao, B.L.; Yang, Q.; Yang, J.; Wang, W.G.; Xie, G.M.; Ma, Z.Y. Enhanced mechanical properties of Mg–Gd–Y–Zr casting via friction stir processing. *J. Alloy. Compd.* **2011**, *509*, 2879–2884.
91. Li, J.; Zhang, D.-T.; Chai, F.; Zhang, W. Microstructures and mechanical properties of WE43 magnesium alloy prepared by friction stir processing. *Rare Met.* **2014**, 1–6.
92. He, L.-Q.; Li, Y.-J.; Li, X.-G.; Ma, M.-L.; Zhang, K.; Wang, X.-W.; Yan, J.-M.; Lin, H.-T. Microstructure and properties of WE93 alloy. *Trans. Nonferrous Metals Soc. China* **2011**, *21*, 790–794.
93. Wang, R.; Dong, J.; Fan, L.-K.; Zhang, P.; Ding, W.-J. Microstructure and mechanical properties of rolled Mg-12Gd-3Y-0.4Zr alloy sheets. *Trans. Nonferrous Metals Soc. China* **2008**, *18*, s189–s193.
94. Xu, C.; Zheng, M.Y.; Xu, S.W.; Wu, K.; Wang, E.D.; Kamado, S.; Wang, G.J.; Lv, X.Y. Microstructure and mechanical properties of rolled sheets of Mg–Gd–Y–Zn–Zr alloy: As-cast *versus* as-homogenized. *J. Alloy. Compd.* **2012**, *528*, 40–44.
95. Xu, C.; Xu, S.W.; Zheng, M.Y.; Wu, K.; Wang, E.D.; Kamado, S.; Wang, G.J.; Lv, X.Y. Microstructures and mechanical properties of high-strength Mg–Gd–Y–Zn–Zr alloy sheets processed by severe hot rolling. *J. Alloy. Compd.* **2012**, *524*, 46–52.
96. Xu, C.; Zheng, M.Y.; Xu, S.W.; Wu, K.; Wang, E.D.; Fan, G.H.; Kamado, S.; Liu, X.D.; Wang, G.J.; Lv, X.Y. Microstructure and mechanical properties of Mg–Gd–Y–Zn–Zr alloy sheets processed by combined processes of extrusion, hot rolling and ageing. *Mater. Sci. Eng. A* **2013**, *559*, 844–851.
97. Freeney, T.A.; Mishra, R.S. Effect of friction stir processing on microstructure and mechanical properties of a cast-magnesium–rare earth alloy. *Metall. Mat. Trans. A* **2010**, *41*, 73–84.
98. Yang, Z.; Li, J.P.; Guo, Y.C.; Liu, T.; Xia, F.; Zeng, Z.W.; Liang, M.X. Precipitation process and effect on mechanical properties of Mg–9Gd–3Y–0.6Zn–0.5Zr alloy. *Mater. Sci. Eng. A* **2007**, *454–455*, 274–280.
99. Al-Samman, T.; Li, X. Sheet texture modification in magnesium-based alloys by selective rare earth alloying. *Mater. Sci. Eng. A* **2011**, *528*, 3809–3822.
100. Stulikova, I.; Smola, B. Mechanical properties and phase composition of potential biodegradable Mg–Zn–Mn–base alloys with addition of rare earth elements. *Mater. Charact.* **2010**, *61*, 952–958.
101. Dobroň, P.; Chmelik, F.; Parfenenko, K.; Letzig, D.; Bohlen, J. On the effect of the extrusion speed on microstructure and plastic deformation of ZE10 and ZEK100 magnesium alloys—An acoustic emission study. *Acta Phys. Polonica A* **2012**, *122*, 593.

102. García, E.M. Influence of alloying elements on the microstructure and mechanical properties of extruded Mg-Zn based alloys. Technischen Universität, Berlin, Germany, 2010. Available online: opus4.kobv.de/opus4-tuberlin/files/2653/mezagarcia_enrique.pdf (accessed on 17 December 2014).
103. Kamrani, S.; Fleck, C. Effects of calcium and rare-earth elements on the microstructure and tension–compression yield asymmetry of ZEK100 alloy. *Mater. Sci. Eng. A* **2014**, *618*, 238–243.
104. Zhang, J.; Li, W.; Zhang, B.; Dou, Y. Influence of Er addition and extrusion temperature on the microstructure and mechanical properties of a Mg–Zn–Zr magnesium alloy. *Mater. Sci. Eng. A* **2011**, *528*, 4740–4746.
105. Zhang, J.; Zhang, X.; Li, W.; Pan, F.; Guo, Z. Partition of Er among the constituent phases and the yield phenomenon in a semi-continuously cast Mg–Zn–Zr alloy. *Scripta Mater.* **2010**, *63*, 367–370.
106. Yu, H.; Yan, H.; Chen, J.; Su, B.; Zheng, Y.; Shen, Y.; Ma, Z. Effects of minor gd addition on microstructures and mechanical properties of the high strain-rate rolled Mg–Zn–Zr alloys. *J. Alloy. Compd.* **2014**, *586*, 757–765.
107. Xiao, W.; Shen, Y.; Wang, L.; Wu, Y.; Cao, Z.; Jia, S.; Wang, L. The influences of rare earth content on the microstructure and mechanical properties of Mg–7Zn–5Al alloy. *Mater. Des.* **2010**, *31*, 3542–3549.
108. Xiao, W.; Wang, J.; Yang, J.; Jia, S.; Wang, L. Microstructure and mechanical properties of Mg–12.3Zn–5.8Y–1.4Al alloy. *Mater. Sci. Eng. A* **2008**, *485*, 55–60.
109. Li, Q.; Wang, Q.; Wang, Y.; Zeng, X.; Ding, W. Effect of nd and y addition on microstructure and mechanical properties of as-cast Mg–Zn–Zr alloy. *J. Alloy. Compd.* **2007**, *427*, 115–123.
110. Yu, H.; Park, S.H.; You, B.S.; Kim, Y.M.; Yu, H.S.; Park, S.S. Effects of extrusion speed on the microstructure and mechanical properties of ZK60 alloys with and without 1 wt% cerium addition. *Mater. Sci. Eng. A* **2013**, *583*, 25–35.
111. Xu, D.K.; Liu, L.; Xu, Y.B.; Han, E.H. The effect of precipitates on the mechanical properties of ZK60-Y alloy. *Mater. Sci. Eng. A* **2006**, *420*, 322–332.
112. Wang, Y.; Guan, S.; Zeng, X.; Ding, W. Effects of Re on the microstructure and mechanical properties of Mg–8Zn–4Al magnesium alloy. *Mater. Sci. Eng. A* **2006**, *416*, 109–118.
113. Wu, A.-R.; Xia, C.-Q. Study of the microstructure and mechanical properties of Mg-rare earth alloys. *Mater. Des.* **2007**, *28*, 1963–1967.
114. Wang, J.; Song, P.; Huang, S.; Pan, F. Effects of heat treatment on the morphology of long-period stacking ordered phase and the corresponding mechanical properties of Mg–9Gd–*x*Er–1.6Zn–0.6Zr magnesium alloys. *Mater. Sci. Eng. A* **2013**, *563*, 36–45.
115. Cheng, W.; Park, S.S.; Tang, W.; You, B.S.; Koo, B.H. Influence of rare earth on the microstructure and age hardening response of indirect-extruded Mg-5Sn-4Zn alloy. *J. Rare Earths* **2010**, *28*, 785–789.

Hot Deformation Behavior of Ti-3.5Al-5Mo-6V-3Cr-2Sn-0.5Fe Alloy in α + β Field

Zhaoxin Du [1,*], Shulong Xiao [2], Jingshun Liu [1], Shufeng Lv [1], Lijuan Xu [2], Fantao Kong [2] and Yuyong Chen [2]

[1] School of Materials Science and Engineering, Inner Mongolia University of Technology, Hohhot 010051, China; E-Mails: Jingshun_Liu@163.com (J.L.); shufenglu@yahoo.com (S.L.)

[2] National Key Laboratory of Science and Technology on Precision Heat Processing of Metals, Harbin Institute of Technology, Harbin 150001, China; E-Mails: xiaoshulong@hit.edu.cn (S.X.); xljuan@hit.edu.cn (L.X.); kft@hit.edu.cn (F.K.); yychen@hit.edu.cn (Y.C.)

* Author to whom correspondence should be addressed; E-Mail: duzhaoxin@163.com

Academic Editor: M. T. Whittaker

Abstract: The deformation behavior of Ti-3.5Al-5Mo-6V-3Cr-2Sn-0.5Fe high strength β titanium alloy is systematically investigated by isothermal compression in α + β field with the deformation temperatures ranging from 1003 K to 1078 K, the strain rates ranging from 0.001 s^{-1} to 1 s^{-1} and the height reduction is around 50%. Essentially, the flow stress-strain curve of isothermal compression in α + β field exhibits a flow softening feature when the strain rate is higher than 0.1 s^{-1} as while it exhibits a steady-state feature as the strain rate is lower than 0.1 s^{-1}. The peak stress increases with a decrease in deformation temperature and the increase of strain rate. The activation energy for deformation in α + β field was calculated and the average activation energy of 271.1 kJ/mol. The microstructure observation reveals that the isothermal deformation in the α + β field of the alloy is mainly controlled by the dynamic recovery mechanism accompanied with the secondary dynamic recrystallizitation of β phase. The α phase shows an obvious pinning effect for the movement of dislocations. During deformation, the α phase was elongated and fragmented.

Keywords: β titanium alloys; activation energy; microstructure evolution; deformation mechanism

---

1. Introduction

Generally, β titanium alloys possess the wide applied range in the applications of aerospace industries due to their excellent mechanical properties [1–3]. The mechanical properties of β titanium alloys are mainly controlled by microstructure, especially the grain size, volume fraction and morphology of α phase (primary, secondary and grain boundary α) [4–7]. For example, Terlinde *et al.* [4] have proven that the Ti-10V-2Fe-3Al with globular α microstructure shows better ductility than elongated α microstructure. However, the disadvantage of β titanium alloys is the poor ductility in high strength level, such as Timetal555 alloy [8]. Fortunately, the mechanical properties of β titanium alloys can be optimized through effectively microstructural controlling during the different stages of thermomechanical processing [9]. Usually, β titanium alloys are forged in the β field to break down the coarse grain microstructure, and then finally forged at the β or α + β field. It should be pointed out that the desired properties of titanium alloys are dependent on the final thermomechanical processing [10–12]. Studies done by Srinivasu *et al.* [13] have proved that rolling Ti-10V-2Fe-3Al alloy in α + β field exhibited better ductility than rolling in β field. Thus, the minimal α + β forging should be carried out if a good ductility is required.

Many works have been conducted on α + β titanium alloys through isothermal compression method to understand deformation mechanisms and microstructure controlling, the majority of these works are mainly focused on the Ti-6Al-4V alloy [14–19]. It is well known that the main deformation mechanism for deformation in β field is dynamic recovery of β titanium alloys, while it is also difficult to determine the deformation mechanism for deformation in α + β field owing to the complex microstructure parameters. Typically, the microstructure of α + β processed β titanium alloys consists of a small sized of α phase, which is a vital factor affecting the flow behavior. There are few works has been done aobut β titanium alloys deformation in α + β field [20–23]. During the deformation of β titanium alloys in α + β field, the α phase could act as a harder particle to pinning the dislocation movement [23] and results in a higher peak stress. Material with high aspect ratio Widmanstäten α platelets produces higher flow stresses than material with globular primary α in subtransus deformation [24]. The flow stress increases with the increasing of volume fraction of α phase at certain strain rate [20]. Technically, the volume fraction of α phase is mainly controlled by deformation temperature, and the strain has little effect on the volume fraction of α phase at the temperatures investigated [25].

Ti-3.5Al-5Mo-6V-3Cr-2Sn-0.5Fe alloy is a newly developed β high strength titanium alloy. The previous work has shown that the alloy exhibits the excellent mechanical property with finally forging in the α + β field and followed by heat treatment [26]. The isothermal compression study has not carried on this alloy yet. This paper aims to investigate the flow behavior and microstructural evolution during isothermal compression in α + β field to accurately determine the hot working process of Ti-3.5Al-5Mo-6V-3Cr-2Sn-0.5Fe alloy.

2. Materials and Methods

The investigations were conducted on Ti-3.5Al-5Mo-6V-3Cr-2Sn-0.5Fe high strength β titanium alloy. The alloy was melted using double vacuum consumable arc melting process to obtain an original ingot. The ingot was pre-forged in the α + β field to a round bars with the diameter of 80 mm. The β transition temperature of the alloy measured by metallographic method is about 1088 K. The microstructure of pre-forged alloy is shown in Figure 1, the α phase showed equiaxed shape and some β subgrains with small size formed in the matrix. The cylindrical specimens with 8.0 mm in diameter and 12.0 mm in height were machined from the pre-forged Ti-3.5Al-5Mo-6V-3Cr-2Sn-0.5Fe alloy bar. The isothermal compression experiment was conducted on a Gleeble-1500dD simulator (Dynamic Systems Inc., Poestenkill, NY, USA) at the deformation temperatures ranging from 1003 K to 1078 K with an interval of 25 K, the strain rates of 0.001 s^{-1}, 0.01 s^{-1}, 0.1 s^{-1}, 1 s^{-1} respectively and the height reduction of 50%. The specimens were heated to compression temperature at a heating rate of 15 Ks^{-1} and held for 3 min before compression. After hot compression, the specimens were cooled in air to room temperature. Microstructure observation of the alloy was characterized by scanning electron microscopy (SEM) and electron backscattered diffraction (EBSD) by using field emission gun scanning electron microscopy Quanta 200FEG (FEI company, Hillsboro, OR, USA). The specimens for SEM and EBSD analysis were roughly polished with 400–2000 grid SiC paper in water, and then electropolished for 150 s at 1 V in reagent of 60% methanol, 30% butyl alcohol and 10% perchlorate at room temperature. The specimens for SEM observation were etched in the Kroll's reagent (Tianjin Yongda Chemical Reagent Co., Ltd., Tianjing, China) (mixed with 10 mL HF, 30 mL HNO_3 and 200 mL H_2O).

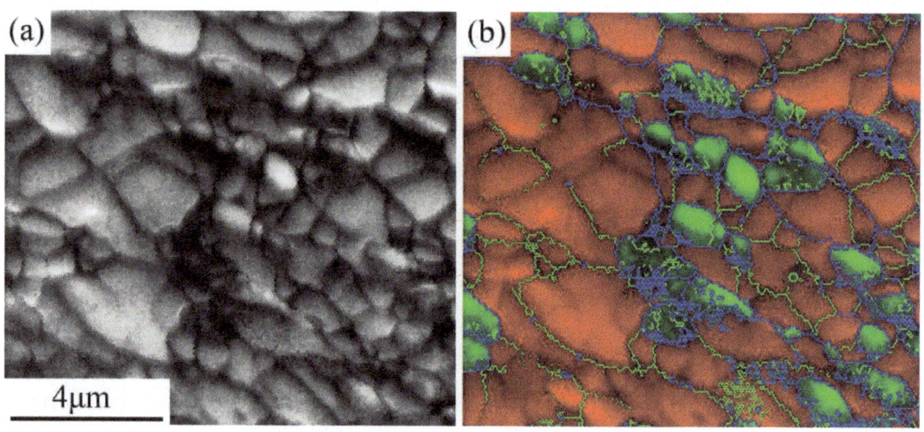

Figure 1. Microstructure of the alloy as-received: (**a**) image quality map; (**b**) grain boundary distribution plus phase map. The red field = β phase, green field = α phase, blue line = high angle grain boundaries (HAGBs, θ ≥ 15°), green line = low angle grain boundaries (LAGBs, 15° ≥ θ ≥ 2°).

3. Results and Discussion

3.1. Flow Stress Behavior

The true stress-strain curves of isothermally compressed deformation of Ti-3.5Al-5Mo-6V-3Cr-2Sn-0.5Fe alloy at different temperatures and strain rates are shown in Figure 2. It can be seen that the flow stress increases sharply with increasing of strain at the beginning stage of deformation. It is just the result of the work hardening behavior resulting from the increasing of dislocation density in the initial stages of deformation. The curves exhibit the steady state flow after an initial sharp peak, except the curves at strain rate of 1 s^{-1} which exhibit dynamic obvious softening behavior. At strain rate of 1 s^{-1} and temperatures of 1003 K, the curve exhibits an increase in true stress after a broad softening, this may be due to the faster dislocation multiplication than dislocation motion resulting from high strain rate and low temperature. It is difficult to determine the deformation mechanisms only according to the shape of the true stress-true strain curves because that the flow softening behavior can be resulted from several reasons, such as dynamic recrystallization, temperature rise of the specimen and texture evolution, or grain size increase during the deformation process [27]. The characteristic of the flow stress tends to a steady value, it is caused by a combination of work hardening and dynamic softening mechanism. After the flow stress reaches the peak value, the effect of work hardening and softening comes to dynamic equilibrium and the flow stress ends to be a steady value.

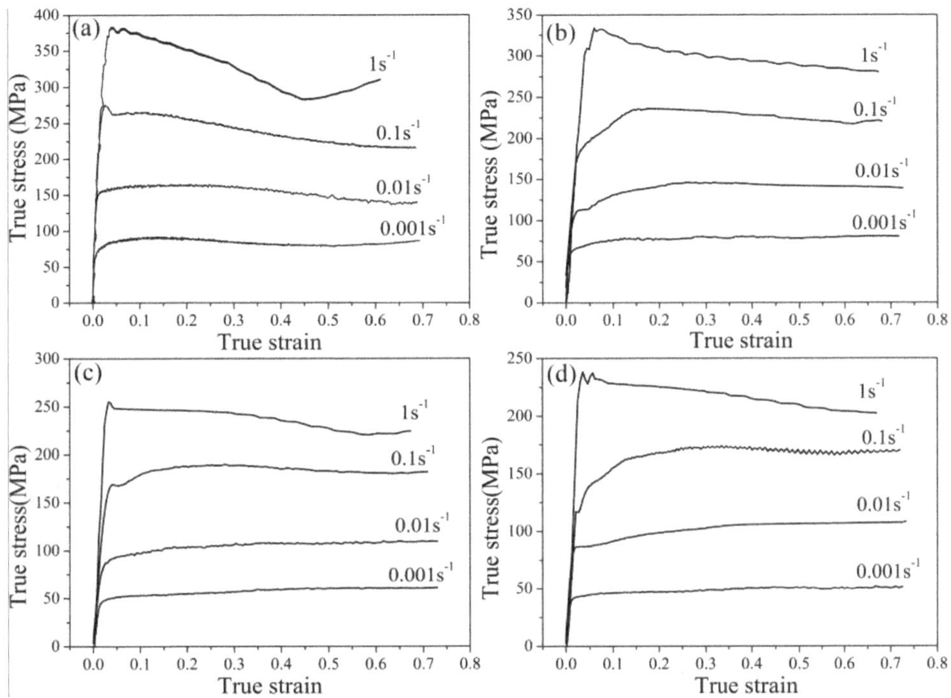

Figure 2. Compressive true stress- strain curves of the alloy under different temperature as follows: **(a)** 1003 K; **(b)** 1028 K; **(c)** 1053 K; **(d)** 1078 K.

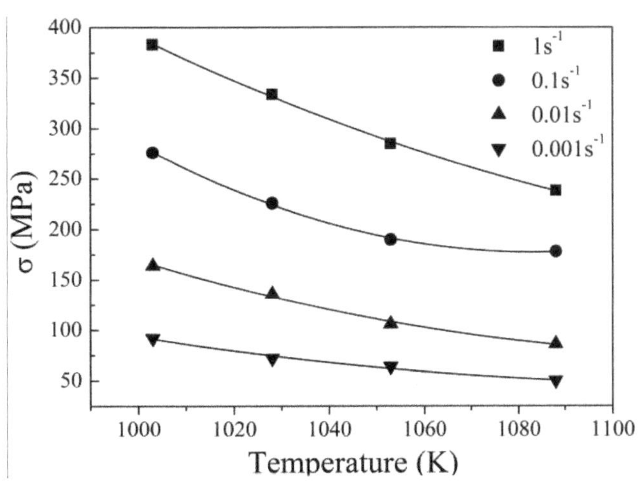

Figure 3. Effect of deformation temperature and strain rate on the peak flow stress.

Figure 3 shows the effects of deformation temperature and strain rate on the peak flow stress in isothermal compression of Ti-3.5Al-5Mo-6V-3Cr-2Sn-0.5Fe alloy. It can be seen that the peak flow stress is so sensitive to the deformation temperature and strain rate. The peak stress increases with the decreasing of deformation temperature and increasing of strain rate. Such as that at a strain rate of 1 s^{-1}, the peak flow stress at 1003 K is higher by about 145 MPa than at a temperature of 1078 K. Moreover, the peak flow stress at strain rate of 1 s^{-1} is higher by about 295 MPa than at strain rate of 0.001 s^{-1} at a same deformation temperature of 1003 K. During the deformation of β titanium alloy in α + β field, the α phase could play a role of a harder particle dispersed in the soft matrix pinning the dislocation movement [23]. However, the volume fraction of α phase decreases with the increasing of deformation temperature. It also leads to a reduction of obstacles for dislocation moving and thus resulted in decrease of flow stress. Oppositely, the high volume fraction of α phase will lead to higher flow stress. There is a support for this position from other studies [20]. In addition, the lower strain rate provides more time to climp and be consumed of the dislocations and with the result that the flow stress increases with the increasing of strain rate.

3.2. Activation Energy

Based on the true flow stress-true strain data from compression test, the Arrhenius type equation which is proposed by Sellars *et al.*, is employed to establish the constitutive equation for alloy and designated as Equation (1) [28]:

$$\dot{\varepsilon}=A\left[\sin h(\alpha\sigma)\right]^{n}\exp\left[-Q/RT\right] \tag{1}$$

where $\dot{\varepsilon}$ is strain rate, A, α, n are constants independent of temperature, Q is activation energy, R is gas constant ($R = 8.3145\ \text{J}\cdot\text{mol}^{-1}\cdot\text{k}^{-1}$), T is the absolute temperature.

The Equation (1) shows different expressions for different ($\alpha\sigma$) values as follows:

$$\dot{\varepsilon} = A_1\sigma^{n_1} \qquad \alpha\sigma< 0.8 \tag{2}$$

$$\dot{\varepsilon} = A_2\exp(\beta\sigma) \qquad \alpha\sigma> 1.2 \tag{3}$$

Assuming that the deformation activation energy Q has no correlation to do with the deformation temperature of T, and the taking logarithm of both sides of Equations (1)–(3), the following equations can be obtained by:

$$\frac{\partial(\ln\dot{\varepsilon})}{\partial\{\ln[\sin h(\alpha\sigma)]\}}\Big|_{1/T}=n \tag{4}$$

$$\frac{\partial(\ln\dot{\varepsilon})}{\partial(\ln\sigma)}\Big|_{1/T}=n_1 \tag{5}$$

$$\frac{\partial(\ln\dot{\varepsilon})}{\partial\sigma}\Big|_{1/T}=\beta \tag{6}$$

The α can then be calculated based on the values of β and n_1, $\beta = \alpha n1$. The material constant α is calculated and the obtained value is 0.00639. The activation energy of Q can be deduced from Equation (1):

$$Q = Rn\{\frac{\partial\ln[\sin h(\alpha\sigma)]}{\partial(1/T)}\}\Big|_{\dot{\varepsilon}}=R\frac{\partial\ln\dot{\varepsilon}}{\partial\ln[\sin h(\alpha\sigma)]}\Big|_{1/T}\cdot\left\{\frac{\partial\ln[\sin h(\alpha\sigma)]}{\partial(1/T)}\right\}\Big|_{\dot{\varepsilon}} \tag{7}$$

The calculated values of Q are 285.7 kJ/mol (1003 K), 283.33 kJ/mol (1028 K), 261 kJ/mol (1053 K), 254.4 kJ/mol (1078 K) respectively. The activation energy represents the energy level to overcome in some atomistic mechanisms such as diffusion. In this study, the activation energy for deformation in α + β field decreased with the increasing of temperature. This is mainly caused by the decreasing of volume fraction of α phase. The average activation energy in α + β deformation field is 271.1 kJ/mol, it is quite similar to Ti-5Al-5Mo-5V-1Cr-1Fe β alloy (291.73 ± 50 kJ mol^{-1}) obtained by Liu [20] and Ti-5Al-5Mo-5V-3Cr-1Zr β alloy (275 $kJmol^{-1}$) obtained by Warchomicka *et al.* [23]. In titanium alloys, the activation energies for self-diffusion in α and β titanium have been reported as 183 kJ mol^{-1} and 153 $kJmol^{-1}$ respectively [29]. It is well known that the activation energy of deformation close to self-diffusion in the β phase suggesting that the deformation is practically dominated by dynamic recovery of β phase. However, the activation energy in present study is higher than self-diffusion of β phase. The reason for higher activation energy in α + β field is possibly resulted from the presence of α phase. McQueen [30] pointed out that the activation energy will rise due to additions of solute, precipitates, dispersoid, reinforcements and inclusions. In titanium alloys, Warchomicka [23] considered that the higher activation energy deformed in α + β field should be related to the energy necessary to rotate, elongate and the α grains. In addition, the α phase shows the pining effect on the dislocation movement as mentioned above. The more α phase resulting from lower temperature creates more obstacles and then the higher activation energy should be provided for dislocation movement. Therefore, a common ground can be deduced from as mentioned above that the presence of the α phase should be the reason for higher activation energy deformed in the α + β field rather than activation energy for self-diffusion of β phase.

3.3. Microstructure Observation

Figure 4 shows the microstructure of Ti-3.5Al-5Mo-6V-3Cr-2Sn-0.5Fe under different deformation temperatures with strain rate of 0.01 s^{-1}. The white area on the SEM image is α phase. The volume fraction of α phase decreased with the increase of deformation temperature, as can be seen from Figure 4. At a lower deformation temperature of 1003 K, the α phase showed three kinds of shape, namely equiaxed α, lamellar α with smaller size and the elongated α with larger size. The smaller α lamellar phase disappeared at temperature 1028 K, and the elongated α phase disappeared up to 1078 K. The volume fraction of equaixed α phase decreased with an increase of temperature, but the size almost is no change at all deformation temperatures. The dynamic recovery phenomenon of α phase can be observed in the present study. It can be observed that some equiaxed α grains exist between elongated α phases in Figure 4a. This suggesting a mechanism of fragmentation of elongated α phases. Margolin and Cohen proposed a model for the globularisation of α phase and the mechanism can be summarized as follows [31,32]. Initially, the recrystallised α grains are formed within the α plate. While the 180° dihedral angle between the α/α boundary can not exist due to the surface tension requirement. Thus, a driving force is provided for the movement of some beta phase into the α/β boundaries and simultaneous rotation of the of the α/β boundaries. Subsequent working will separate the neighbouring recrystallizated α phases and leads to equiaxed shape of the α phases. The mechanism of fragmentation of α phases is not only good applicable to the deformation of α + β titanium alloys with lamellar microstructure [17,31,33,34], but also to β titanium alloys [13,24].

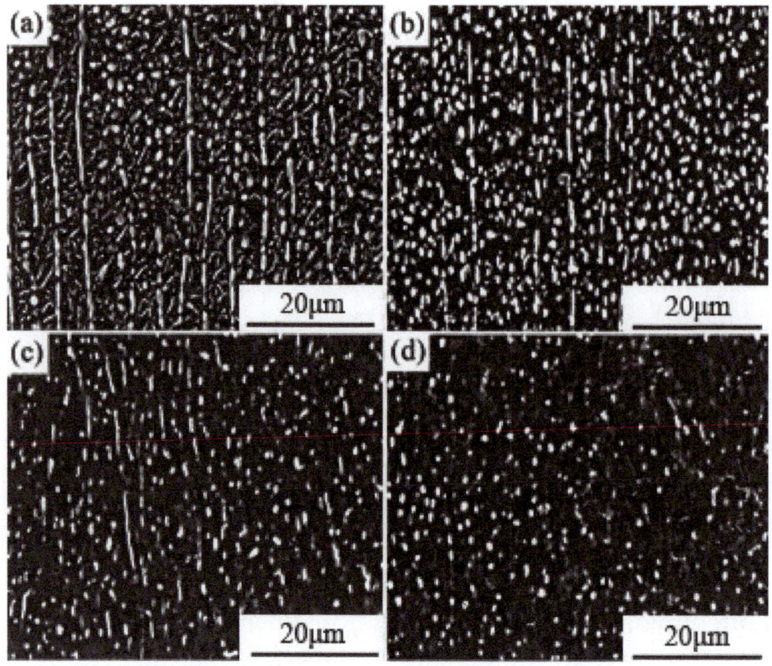

Figure 4. Morphology of the alloy deformed at different temperatures with strain rate of 0.01 s^{-1}: (**a**) 1003 K; (**b**) 1028 K; (**c**) 1053 K; (**d**) 1078 K.

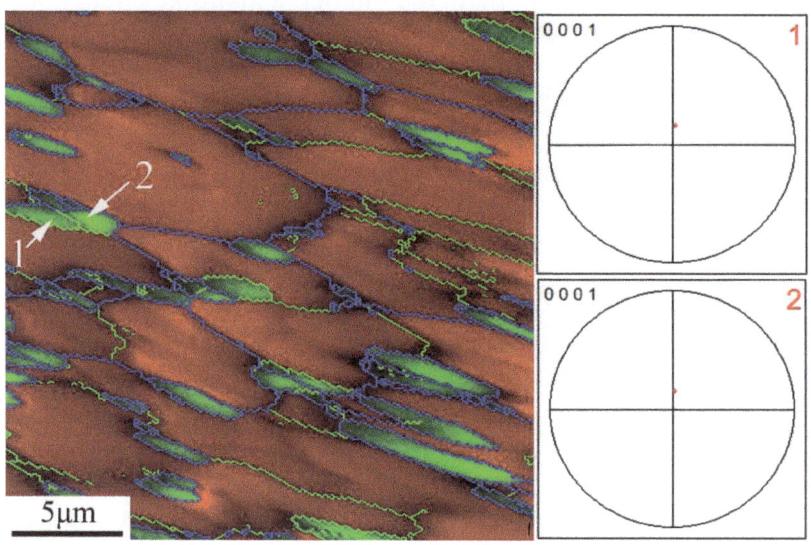

Figure 5. Phase composition and grain boundary map of the alloy deformed at 1028 K with the strain rate of 0.001 s^{-1}. (The red field = β phase, green field = α phase, blue line = HAGBs ($\theta \geq 15°$), green line = LAGBs ($15° \geq \theta \geq 2°$)).

The direct evidence for fragment of α phase displays in Figure 5. Figure 5 shows phase composition and grain boundary map of the alloy deformed at 1028 K with the strain rate of 0.001 s^{-1}. It is obvious that the α phase was also deformed during isothermal compression and elongated perpendicular to the compression direction. As indicated by white uni-directional arrows in Figure 5, a subboundary with low angle developed inside α phase, and the right panels of Figure 5 shows that each part of this α grain exhibits the same orientation. It can confirms that the dynamic recovery phenomenon of the α phase. Subsequent compression deformation can provide a driving force for penetration of β phase along the α/α subboundaries and then resulting in equiaxed morphology α phase. However, the low angle grain boundaries are mainly distributed in β grains. It indicates that the β phase is mainly deformed. As previouly work reported by Jones *et al.* [25] indicated that the majority of the flow occurs within the β phase, but little load transfers to the α phase. The α phase plays a vital role of hindering effect of dislocation climp during hot processing in α + β field, taht it could act as a harder particle dispersed in the soft matrix pinning the dislocation movement and hence the β subgrains cannot grow beyond the α grains [23]. As shown in Figure 6, the α grains are mainly distributed in grain or subgrain boundary triple points, and the low angle grain boundaries terminated at α phase.

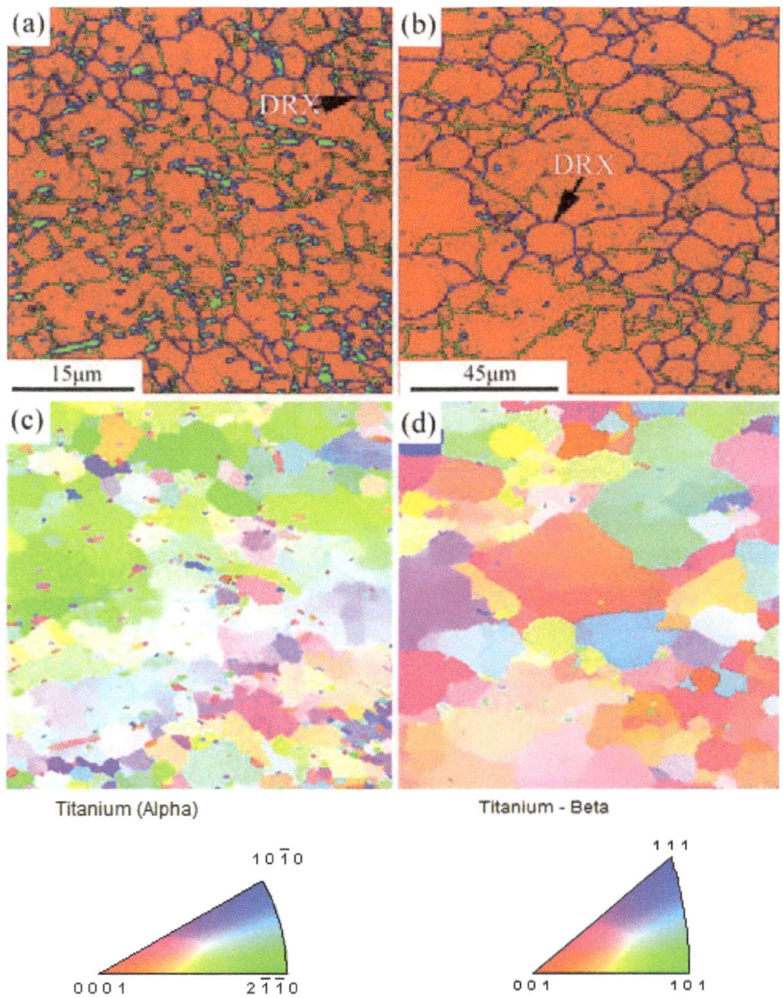

Figure 6. Electron backscattered diffraction (EBSD) images showing microstructure of the alloy deformed at 1003 K (**a**) and (**c**) 1078 K (**b**) and (**d**) with the strain rate of 0.001 s^{-1}. (In **a** and **b**), the red field = β phase, green field = α phase, blue line = HAGBs, green line = LAGBs, DRX = dynamic recrystallization).

Figure 6 shows the EBSD images of the alloy deformed at 1003 K and 1078 K with strain rate of 0.001 s^{-1}, respectively. The grain size deformated at 1003 K is quite smaller than high deformation temperature of 1078 K as can be seen from Figure 6a,b. The smaller grain size is attributed to the volume fraction of α phase. As mentioned above, the α phase acts as a harder particle dispersed in the soft matrix to pinning the moving dislocation and hence the β subgrains cannot grow beyond the α grains. Therefore, the more content of α phase obtained at lower deformation temperature resulting in small β grain or subgrain size. Some new formed equaixed β grains with large angle grain boundaries can be observed in

Figure 6a,b. However, most of the β grains with large angle grain boundaries in Figure 6a exhibit similar color with neighbor grains, as shown in inverse pole figure map as shown in Figure 6c. It means that these new β grains are obtained by rotation and growth of subgrains, and it can be conclued that the dynamic recovery is the dominant mechanism during deformation at lower temperature since there are also few in number of dynamic recrystallized β grains deformated at 1003 K. The dynamic recovery still is the dominant mechanism at higher deformation temperature of 1078 K, but the dynamic recrystallized β grains are more and larger than lower deformation temperature of 1003 K.

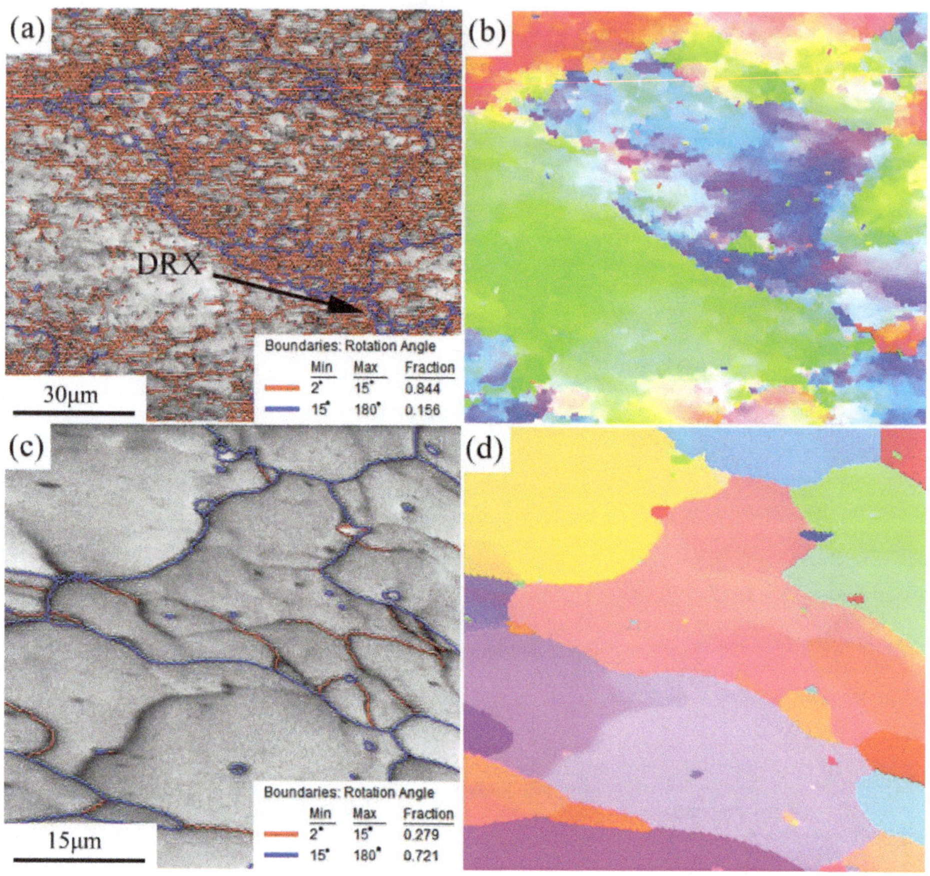

Figure 7. Grain boundary maps and inverse pole figure maps of the alloy deformed at 1078 K with the strain rate of 1 s^{-1} (**a**), (**b**) and 0.001 s^{-1} (In **a** and **c**), blue line = HAGBs, green line = LAGBs, DRX = dynamic recrystallization).

Figure 7 reveals the EBSD imaged of the alloy deformed at 1078 K with the strain rate of 1 s^{-1} and 0.001 s^{-1}. The microstructure of the alloy is very sensitive to the deformation strain rate. The higher strain rate of 1 s^{-1} resulted in large number fraction of low angle grain boundaries (Figure 7a) and formation of lots of sub-structures with small misorientation inside β grains (Figure 7b). However, at lower deformation strain rate of 0.001 s^{-1} the number fraction of low angle grain boundaries is much less than the high strain rate of 1 s^{-1} (compared with Figure 7a,c), and the sub-structures show larger size and less number (compared with Figure 7b,d). The alloy deformated at 1078 K and strain rate of 1 s^{-1} shows a serrated grain boundary, while the uniform grain boundary is obtained at the strain rate of 0.001 s^{-1}. The serrated grain boundary in large strain rate conditon caused by the existence of some new fine recrystallized β grains with high angle boundaries in the vicinity of boundaries as shown in Figure 7a (black arrow), but there is no appearance at the lower strain rate of 0.001 s^{-1}. The recrystallized β grains are usually observed at the borders of β grains [35]. The content of low and high angle boundaries is relevant to deformation time. A massive dislocations produced during the process of isothermal deformation, but it seems no sufficient time to climp and be consumed because the high strain rate, and then reuslted in lots of number low angle grain boundaries inside β grains. On the contrary, the dislocations have sufficient time to climp and be consumed at lower strain rate conditions. Meanwhile, the recrystallized grains can also have more time to nucleation and growth at lowe strain rate. The evidence can be represented by comparision of Figures 6b and 7b, the recrystallized grains with lower strain rate shows a larger size than higher strain rate conditions.

4. Conclusions

The flow behavior and microstructural evolution of Ti-3.5Al-5Mo-6V-3Cr-2Sn-0.5Fe alloy during isothermal compression in α + β field has been investigated. The following principal conclusions can be drawn from the present study:

(1) The flow stress–strain curves of Ti-3.5Al-5Mo-6V-3Cr-2Sn-0.5Fe alloy isothermally compressed in α + β field exhibit a flow softening feature as the strain rate is higher than 0.1 s^{-1} while it exhibits a steady-state feature as the strain rate is lower than 0.1 s^{-1}. The peak stress increases with the decrease of deformation temperature and an increase of strain rate.

(2) The activation energy of Ti-3.5Al-5Mo-6V-3Cr-2Sn-0.5Fe alloy for deformation in α + β field was calculated. The average activation energy in α + β field deformation is 271.1 kJ/mol, it is higher than that of self-diffusion of β phase resulting from presence of α phase.

(3) The α phase shows an obvious pinning effect for the movement of dislocations deformed at α + β field. The high volume fraction of α phase is responsible for higher flow stress at lower deformation temperature.

(4) The deforamtion mechanism of Ti-3.5Al-5Mo-6V-3Cr-2Sn-0.5Fe alloy in α + β field is mainly attributed to dynamic recovery accompanied with dynamic recrystallizitation of β phase, but the dynamic recovery is dominant.

Acknowledgments

J.S.L. acknowledges the support from Scientific Research Foundation of Inner Mongolia University of Technology under grant Nos. ZD201405.

Author Contributions

Zhaoxin Du, Shulong Xiao, Lijuan Xu, Fantao Kong and Yuyong Chen conceived and designed the experiments; Zhaoxin Du and Shulong Xiao performed the experiments; Zhaoxin Du and Shulong Xiao, Jingshun Liu and Shufeng Lv analyzed the date; Zhaoxin Du wrote the paper.

Conflicts of Interest

The authors declare no conflict of interest.

References

1. Boyer, R.R.; Briggs, R.D. The use of β titanium alloys in the aerospace industry. *J. Mater. Eng. Perform.* **2005**, *14*, 681–685.
2. Boyer, R.R. An overview on the use of titanium in the aerospace industry. *Mater. Sci. Eng. A* **1996**, *213*, 103–114.
3. Boyer, R.R. Titanium for aerospace Rationale and applications. *Adv. Perform. Mater.* **1995**, *2*, 349–368.
4. Terlinde, G.T.; Duerig, T.W.; Williams, J.C. Microstructure, tensile deformation, and fracture in aged Ti-10V-2Fe-3Al. *Metall. Trans. A* **1983**, *14A*, 2101–2115.
5. Bhattacharjee, A.; Varma, V.K.; Kamat, S.V.; Gogia, A.K.; Bhargava, S. Influence of β grain size on tensile behavior and ductile fracture toughness of titanium alloy Ti-10V-2Fe-3Al. *Metall. Trans. A* **2006**, *37*, 1423–1433.
6. Ghosh, A.; Sivaprasad, S.; Bhattacharjee, A.; Kar, S.K. Microstructure-fracture toughness correlation in an aircraft structural component alloy Ti-5Al-5V-5Mo-3Cr. *Mater. Sci. Eng. A* **2013**, *568*, 61–67.
7. Kar, S.K.; Suman, S.; Shivaprasad, S.; Chaudhuri, A.; Bhattacharjee, A. Processing-microstructure-yield strength correlation in a near β Ti alloy, Ti-5Al-5Mo-5V-3Cr. *Mater. Sci. Eng. A* **2014**, *610*, 171–180.
8. Nyakana, S.; Fanning, J.; Boyer, R.R. Quick reference guide for β titanium alloys in the 00 s. *J. Mater. Eng. Perform.* **2005**, *14*, 799–811.
9. Weiss, I.; Semiatin, S.L. Thermomechanical processing of beta titanium alloys—An overview. *Mater. Sci. Eng. A* **1998**, *243*, 46–65.
10. Boyer, R.R. Design properties of a high-strength titanium alloy, Ti-10V-2F-3Al. *JOM* **1980**, *32*, 61–65.
11. Chen, C.C.; Boyer, R.R. Practical considerations for manufacturing high strength Ti-10V-2Fe-3Al forgings. *JOM* **1979**, *31*, 33–39.
12. Jones, N.G.; Jackson, M. On mechanism of flow softening in Ti-5Al-5Mo-5V-3Cr. *Mater. Sci. Technol.* **2011**, *27*, 1025–1032.
13. Srinivasu, G.; Natraj, Y.; Bhattacharjee, A.; Nandy, T.K.; Nageswara Rao, G.V.S. Tensile and fracture toughness of high strength β Titanium alloy, Ti-10V-2Fe-3Al, as a function of rolling and solution treatment temperatures. *Mater. Des.* **2013**, *47*, 323–330.
14. Zong, Y.Y.; Shan, D.B.; Xu, M.; Lv, Y. Flow softening and microstructural evolution of TC11 titanium alloy during hot deformation. *J. Mater. Process. Technol.* **2009**, *209*, 1988–1994.
15. Li, H.; Li, M.Q.; Han, T.; Liu, H.B. The deformation behavior of isothermally compressed Ti-17 titanium alloy in alpha plus beta field. *Mater. Sci. Eng. A* **2012**, *546*, 40–45.
16. Quan, G.Z.; Luo, G.C.; Liang, J.T.; Wu, D.S.; Mao, A.; Liu, Q. Modelling for the dynamic recrystallization evolution of Ti-6Al-4V alloy in two-phase temperature range and a wide strain rate range. *Comput. Mater. Sci.* **2015**, *97*, 136–147.
17. Zherebtsov, S.; Murzinova, M.; Salishchev, G.; Semiatin, S.L. Spheroidization of the lamellar microstructure in Ti-6Al-4V alloy during warm deformation and annealing. *Acta Mater.* **2011**, *59*, 4138–4150.

18. Zhao, J.; Ding, H.; Zhao, W.; Jiang, Z. Effects of hydrogen on the hot deformation behaviour of Ti-6Al-4V alloy: Experimental and constitutive model studies. *J. Alloy. Compd.* **2013**, *574*, 407–414.
19. Kim, J.H.; Semiatin, S.L.; Lee, C.S. Constitutive analysis of the high-temperature deformation mechanisms of Ti-6Al-4V and Ti-6.85Al-1.6V alloys. *Mater. Sci. Eng. A* **2005**, *394*, 366–375.
20. Liu, S.F.; Li, M.Q.; Luo, J.; Yang, Z. Deformation behavior in the isothermal compression of Ti-5Al-5Mo-5V-1Cr-1Fe alloy. *Mater. Sci. Eng. A* **2014**, *589*, 15–22.
21. Liang, H.; Guo, H.; Ning, Y.; Peng, X.; Qin, C.; Shi, Z. Dynamic recrystallization behavior of Ti-5Al-5Mo-5V-1Cr-1Fe alloy. *Mater. Des.* **2014**, *63*, 798–804.
22. Li, C.; Zhang, X.Y.; Zhou, K.C.; Peng, C.Q. Relationship between lamellar α evolution and flow behavior during isothermal deformation of Ti-5Al-5Mo-5V-1Cr-1Fe near β titanium alloy. *Mater. Sci. Eng. A* **2012**, *558*, 668–674.
23. Warchomicka, F.; Poletti, C.; Stockinger, M. Study of the hot deformation behaviour in Ti-5Al-5Mo-5V-3Cr-1Zr. *Mater. Sci. Eng. A* **2011**, *528*, 8277–8285.
24. Jackson, M.; Jones, N.G.; Dye, D.; Dashwood, R.J. Effect of initial microstructure on plastic flow behaviour during isothermal forging of Ti-10V-2Fe-3Al. *J. Mater. Sci. Eng. A* **2009**, *501*, 248–254.
25. Jones, N.G.; Dashwood, R.J.; Dye, D.; Jackson, M. Thermomechanical processing of Ti-5Al-5Mo-5V-3Cr. *Mater. Sci. Eng. A* **2008**, *490*, 369–377.
26. Du, Z.X.; Xiao, S.L.; Xu, L.J.; Tian, J.; Kong, F.T.; Chen, Y.Y. Effect of heat treatment on microstructure and mechanical properties of a new β high strength titanium alloy. *Mater. Des.* **2014**, *55*, 183–190.
27. Li, L.X.; Lou, Y.; Yang, L.B.; Peng, D.S.; Rao, K.P. Flow stress behavior and deformation characteristics of Ti-3Al-5V-5Mo compressed at elevated temperatures. *Mater. Des.* **2002**, *23*, 451–457.
28. Sellars, C.M.; McTegart, W.J. On the mechanism of hot deformation. *Acta Metall.* **1966**, *14*, 1136–1138.
29. Dyment, F.; Libanati, C.M. Self-diffusion of Ti, Zr, and Hf in their HCP phases, and diffusion of Nb95 in HCP Zr. *J. Mater. Sci.* **1968**, *3*, 349–359.
30. McQueen, H.J.; Ryan, N.D. Constitutive analysis in hot working. *Mater. Sci. Eng. A* **2002**, *322*, 43–63.
31. Margolin, H.; Cohen, P. Evolution of the equiaxed morphology of phases in Ti-6Al-4V. In *Titanium 80: Science and Technology*; Warrendale, TMS-AIME: Kyoto, Japan, 1980; pp. 1555–1561.
32. Margolin, H.; Cohen, P. Kinetics of Evolution of Alpha in Ti-6Al-4V. In *Titanium 80: Science and Technology*; Warrendale, TMS-AIME: Kyoto, Japan, 1980; pp. 2991–2997.
33. Wang, K.X.; Zeng, W.D.; Zhao, Y.Q.; Lai, Y.J.; Zhang, X.M.; Zhou, Y.G. Flow behaviour and microstructural evolution of Ti-17 alloy with lamellar microstructure during hot deformation in alpha plus beta phase field. *Mater. Sci. Technol.* **2011**, *27*, 21–28.
34. Weiss, I.; Froes, F.H.; Eylon, D.; Welsch, G.E. Modification of alpha morphology in Ti-6Al-4V by thermomechanical processing. *Metall. Trans. A* **1986**, *17A*, 1935–1947.
35. OuYang, D.L.; Fu, M.W.; Lu, S.Q. Study on the dynamic recrystallization behavior of Ti-alloy Ti-10V-2Fe–3Al in β processing via experiment and simulation. *Mater. Sci. Eng. A* **2014**, *619*, 26–34.

3

Temperature-Induced Liquid-Liquid Transition in Metallic Melts: A Brief Review on the New Physical Phenomenon

Fang-Qiu Zu

Liquid/Solid Metal Processing Institute, School of Materials Science & Engineering, Hefei University of Technology, Hefei 230009, China; E-Mail: fangqiuzu@hotmail.com;

Academic Editor: Enrique Louis

Abstract: Understanding the nature of liquid structures and properties remains an open problem for many fundamental and applied fields. It is well known that there is no other defined phase line above *liquidus* (T_L) in phase diagrams of ordinary alloys. However, via resorts of internal friction, electric resistivity, thermal analysis, X-ray diffraction, solidification, *etc.*, the results of our research on lots of single- and multiple-component melts show a novel physical image: temperature induced liquid-liquid structure transition (TI-LLST) can occur above T_L. Moreover, the solidification behaviors and structures out of the melts that experienced TI-LLST are distinct from those out of the melts before TI-LLST. In this paper, some typical examples of TI-LLST and characteristic aspects of the TI-LLST are briefly reviewed, in which the main contents are limited in our own achievements, although other groups have also observed similar phenomena using different methods. In the sense of phenomenology, TI-LLST reported here is quite different from other recognized liquid transitions, *i.e.*, there are only a few convincing cases of liquid P, Si, C, H_2O, Al_2O_3-Y_2O_3, *etc.* in which the transition occurs, either induced by pressure or at a supercooled state and near *liquidus*.

Keywords: metals; alloys; liquid structures; properties of melts; solidification behaviors

---

1. Introduction

It is more and more clear that polymorphism in liquid states [1–3] is one of the most promising new fields, in which there are lots of puzzling phenomena and nature for experimentalists and theoreticians to explore and to find their way to its applications, e.g., critical points between stable liquids and/or metastable ones [3–6]. Growing evidence suggests that the pressure-induced discontinuous phase changes might happen in some one-component liquids, e.g., SiO_2, H_2O, C, Si, Cs, Ga, Bi, I_2, and Se [1,6–12]. Such liquid state polyamorphism has been verified by an *in situ* X-ray study in liquid phosphorus, with an abrupt, pressure-induced structural change between two distinct forms [13]. Moreover, a temperature-induced phase transition was observed in the supercooled liquid system Al_2O_3-Y_2O_3 [14]. In addition, some liquid alloys, such as S-Te [15], Se-Te [16], and Cu-Ge [17], have been added to the systems in which the well-known conventional liquid phase separations can occur in the region near the *liquidus* with the chemical composition as the changing parameter. Because of the versatility of matter and the versatile complicated liquid structural nature, it is believed that novel phenomena and new knowledge will be uncovered as we probe liquid behaviors more deeply [6]. Accordingly, it is meaningful to explore if a temperature-induced structural change could discontinuously occur at ordinary pressure and at temperature far above melting point, or T_L, in single- or multiple-component liquids with a constant chemical composition

On the other side, liquid to solid transitions could be involved in most cases of the production or development of materials. It is a common phenomenon that structures and properties of many materials are related to the thermal history of their original melts, which has been intensively investigated in recent years. In developing semiconductors, the melt thermal history was shown to have a decisive influence on defect formation (twins) during crystallization and quality of ZnSe-based blue light-emitting laser diodes [18,19]. The microstructure of amorphous alloys is found to be much more homogeneous if precursor melts are much over-heated [20,21]. In order to refine solidified crystals, various techniques, like melt overheating and melt thermal treatment, have been explored and sometimes proved to be quite effective [22–26]. Although it is believed that the correlation between solidified microstructures and the melt thermal history would be attributed to different melt microstructures, the underlying nature and rules are unclear to date.

Unfortunately, knowledge about the liquid state has been much retarded in contrast to that of solid and gaseous states. On one hand, near *liquidus*, there is a general view that some kinds of residual micro-crystals still remain in the melts after fusion of crystal materials [1,27], *i.e.*, topological and/or chemical short-range ordering domains. On the other hand, near the liquid-gas critical temperature, the liquid structure of matter is hard to distinguish from its gaseous structure. Between the two typical cases, however, the change pattern of liquid structures and properties has been rarely concerned. Well what then is the underlying nature for the liquid-solid correlations and what rules are there for controlling solidification more effectively with less blindness by manipulating melt thermal history? Taking into account the abovementioned effects of liquids on solids and enlightened by liquid polymorphism [1–12,28,29], in the last ten years, our works suggest TI-LLST occurs well above the *liquidus* at ambient pressure in some ordinary alloys. In addition, we confirmed it is TI-LLST that decides liquid-solid effects at constant pressure and gives rise to significant alterations to solidification behaviors and structures.

A widely accepted conventional view about liquids is that the structures and properties change gradually with temperature or pressure [6,27]. It is worth noting that there is no other defined phase line above the *liquidus* in the phase diagrams of alloys. However, with the revised internal friction method, resistivity, θ-θ X-ray diffraction, DSC, DTA, *etc.*, a new physical phenomenon, *i.e.*, TI-LLST, was experimentally observed in various kinds of alloy systems and pure elements, including:

- Simple eutectic systems, such as Pb-Sn, Sn-Bi, Pb-Sb, *etc.*
- Infinitely mutual soluble alloy, like Bi-Sb.
- Complex systems with peritectic reaction, e.g., Pb-Bi, In-Sn, Sn-Sb, Bi-Te, *etc.*
- Complex systems with compounds like Bi-In, Cu-Sn, Cu-Sb, In-Sb, Pb-In, *etc.*
- Pure elements like Sn, Bi, Sb, *etc.*
- Some multiple component systems.

Some typical examples of TI-LLST and its characteristic aspects are briefly reviewed as follows.

2. The Behaviors of Internal Friction with Changing Temperature in Some Liquid Alloys

Internal friction is a structure sensitive physical property and as a powerful technique widely used for studying crystal structures, defects and phase transitions in solids [30]. In our work, an improved torsion pendulum internal friction apparatus was, for the first time, used to study liquids, and it was verified to also be valid and sensitive to liquid structures [31–35]. The prior-prepared samples were re-melted and held for 60 min before internal friction experiments and shielded with 5N pure argon, and all experimental conditions and data collection were automatically controlled by a computer, for details of which please refer to [31–33].

Figure 1 shows the internal friction experiment results of the liquid samples with eutectic composition (PbSn61.9 wt%). Surprisingly, there is a notable peak in each Q^{-1}-T curves, around 670 °C at 2.5 °C/min (Figure 1a) and 712 °C for 6.0 °C/min (Figure 1b), hundreds of degrees above T_L (183 °C). The peak position (T_P) moves to higher temperature when dT/dt rises, which is consistent with the kinetic nature of phase transitions. In order to detect what the Q^{-1}-T peak means for the Pb-Sn melt, the characteristics of the peak were analyzed. Figure 1 indicates clearly that T_P does not change with torsion frequencies (f), which resembles the internal friction feature in solid transitions. On the other hand, the magnitude of the peaks, *i.e.*, $Q_P^{-1} = Q^{-1}{}_{max} - Q^{-1}{}_{min}$, drops with increasing frequency (see Figure 1), but rises with the heating rate dT/dt; for example, at f = 0.5 Hz, Q_P^{-1} = 0.0125 for 2.5 °C/min, while Q_P^{-1} = 0.0143 for 6 °C/min. These features are perfectly in accordance with those of the first order phase transitions in solid internal friction (Delorme and Belko's Model: $Q_P^{-1} = A(dT/dt)/f$).

The internal friction characteristics in the eutectic Pb-Sn melt can be well repeated in other Pb-Sn melts with different compositions. However, peak position and magnitude change with Sn contents at the same heating rate and torsion frequency. The relevant data of different compositions for T_P at 2.5 °C/min and 0.5 Hz are shown in the Pb-Sn phase diagram (Figure 2). We assume that, at a lower temperature, there are residual orders, *i.e.*, the phases L' (rich in Pb) and L" (rich in Sn), which are metastable and prone to homogenization. As temperature elevated, the atoms within the residual orders gain higher and higher energy. Once the critical temperature is reached, the energy is high enough to overcome the energy barrier so that the Pb-Pb and Sn-Sn atomic bonds in original phases are broken continuously. At the same time, the new Pb-Sn atomic bonds build up, which makes relatively homogeneous liquid phase (L) nucleates and they grow. That is, the phase transition

L'(Pb) + L''(Sn)→L(Pb-Sn) takes place. Similar patterns of internal friction with temperature were also observed in other liquid alloys, e.g., In-Sn80, In-Bi62 shown in Figures 3 and 4.

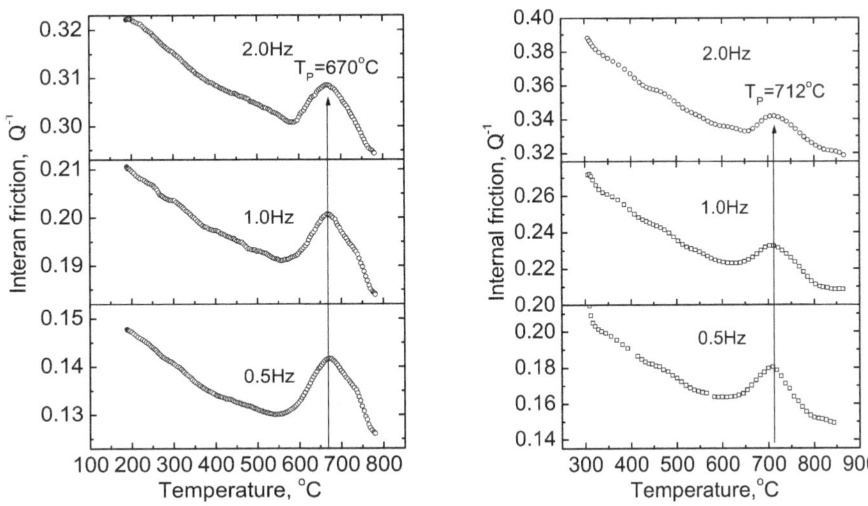

(**a**) $dT/dt = 2.5$ °C/min (**b**) $dT/dt = 6.0$ °C/min

Figure 1. Internal friction behavior of the melt Pb-Sn61.9.

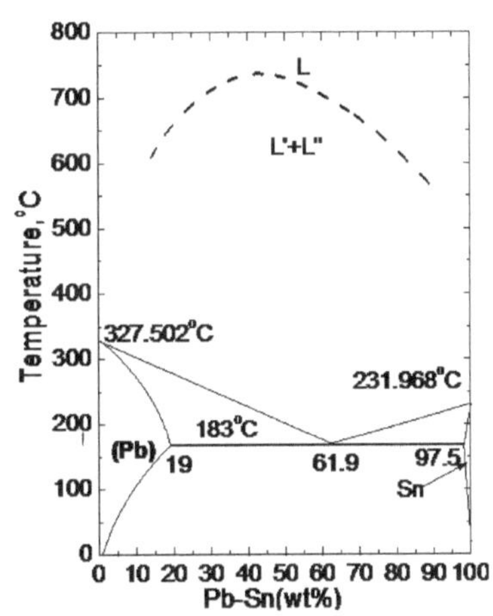

Figure 2. The temperature T_P of different compositions of Pb-Sn during heating process.

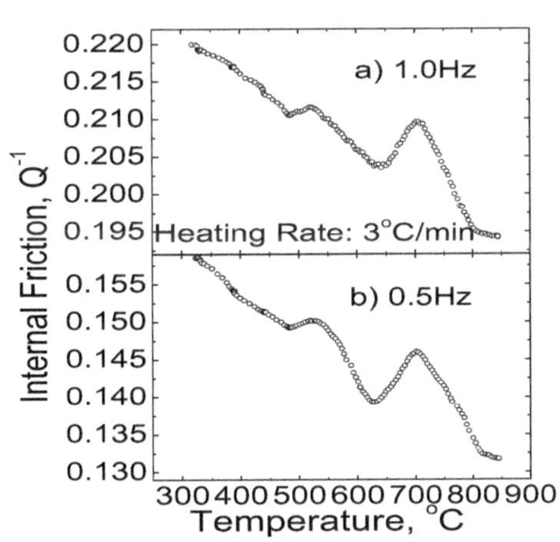

Figure 3. Q^{-1}-T behavior of the melt In-Sn80. (**a**) at frequency 1.0 Hz; (**b**) at frequency 0.5 Hz.

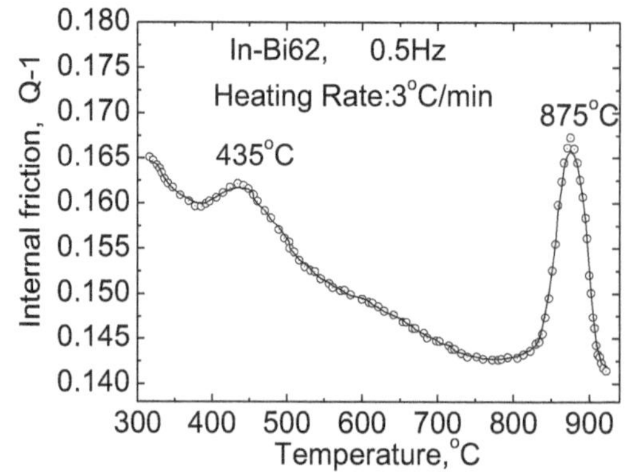

Figure 4. Q^{-1}-T behavior of the melt In-Bi62.

3. The Physical Characteristics of TI-LLST Confirmed by X-ray Diffraction

The results of internal friction experiments implied that TI-LLST occurred hundreds of degrees above T_L in some liquid alloys. However, it should be further confirmed if the results reflect a structure change and be found out what and how this happens during the process. To this goal, we performed X-ray diffraction measurements on In-Sn alloys. The reason for selecting In-Sn in the diffraction study is that the vapor pressure of both elements is quite low, so as not to stain the apparatus. The X-ray diffraction measurements were carried out using a θ-θ diffractometer with Mo-Kα-radiation; graphite monochromator; tantalum heater; corundum crucible; the sample chamber was evacuated beforehand to 10^{-6} mm Hg for inhibiting possibilities of contamination; and then filled with high pure Ar (4N) at 1.3 atm (see Refs [34,35] for detailed experiment procedures).

Based on the results of the diffraction experiment, through the use of correcting, normalization (Krogh-Moe-Norman Method) and inverse Fourier transform, we got the pair distribution functions g(r), shown in Figure 5, for the melt In-Sn80 at each temperature. Figure 6 shows the mean nearest neighbor distance (r_1) and the coordination number (N_1) derived from the pair distribution function g(r) at each temperature, respectively. For comparison, the internal friction pattern Q^{-1}-T of the melt In-Sn80 is also shown in the same figure, in which the prominent peak, around 700 °C, was focused for selecting temperatures in X-ray diffraction measurements. Corresponding to the prominent peak of the internal friction in the curve Q^{-1}-T, as can be seen in Figure 6, a remarkable valley appears abnormally around about 700 °C for both r_1 and N_1, demonstrating evidently a discontinuous configuration reconstruction in the melt, while r_1 and N_1 vary only slightly with temperature before and after the valley. It should be noted that the values of both r_1 and N_1 after the valley are not much different from those before it (Figure 6). We assume that this is because the atomic radii of indium and tin are quite similar (the difference is less than 6%). However, the liquid structures before and after the configuration change are unlike. Before the change, there are the residual covalent bonds of solid tin in the melt, *i.e.*, the short-range ordering of tetrahedrons still exists. We suppose, in the melt, the atoms of indium do not distribute uniformly with those of tin, whereas they aggregate to some extent in microscopic zones surrounded by the domains composed mainly of tin atoms, showing a kind of metastable local ordering. During the transition, the residual covalent bonds are broken around 700 °C, at the same time, new atomic bonds build up, with a relatively uniform melt forming. Since the melt is mainly composed of tin, after the reconstruction, a reference Sn atom is surrounded by the nearest shell made up of about five or six atoms of tin and one or two atoms of indium (N_1 = 7.66 at 800 °C).

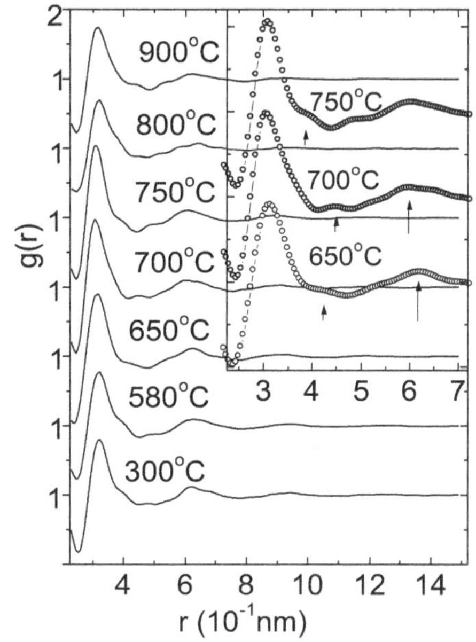

Figure 5. g(*r*) patterns of the melt In-Sn80 at each temperature.

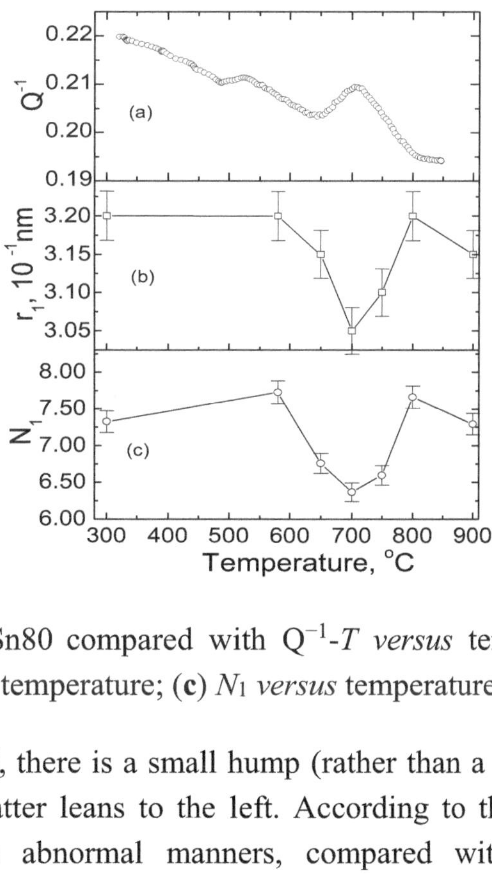

Figure 6. r_1, N_1 of In-Sn80 compared with Q^{-1}-*T* *versus* temperature. (**a**) Q^{-1} *versus* temperature; (**b**) r_1 *versus* temperature; (**c**) N_1 *versus* temperature.

In Figure 5, at about 700 °C, there is a small hump (rather than a shoulder) between the first peak and the second one, and the latter leans to the left. According to the physical meaning of the pair correlation function g(*r*), the abnormal manners, compared with those at lower and higher temperatures, demonstrate that some atoms move amid the nearest atom shell and the next one around the reference atom, for example, the atoms of tin outwards and those of indium inwards. Of course, the same move takes place between further shells.

Using a semi-empirical approach [36], we got the functions of both r_C (the averaged ordering domain radius) and N_C (the atom number within the domains) *versus* temperature for the melt In-Sn80, as illustrated in Figure 7. From 750 °C to 800 °C, r_C drops from 10.75 to 9.55, and N_C from 192 to 141, which indicate obviously that the ordering range becomes much smaller abruptly at the end of the structure change. In addition, Figure 7 also shows the parameter ζ (=r_C/r_1) whose pattern signifies that the ordering degree of the melt In-Sn80 decreases evidently along with the change.

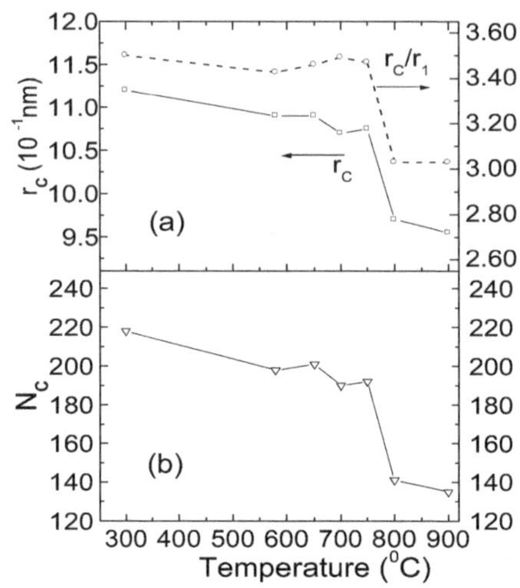

Figure 7. r_C, N_C and r_C/r_1 of In-Sn80 *versus* temperature. (**a**) r_C and r_C/r_1 *versus* temperature; (**b**) N_C *versus* temperature.

As is well known, r_1, N_1, r_C, N_C and r_C/r_1 are all important structural parameters for liquid matter, the peculiar behaviors confirm the occurrence of a discontinuous structural change in the melt In-Sn80. More importantly, they demonstrate explicitly the concrete physical meaning and characteristics of TI-LLST: during the liquid transition, through the adjustment of atomic bonds and the structural rearrangement, the size of short-range orders becomes smaller, and liquid structures are distinct from the original ones and reach a more homogeneous level. Sİmilar phenomenon was also revealed in In-Sn20 melt, see Ref. [35].

4. Electrical Resistivity Behaviors of Liquid Alloys *versus* Temperature

In our investigations, the electrical resistivity method was used to study liquid alloys, too, and proved to be valid and sensitive to explore liquid structural changes because the results were in good accordance with those of other methods, such as internal friction, X-ray diffraction, DTA, and so on. In Figure 9, the abnormal behavior on resistivity-temperature curve (ρ-T) is in good accordance with the prominent peak of Q^{-1} and the valley of r_1, N_1 in Figure 6, although there is no sign of agreement with the small pre-peak of Q^{-1}, due to the difference of the methods. The detailed method of the electrical resistivity was demonstrated in Refs. [37–41].

The resistivity experiment results display abundant physical images of liquid matter in response to temperature. For examples, Figures 8–10 show some resistivity behaviors of several melts during heating and cooling processes, exhibiting different ρ-T patterns of liquid state and characteristics of TI-LLST. In some liquid alloys, two or more TI-LLST happen during heating process, e.g., Cu-Sn30 melt shown in Figure 8, while in some others only one TI-LLST happens, e.g., BiSb10 and Sn-Sb5 melts in Figures 10 and 11.

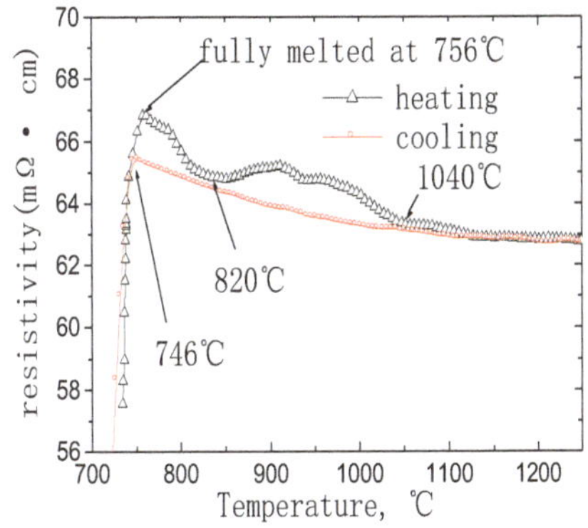

Figure 8. ρ-*T* curves of Cu-Sn30.

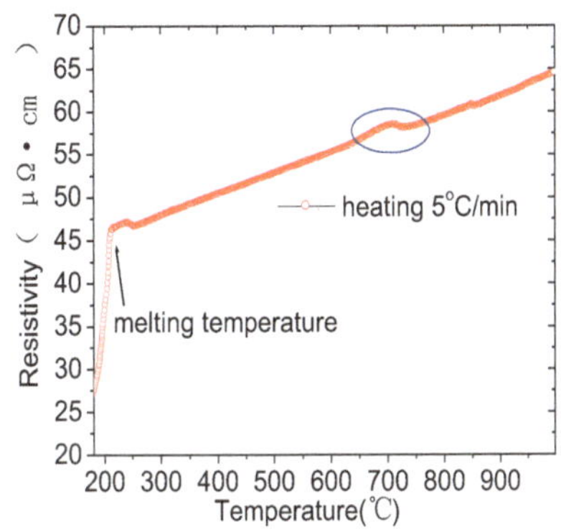

Figure 9. ρ-*T* curves of In-Sn80.

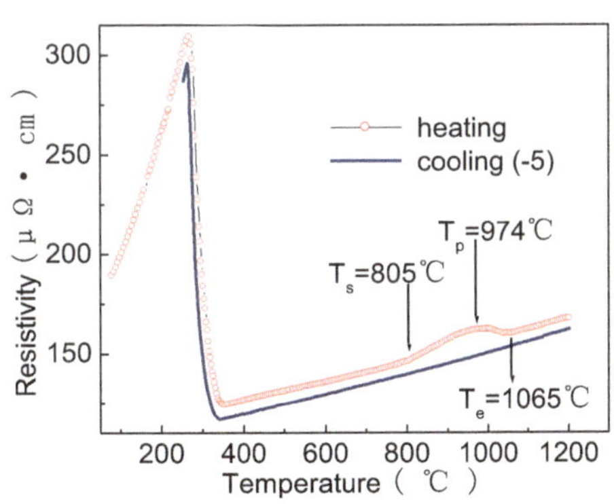

Figure 10. ρ-*T* curves of BiSb10.

Figure 12 shows the isothermal experiments of Pb-Sn61.9 holding at 888 K with the methods of resistivity and DTA (differential thermal analysis) [40]. During TI-LLST of Pb-Sn61.9, resistivity of the melt rises with a long "S" pattern (Figure 12). However, there are other resistivity patterns displaying TI-LLST during the heating process, such as a peak on ρ-*T* curve (Figures 9 and 10), a "Z" pattern (Figure 11), or a more complex pattern (Figure 8). In agreement with the ρ-*T* curve of Pb-Sn61.9 melt, there is an endothermic peak on the DTA curve, which suggests the entropy-driven mechanism and the discontinuous characteristic of TI-LLST.

Because of the sensitivity and the relative convenience of the resistivity method, up to now, the ρ-*T* behaviors of over one hundred liquid systems (alloys with different compositions or a single component) have been explored, whose results are listed in the table "Summary of TI-LLST suggested by the resistivity phenomena" as a Supplementary Material.

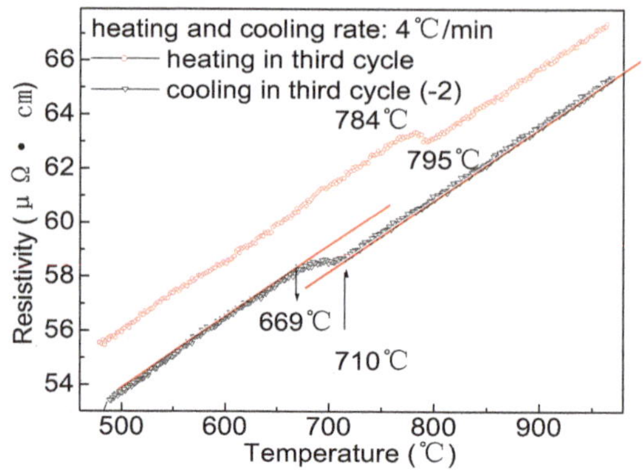

Figure 11. ρ-*T* curves of Sn-Sb5wt% melt.

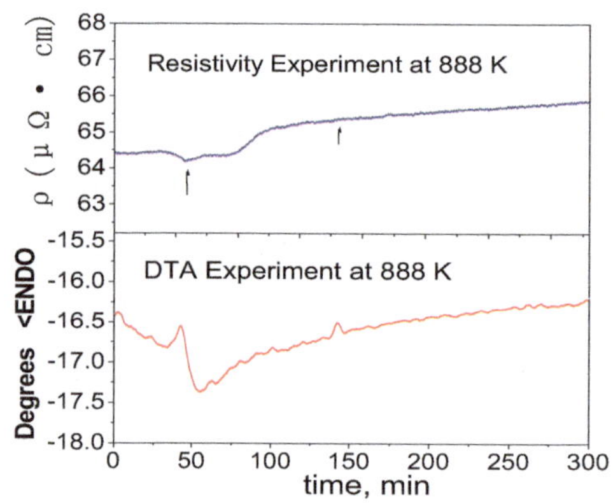

Figure 12. Isothermal experiments holding Pb-Sn61.9.

5. Primary Discussions about Temperature-Induced Liquid-Liquid Transition

Based on the abovementioned content, and the table "Summary of TI-LLST suggested by the resistivity phenomena", some aspects of TI-LLST are summarized as following.

5.1. If and How the Temperature-Induced Liquid-Liquid Transition Happens?

Because of the easy oxidation tendency of liquid metals and the various technical difficulties at high temperature, in our experiments, the temperature scope for the internal friction method (technically being only valid in heating process for liquids at present) is lower than 900 °C, and that of the resistivity method is around 1200 °C (recently ameliorated up to 1400 °C or so) at most. So the alloys or pure components in our explorations are primarily limited in those whose *liquidus* or melting temperature and the oxidation tendency are as low as possible, specifically; chiefly limited in part of the IIIA-VIA elements (No. 49–52: In, Sn, Sb, Te and No. 82–83: Pb, Bi) and part of IB-IIB elements (No. 29–30: Cu, Zn and No. 47: Ag), and their alloys.

As can be seen in the Supplementary Materials, for all seven investigated pure elements, there is no sign of TI-LLST in liquid In, Pb, Te and Zn during the heating and cooling processes, while TI-LLST is suggested by repeated experiment results to occur in liquid Sn, Sb and Bi. Intriguingly, for the candidates of LLST in the supercooled liquid state, Poole *et al.* [1] prophesied that liquids with a locally open tetrahedral molecular structure are the most promising, such as Si, Ge, C, SiO_2, GeO_2, and H_2O. It has been confirmed with a marked shoulder at the right side of the first peak of g(r) and S(Q) that, in liquid Sn, Sb and Bi, there are residual covalent bonds of the solid with short-range ordering of tetrahedrons. The observations of TI-LLST here in liquid Sn, Sb and Bi widen Poole's theoretical prediction (liquids with open tetrahedral structure being the promising candidates of LLST) from the supercooled state to a much higher temperature scope, well above T_m. Another character of liquid Sn, Sb and Bi is that their ratio of Q_2/Q_1, Q_2 and Q_1 denoting the structure factor's position of second and first peaks, is much higher than that of the hard-sphere structure ($Q_2/Q_1 = 1.86$), being 1.96, 1.96 and 1.9, respectively [36].

For the investigated liquid alloy systems consisting of the abovementioned seven elements, except for Ag, Cu, Al, Si and Zr, phenomena of TI-LLST are observed in all the ternary liquid systems and most of the binary liquid systems, at least during the first heating process, while there is no sign of TI-LLST in some of the binary liquid systems. The exceptions include some Pb-Te melts with Te content higher than 75 weight percent (wt%), some Sn-Sb melts with Sb higher than 30 wt%, most of the In-Sb melts with Sb lower than 90 wt% and all the binary melts containing Zn. From the data accumulated to date, it is hard to conclude in what kinds of binary-component liquid systems TI-LLST can or cannot happen. We have attempted to understand the possibility of TI-LLST in binary melts from the viewpoint of the mixing enthalpies (ΔH_{mix}) indicating the inter-atomic force of unlike elements to some extent, but have come to no conclusion. We have also tried to understand from the kinds of constituent element in binary melts, but the only element that could be confirmed is Zn, with which alloys there is no TI-LLST. Indeed, Zn is a special metal element in respect of its liquid structure, whose first peak of structure factor S(Q) is quite asymmetrical, leaning towards the right, and is almost independent of temperature. Furthermore, the ratio of Q_2/Q_1 for Zn is the smallest (1.75) [36] in all the liquid metals, being considerably below the value of the hard-sphere structure ($Q_2/Q_1 = 1.86$).

The other fascinating fact is that, for the liquid systems in which TI-LLST is suggested to occur during the first heating process, for some melts there is no abnormality on the ρ-T curve during the cooling process, while for some others anomalies do occur during the first cooling process and even in further heating and cooling processes (in the second and third running). For the former cases, the TI-LLST is an irreversible liquid change, as shown in Figures 13a–c and 10. For the latter cases, the situations are relatively complicated. During the first cooling process, the abnormality of ρ-T is evidently corresponding to that or one of those during the first heating process for some melts. As can be seen in Figure 13d, for the melt Sn97Ag2.5Cu0.5, the second liquid change is reversible, while the first liquid change is irreversible. It should be noted that in a majority of the latter cases, it is hard to identify if the ρ-T abnormality on the first cooling marks the reversed process to that of the first heating, because the temperature difference between them is quite large, and the patterns of them are sometimes also distinct, as shown in Figure 13e,f. Intriguingly, as long as there is a sign of TI-LLST on the first cooling, the reversibility of TI-LLST will show quite well in the second and third running, see also Figure 13e,f.

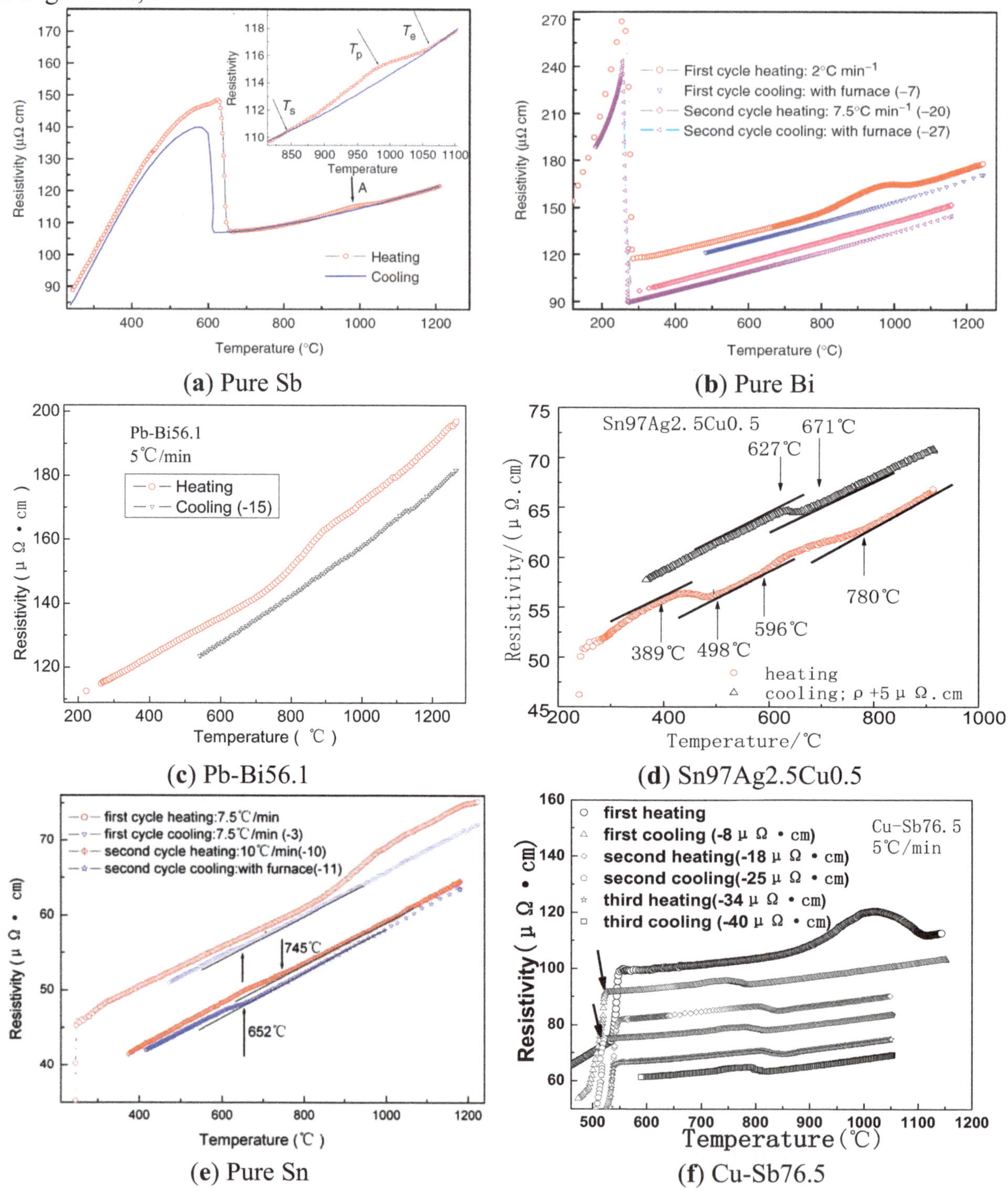

Figure 13. Is the TI-LLST reversible? Different behaviors of ρ-T in the first and further heating and cooling cycles.

5.2. What Happens During the Temperature-Induced Liquid-Liquid Transition?

With regard to what happens for TI-LLST during the first heating process, generally, we assume that it is a transition from an inhomogeneous liquid state to a more uniformed liquid state.

We think that there are topological short-range orders for pure liquids and the molten alloys with the positive or zero excess enthalpy, for example, different minor domains (rich in Pb and rich in Sn, as mentioned before) coexist heterogeneously in molten Pb-Sn alloys when temperature is not quite high. On the other hand, both the chemical and topological short-range orders exist in the liquid alloys with the negative excess enthalpy, for instance, both chemical short-range domains with the micro-structure similar to Ag_3Sn and the residual short-range ordering of tetrahedron of solid Sn coexist in molten AgSn alloys. Nevertheless, these minor domains dissipate and engender always with time and space as the fluctuation of energy, but they do not vanish with their statistical equilibrium structures, sizes and constituents, and change continuously with temperature elevated. However, as long as temperature reaches the critical scope, the inter-atomic bonds in original domains (the topological and/or chemical short-range orders) are broken, and the old domains (or minor phases) are reduced little by little, at the same time, new domains form with a relatively more uniformed liquid structure state.

The critical temperature range of TI-LLST depends on how strong the atomic bonds in the original domains are because the microstructures and the kinds of short-range orders are diversified in different pure liquids and molten alloys. For example, with the experimental results of scattering techniques, it was proposed that there are domains of both spherical close packing and layer lattice in the melts of Au, Ag, Pb, Tl and the alkali metals [27]. For liquid Sn, Bi, Sb, Ga and Si, shoulders appear at the right side of the first peak on the scattering curves after fusion, which is attributed to the short range ordering of tetrahedrons with the residual covalent bonds of solids in the liquids. It has been proved that there are regular icosahedrons and other clusters in liquid Pb [42] and many other simple liquids. Of course, the inter-atomic bonds in chemical short-range orders are also various, depending on liquid alloy systems and compositions. Generally, the stronger the inter-atomic bonds in old short-range orders, the higher the temperature range of TI-LLST. If there are species of short range orders with variety of bonding strength, which are able to change to others by TI-LLST, more than one liquid-liquid transitions will occur in different temperature ranges during the heating process, see the ρ-T patterns in Figures 8 and 13d. Typically, when the chemical short-range orders with strong bonds (*i.e.*, for the alloy with a large value of negative excess enthalpy) exist in the old liquid state, the temperature range of TI-LLST will be high. For example, for Cu-Zr alloy $\Delta H_{mix} = -23$ kJ/mol, the TI-LLSTs of the molten alloy Cu50Zr50 ($T_L = 935$ °C) happen around 1230 °C and 1360 °C, with two turning zones, see the table in Supplementary Materials.

As for the TI-LLST suggested by the anomalies of resistivity and other resorts during the first cooling process and even in further heating and cooling processes, based on their characteristics we assume that they could be divided into two kinds of TI-LLSTs. One is the reversed liquid-liquid transition corresponding to that during the first heating process. For instance, in liquid tin and some of its molten alloys, the domains of topological short-range ordering of tetrahedron of Sn dissipate during heating and recover during cooling. Similar cases can also happen for some chemical short-range orders in molten alloys with negative excess enthalpy. Another one is just such a kind of TI-LLST that is quite a distinct one from the TI-LLST during the first heating process because the newly generated short-range orders could not be recovered to those before the TI-LLST during the first heating. In this kind of TI-LLST, during the first cooling process and the further thermal runs, the micro-structures can be reversely transformed between different liquid states, both of which is dissimilar to that before the TI-LLST during the first heating.

5.3. Kinetics and Thermodynamics of Temperature-Induced Liquid-Liquid Transition

The investigation of kinetics and thermodynamics on TI-LLST reveal that the TI-LLST is accordance with kinetics characteristic and can be ranged into entropy driven type.

In terms of the ordering degree of a system, it is thermodynamically expressed by entropy. The pair correlation entropy S_2 (or called two-body excess entropy), defined by $S_2 = -2\pi\rho\int_0^\infty [g(r)\ln g(r) - h(r)]r^2 dr$ for amorphous systems, contributes at least 85% of the excess entropy S^E for liquids [43]. S^E signifies the difference between the thermodynamic entropy S^R of a real system and that of the equivalent ideal gas S^{ID}, *i.e.*, $S^E = S^R - S^{ID}$. Because entropy is always positive and the entropy of ideal gas S^{ID} larger than that of a real liquid S^R, the excess entropy S^E as well as the pair correlation entropy S_2 is negative forever for liquids. The smaller the absolute value of S_2 is, the less the ordering degree is in liquids. For convenience, we define the integration function of S_2 as $S(r) = -2\pi\rho[g(r)\ln g(r) - h(r)]r^2$. Here, $h(r) = g(r) - 1$, and is called the total pair distribution function, representing deviations from randomness, and ρ is the number density. It is reasonable to infer that the function $S(r)$ can give an intrinsic picture of the ordering degree. Figure 14 illustrates the patterns of the function $S(r)$ curves for the melt In-Sn20 as well as their integration results, the pair correlation entropy S_2, at different temperatures.

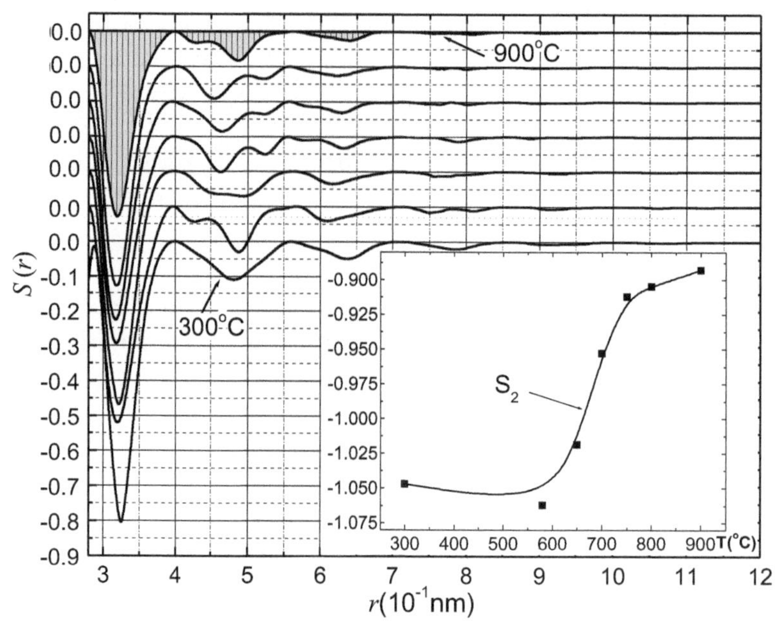

Figure 14. Patterns of the function $S(r)$ curves for the melt In-Sn20 as well as the pair correlation entropy S_2 (Inset) at different temperatures [35].

It is interesting to note that the function $S(r)$ is better than $g(r)$ in revealing the short-range ordering degree of liquid. From Figure 14, we clearly see that the magnitudes of the first, second, third and fourth inversed peaks of $S(r)$ decline quickly from 580 °C to 750 °C. Up to high temperatures (including 750 °C, 800 °C, 900 °C), the fourth inversed peak blurs out nearly completely, indicating the size of local short-range ordering reduces abruptly. Meanwhile, even more clearly, the absolute value of S_2, *i.e.*, the dark area surrounded by the curve $S(r)$, becomes smaller very fast from 580 °C to 750 °C (by about 15% between two liquid states), indicating a discontinuous drop of the ordering degree in the melt In-Sn20 during the configuration reconstruction. From the standpoint of thermodynamics, the steep change of entropy manifests the discontinuous characteristic of the L-L structure change. The entropy driven type characteristic of TI-LLST has also been verified by the thermal effect on DSC (differential scanning calorimetry) and DTA curves for lots of liquid systems, for instance, illustrated in Figures 12 and 15.

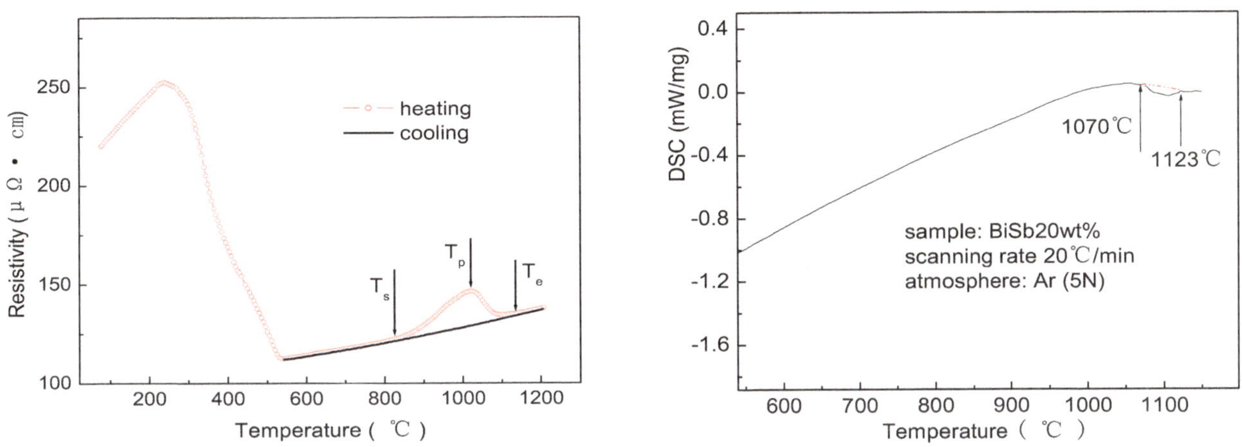

(**a**) Resistivity-temperature pattern (5 °C/min) (**b**) DSC analysis in heating process (20 °C/min)

Figure 15. Results of resistivity and DSC analysis for BiSb20 wt% alloy [41].

The kinetic behavior in TI-LLST process of those alloys has been investigated through continuous heating experiments and isothermal experiments. Using kinetic theory of phase transition, we found that the kinetic transition process is nucleation-growth type, and the nucleation rate is the dominant controlling factor of the TI-LLST process. There are two different kinds of transition modes for TI-LLST: Johnson-Mehl-Avrami transition mode and self-catalysis transition mode, whose microcosmic mechanism of the transition process has been discussed in Ref. [40]. Two characteristics of TI-LLST should be stressed here. At first, the transition speed of TI-LLST is much slower than that of normal solid state phase transitions, which can be seen from the large temperature width of TI-LLST, particularly for TI-LLST during the first heating process, normally between 100–200 °C or even wider at moderate speeds. Secondly, because both phases are liquid, the changes in energy, entropy, and density, *etc.* across the transition will be much smaller than those for the solid-solid, liquid-solid and liquid-gas transitions, which was also pointed out for liquid-liquid phase transitions in the supercooled liquid [1]. So it is not strange that the difference of S_2 before and after TI-LLST in Figure 14 is only 0.15, and that the thermal effect shown in Figure 15b is much less prominent than that of normal phase transitions. Based on the results of heating experiments, the apparent activation energy E_k of TI-LLST has been calculated using Kissinger equation for some liquid alloys [40,44]. E_k is the energy needed to overcome the energy barrier of the structure transition, and could reflect the difficulty level of TI-LLST of different alloy melts, for example InSn80 (1.60 ev) > InPb80 (1.45 ev) > PbSn61.9 (1.22 ev). Furthermore, compared with normal solid-state phase transition, the E_k value of TI-LLST is much smaller.

Based on the results of electronic transport properties and the calculated mean nearest neighbor atomic distance, the dependence of the density of electronic states $N(E_F)$ and its gradient value d$N(E_F)$/dE on temperature were deduced with the help of Faber-Ziman theory, reflecting that the electronic structure of the melt at Fermi level also changes during the transition, e.g., illustrated in Figure 16 for In-Sn20 melt.

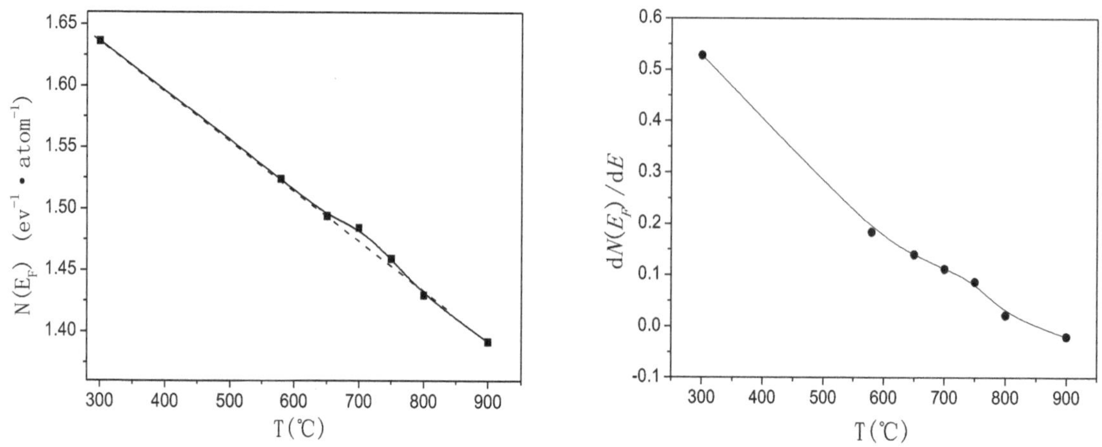

(**a**) $N(E_F)$-T of liquid InSn20 (**b**) d$N(E_F)$/dE of liquid InSn20

Figure 16. The density of state $N(E_F)$ and The d$N(E_F)$/dE of liquid InSn20 *versus* temperature [45].

Restricted to the paper space, the other specific characteristics of TI-LLST could not be illustrated here in detail; for further information please refers to Refs. [31–35,44–48].

It should be pointed out that, although the abovementioned results of TI-LLST are limited in the investigations of ourselves, the similar phenomena have been observed by other groups with different resorts like viscosity [49,50], density and sound velocity methods [51,52], as well as computer simulations [53,54]. For example, by high-resolution measurements of the density, the heat capacity and the static structure factor, scientists out of Israel, USA and France found that a temperature-driven liquid-liquid structural transformation takes place quite above *liquidus* of Bi [55]. Also, with high precision measurements, Greenberg *et al.* investigated the sound velocity-temperature behaviors in pure liquid tin and the results showed that at approximately 500 °C and 850 °C the data deviates by three times the experimental error from the linear fit [56]. By neutron diffraction and small-angle neutron scattering, a collaborative study on molten Al-Si alloys were carried out by University of Rouen (France), Ames Laboratory and Oak Ridge National Laboratory, and the results unambiguously show that for the eutectic Al-Si melt, the microstructure changes with increasing temperature in a partly reversible way, while for the hypereutectic (Al-Si20at%) melt, the change is almost completely irreversible [57]. A similar case was also found in a liquid Bi–Ga System [58]. Also, with the temperature dependence of the density, the viscosity and other properties of the Pd-18 at.% Si melt measured, Sivkov (Russia) *et al.* observed the existence of micro heterogeneities in the melt and a transformation of its microstructure at 1380-1430 K during a temperature cycle of heating and cooling [59]. In China, with experimental procedures like X-ray diffraction, internal friction, viscosity, resistivity, *etc.* as well as theoretical simulation methods, several other groups in Institute of Solid State Physics of Chinese Academy of Sciences [54,60,61], Shandong University [49,62,63], Jinan University [64] and Fudan University [53] observed lots of TI-LLST phenomena in liquid metals and alloys.

6. The Effects of TI-LLST on Solidification Behaviors and Structures

Based on the new physical phenomenon of TI-LLST, in recent years, the effects on solidification behaviors and structures resulted from different liquid structure states before and after TI-LLST have been explored. By comparison, for instance, two different melt treating procedures were used for carrying out solidification experiments, *i.e.*, the samples were melted at tempratures below and above the range of TI-LLST and held for one hour or longer, respectively. For the former, the melted samples were taken out of the furnace directly and cooled with the crucible in air (procedure A). For the latter, the melted samples were rapidly cooled to the same temperature as procedure A and held for 30 min, then taken out of the furnace and cooled with the crucible in air, too (procedure B), thus to obtain the same cooling condition as the procedure A. All samples were prepared from high pure elements (4~5 N) and shielded by flux, whose cooling curves *T-t* (Temperature-time) of solidification were automatically recorded by a system with the error less than 0.1 °C. The effects of TI-LLST on solidification behaviors and structures are briefly summarized as follows.

Illustratively, the results of Bi-Sb 10 alloy are shown in Figures 17–19.

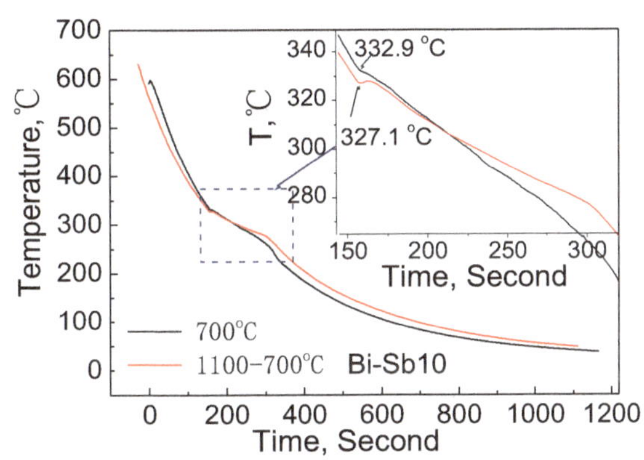

Figure 17. Bi-Sb10 T-t curves of different melts.

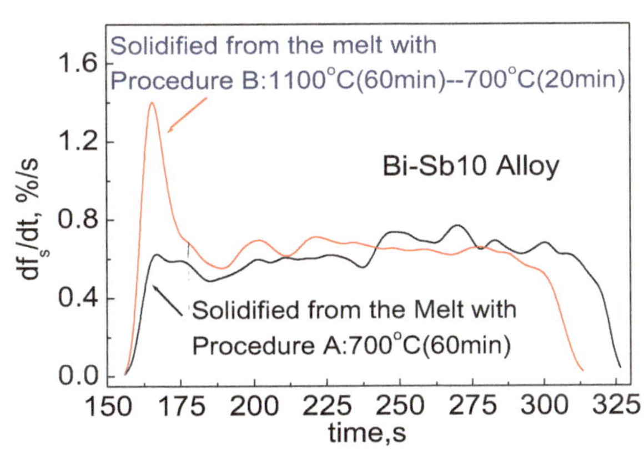

Figure 18. Solidification rate obtained by NTA method.

At first, it is evident that, for the liquid alloys in which TI-LLST occurs and is not reversible, the under-cooling of the samples treated with procedure B is larger than that of those with procedure A, the normal enlargement of the under-cooling being about 5–20 °C. As shown in Figure 17 for Bi-Sb10 alloy [65], as the T_L of Bi-Sb10 is 346 °C, the nucleation under-cooling increases from 13.1 °C of the procedure A to 18.9 °C of the procedure B. Figure 18 shows the solidification rate curves of Bi-Sb10, d*f*s/d*t*-*t*, obtained by Newton Thermal Analysis (NTA) method [66], here dfs/dt indicates the solid fraction *f*s increased in unit time. According to Kurz' theory [67], the altitude of the first peak on d*f*s/d*t*-*t* curve betokens the level of nucleation rate, the apparently enhanced peak of d*f*s/d*t*-*t* for the sample with procedure B signifies a quite higher nucleation rate. The enlarged under-cooling brings about the higher nucleation rates and evident refined solidification structures, which can be seen in Figure 19.

Similar effects of refined solidification structures were also observed in dozens of different alloys, such as Babbitt alloys (Sn-Cu-Sb), piston alloys (hyper-eutectic Al-Si), Pb-free solder alloys, thermoelectric alloys (Bi-Te/Sb), *etc*. As shown in Figure 20, for example, in hyper-eutectic Al-Si20 wt% alloy, which can be used for pistons of engines and other wear-resisting parts, the crystals of primary silicon are much refined with a more uniformed distribution with the procedure B process.

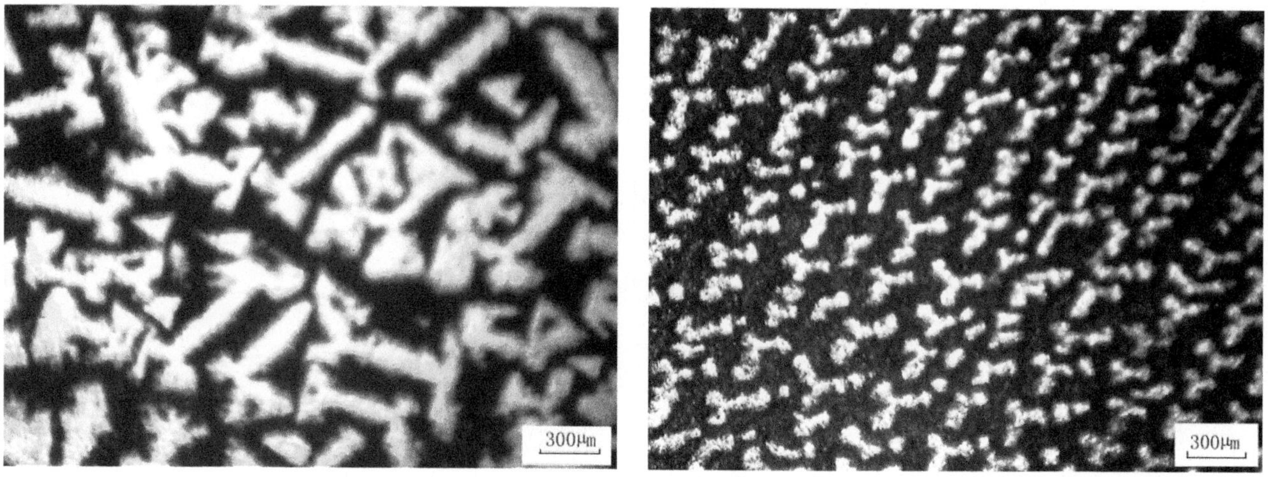

(**a**) procedure A: 700 °C for 60 min (**b**) procedure B: 1100 °C for 60 min + 700 °C for 30 min

Figure 19. Solidified structures of the Bi-Sb10 (wt %) alloy from different liquid states.

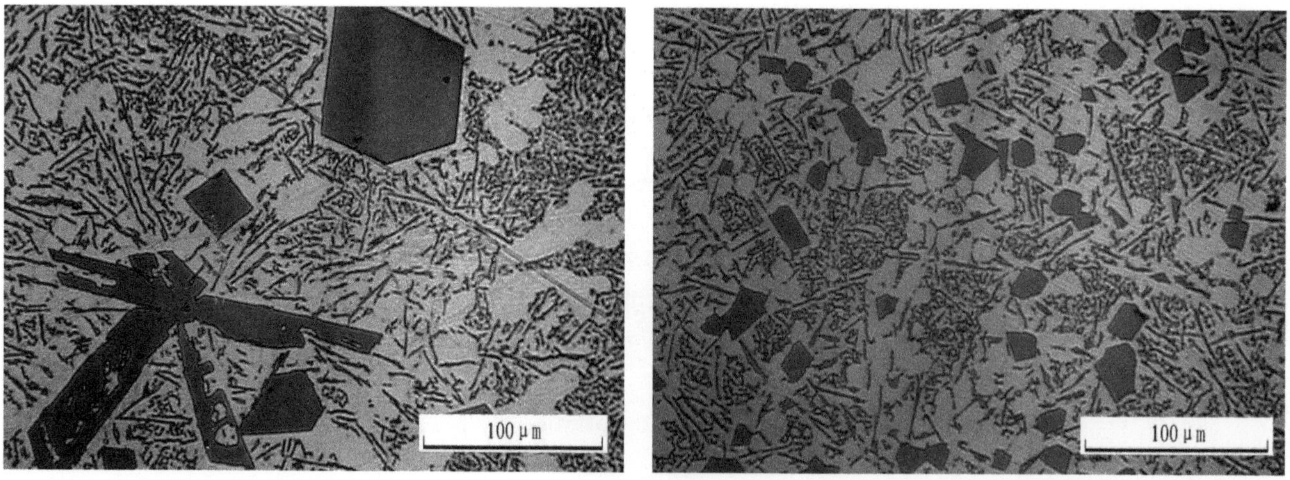

(**a**) procedure A: 760 °C for 90 min (**b**) procedure B: mixing the melt at 1150 °C with the melt at 660 °C (1:1), and then hold in the furnace at 760 °C for 5 min

Figure 20. Solidified structures of the Al-Si20 (wt%) alloy from different liquid states.

Because of the changes of solidified structures of alloys with TI-LLST, some of the engineering properties can be evidently improved. For instance, the wettability (Figure 21) and the shear strength (Figure 22) of the Pb-free solder Sn-3.5Ag-*x*Bi/Cu joint are ameliorated.

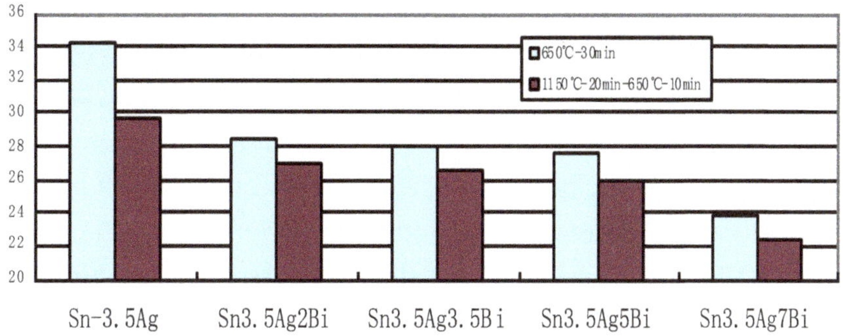

Figure 21. The effect of melt state on the wettability (wetting angle θ°) of Pb-free solder Sn-3.5Ag-*x*Bi/Cu joint [68].

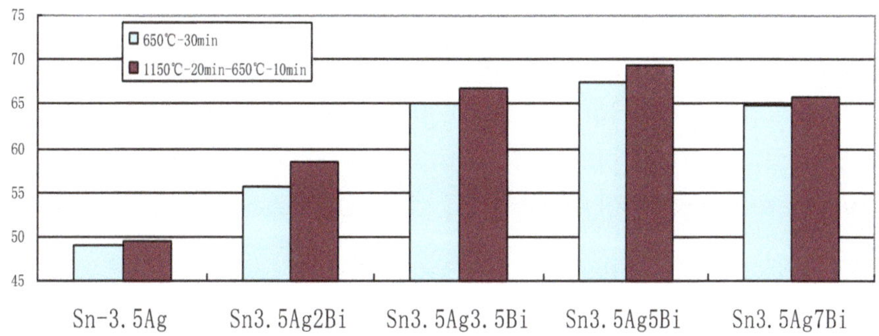

Figure 22. The effect of melt state on the shear strength (MPa) of Sn-3.5Ag-*x*Bi/Cu joint [68].

On the other hand, the effective partition coefficient $K_E(C_S/C_L)$ becomes smaller because the melt experienced TI-LLST, thus leading to changes of the solute redistribution process during solidification and the morphology of the solidified structures.

It should be pointed out that the liquid thermal history has no influence on solidification because it results in no different liquid state if a liquid system without TI-LLST, *i.e.*, for the alloy Sn-Sb42 without TI-LLST detected, no apparent difference of solidified structures can be observed between lower and higher melt processing temperature. Even though liquid structure transition could occur, it is not always effectual to change solidification behaviors by melt thermal resorts if TI-LLST is reversible in the system. Whether it is effective to control solidification depends on how to manipulate the melt thermal history for curbing the structural reverse process, too.

7. Summary

1. As opposed to the conventional view of liquids that liquid structures and properties change gradually with temperature, with different resorts, temperature induced discontinuous liquid structure transitions were observed to occur in some alloy systems and pure elements.
2. Results show that the averaged ordering domain radius (r_C) and the atom number within the domains (N_C) drop abruptly during TI-LLST, and the ordering degree parameter $\zeta(=r_C/r_1)$ decreases evidently

along with the liquid change. Analyses of the results suggest the entropy-driven mechanism for TI-LLST. The temperature dependence of density of electronic states $N(E_F)$ and its gradient value $dN(E_F)/dE$ reflects that the electronic structure of the melt at Fermi level also changes during TI-LLST.

3. The kinetic process of TI-LLST belongs to nucleation-growth type, and the nucleation rate is the dominant controlling factor of the TI-LLST process. There are two different kinds of transition modes for TI-LLST: Johnson-Mehl-Avrami transition mode and self-catalysis transition mode. Through the adjustment of atomic bonds, the size of short-range orders becomes smaller, and liquid structures are distinct from the original ones and reach a more homogeneous level.
4. If an irreversible TI-LLST occurs, it brings about different liquid states and results in enlarged undercooling and an altered solute redistribution process during solidification, which gives rise to refined grains and improved configurations. However, in a system without TI-LLST, the liquid thermal history has no influence on solidification owing to no difference in liquid state. Furthermore, if TI-LLST is reversible, then effects of the melt manipulation are dependent on whether the reverse process could be prevented after TI-LLST.
5. Because of the present technical and theoretical limits, and the restricted scope of liquids already investigated to date, there is a long way to go to understand the nature of TI-LLST. However, the already disclosed abundant phenomena of TI-LLST open a new window for experimentalists and theoreticians to recognize polymorphism in liquid states, to explore it further and to find their way to its applications.

Acknowledgments

The author is indebted to all the colleagues and students who partly contributed to the relevant contents in this paper, and to Zhu Z.G. and Jiang J.Z. for their support and valuable discussions. The works have been financially supported by the National Natural Science Foundation of China Grant No.50571033 and No.50371024, the National Basic Research Program of China (973 Program) (No.2012CB825702), the key project of Chinese Ministry of Education No.104106, the Natural Science Foundation of Anhui Province No.070414178, and the Grant Project of Hefei University of Technology RC-2002-01.

Author Contributions

The author is the principal researcher and the director of all the above listed projects except for No.2012CB825702, and the writer of this paper.

Conflicts of Interest

The author declares no conflict of interest.

References

1. Poole, P.H.; Grande, T.; Angell, C.A.; McMillan, P.F. Polymorphic phase transitions in liquids and glasses. *Science* **1997**, *275*, 322–323.

2. Rao, K.R. Phase transitions in liquids. *Curr. Sci.* **2001**, *80*, 1098–1100.
3. Yarger, J.L.; Wolf, G.H. Polymorphism in Liquids. *Science* **2004**, *306*, 220–221.
4. Kurita, R.; Tanaka, H. Critical-like phenomena associated with liquid-liquid transition in a molecular liquid. *Science* **2004**, *306*, 845–848.
5. White, J.A. Multiple critical points for square-well potential with repulsive shoulder. *Physica A* **2005**, *346*, 347–371.
6. McMillan, P. Jumping between liquid states. *Nature* **2000**, *403*, 151–152.
7. Saika-Voivod, I.; Sciortino, F.; Chester, P.H. Computer simulations of liquid silica: Equation of state of liquid-liquid transition. *Phys. Rev. E* **2001**, *63*, 011202.
8. Lacks, D.J. First-order amorphous-amorphous transformation in silica. *Phys. Rev. Lett.* **2000**, *84*, 2881–2884.
9. Mishima, O.; Stanley, H.E. Decompresion-induced melting of ice IV and the liquid-liquid transition in water. *Nature* **1998**, *392*, 164–168.
10. Koga, K.; Tanaka, H.; Zeng, X.C. First-order transition in confined water between high-density liquid and low-density amorphous phases. *Nature* **2000**, *408*, 564–567.
11. Soper, A.K.; Ricci, M.A. Structures of high-density and low-density water. *Phys. Rev. Lett.* **2000**, *84*, 2881–2884.
12. Glosli, J.N.; Ree, F.H. Liquid-liquid phase transition in carbon. *Phys. Rev. Lett.* **1999**, *82*, 4659–4662.
13. Katayama, Y.; Mizutani, T.; Utsumi, W.; Shimomura, O.; Yamakata, M.; Funakoshi, K. A first-order liquid-liquid phase transition in phosphorous. *Nature* **2000**, *403*, 170–172.
14. Aasland, S.; MaMillan, P.F. Density-driven liquid-liquid phase separation in the system Al_2O_3-Y_2O_3. *Nature* **1994**, *369*, 633–636.
15. Coulet, M.-V.; Bergman, C.; Bellissent, R.; Bichara, C. Local order and phase separation in sulphur–tellurium melts: A neutron scattering study. *J. Non-Crystal. Solids* **1999**, *250–252*, 463–467.
16. Tsuchiya, Y.; Kakinuma, F.; Bergman, C. Structural changes and concentration fluctuations in the liquid Se-Te system. *J. Non-Crystal. Solids* **1996**, *205–207*, 143–146.
17. Kaban, I.; Halm, T.; Honyer, W. Structure of molten copper-germanium alloys. *J. Non-Crystal. Solids* **2001**, *288*, 96–102.
18. Rudolph, P.; Schaefer, N.; Fukuda, T. Crystal growth of ZnSe from the melt. *Mater. Sci. Eng. R Rep.* **1995**, *15*, 85–133.
19. Wang, J.F.; Omino, A.; Isshiki, M. Bridgman growth of twin-free ZnSe single crystals. *Mater. Sci. Eng. B* **2001**, *83*, 185–191.
20. Dahlborg, U.; Calvo-Dahlborg, M. Influence of the production conditions on the structure and the microstructure of metallic glasses studied by neutron scattering techniques. *Mater. Sci. Eng. A* **2000**, *283*, 153–163.
21. Popel, P.S.; Sidorov, V.E. Microheterogeneity of liquid metallic solutions and its influence on the structure and properties of rapidly quenched alloys. *Mater. Sci. Eng. A* **1997**, *226–228*, 237–244.
22. Enisz, M.; Kristof-Mako, E.; Oravetz, D. Phase transformation in doped Y-Ba-Cu-O superconductors obtained by different melt processing techniques. *J. Eur. Ceram. Soc.* **2007**, *27*, 1105–1111.
23. Koh, H.J.; Rudolph, P.; Schaefer, N.; Umestsu, K.; Fukuda, T. The effect of various thermal treatments on supercooling of Pb-Te melts. *Mater. Sci. Eng. B* **1995**, *34*, 199–203.

24. Li, P.; Nikitin, V.I.; Kandalova, E.G.; Nikitin, K.V. Effect of melt overheating, cooling and solidification rates on Al–16wt.%Si alloy structure. *Mater. Sci. Eng. A* **2002**, *332*, 371–374.
25. Sidorov, V.; Popel, P.; Calvo-Dahlborg, M.; Dahlborg, U.; Manov, V. Heat treatment of iron based melts before quenching. *Mater. Sci. Eng. A* **2001**, *304–306*, 480–486.
26. Wang, J.; He, S.H.; Sun, B.D.; Guo, Q.X.; Nishio, M. Grain refinement of Al–Si alloy (A356) by melt thermal treatment. *J. Mater. Process. Technol.* **2003**, *141*, 29–34.
27. Ubbelohde, A.R. *The Molten State of Matter: Melting and Crystal Structure*; Wiley: New York, NY, USA, 1978.
28. Franzese, G.; Malescio, G.; Skibinsky, A.; Buldyrev, S.V.; Stanley, H.E. Generic mechanism for generating a liquid-liquid phase transition. *Nature* **2001**, *409*, 692–695.
29. Tanaka, H. General view of a liquid-liquid phase transition. *Phys. Rev. E* **2000**, *62*, 6968–6976.
30. Nowic, A.S.; Berry, B.S. *Anelastic Relaxation in Crystalline Solids*; Academic Press: New York, NY, USA, 1972.
31. Zhu, Z.G.; Zu, F.Q.; Guo, L.J.; Zhang, B. Internal friction method: suitable also for structural changes of liquids. *Mater. Sci. Eng. A* **2004**, *370*, 427–430.
32. Zu, F.Q.; Zhu, Z.G.; Guo, L.J.; Zhang, B.; Shui, J.P.; Liu, C.S. Liquid-liquid transition in Pb-Sn melts. *Phys. Rev. B* **2001**, *64*, 180203.
33. Zu, F.Q.; Zhu, Z.G.; Zhang, B.; Feng, Y.; Shui, J.P. Post-melting anomaly of Pb-Bi melts observed by internal friction technique. *J. Phys.* **2001**, *13*, 11435–11441.
34. Zu, F.Q.; Zhu, Z.G.; Guo, L.J.; Qin, X.B.; Yang, H.; Shan, W.J. Observation of an anomalous discontinuous liquid-structural change with temperature. *Phys. Rev. Lett.* **2002**, *89*, 125505.
35. Zu, F.Q.; Li, X.F.; Guo, L.J.; Yang, H.; Qin, X.B.; Zhu, Z.G. Temperature dependence of liquid structures in In–Sn20: Diffraction experimental evidence. *Phys. Lett. A* **2004**, *324*, 472–478.
36. Waseda, Y. *The Structure of Non-Crystalline Materials*; McGraw-Hill: New York, NY, USA, 1980.
37. Li, X.F.; Zu, F.Q.; Ding, H.F.; Yu, J.; Liu, L.J.; Li, Q.; Xi, Y. Anomalous change of electrical resistivity with temperature in liquid Pb-Sn alloys. *Phys. B Condens. Matter* **2005**, *358*, 126–131.
38. Xi, Y.; Zu, F.Q.; Li, X.F.; Yu, J.; Liu, L.J.; Li, Q.; Chen, Z.H. High-temperature abnormal behavior of resistivities for Bi-In melts. *Phys. Lett.* **2004**, *329*, 221–225.
39. Li, Q.; Zu, F.Q.; Li, X.F.; Xi, Y. The electrical resistivity of liquid Pb-Bi alloy. *Morden Phys. Lett. B* **2006**, *20*, 151–158.
40. Chen, Z.H.; Zu, F.Q.; Li, X.F.; Yu, J.; Xi, Y.; Shen, R.R. Temperature-induced liquid-liquid transition process in eutectic Pb–Sn melt explored from kinetic viewpoint. *J. Phys. Condens. Matter* **2007**, *19*, 116106.
41. Li, X.F.; Zu, F.Q.; Ding, H.F.; Yu, J.; Liu, L.J.; Xi, Y. High-temperature liquid-liquid structure transition in liquid Sn-Bi alloys: Experimental evidence by electrical resistivity method. *Phys. Lett. A* **2006**, *354*, 325–329.
42. Spaepen, F. Five-fold symmetry in liquids. *Nature* **2000**, *408*, 781–785.
43. Baranyai, A.; Evans, D.J. Direct entropy calculation from computer simulation of liquids. *Phys. Rev. A* **1989**, *40*, 3817–3822.
44. Zu, F.Q.; Chen, Z.H.; Zou, L. Kinetics of liquid–structure change of In–Sn and In–Pb melt. *Phys. Chem. Liquids* **2008**, *46*, 433–441.

45. Zou, L.; Zu, F.Q.; Li, X.F.; Chen, Z.H.; Chen, H.S.; Guo, L.L.; Sun, Q.Q. Electronic transport properties of liquid InSn20 alloy. *Phase Transit.* **2008**, *81*, 889–896.
46. Zu, F.Q.; Shen, R.R.; Xi, Y. Electrical resistivity of liquid Sn-Sb alloy. *J. Phys. Condens. Matter* **2006**, *18*, 2817–2823.
47. Zu, F.Q.; Xi, Y.; Shen, R.R.; Li, X.F.; Zhang, Y.; Chen, Z.H. Effects of compound formation on liquid structure in Cu-Sn melts as a function of temperature. *Phys. Chem. Liquids* **2006**, *44*, 543–550.
48. Shen, R.R.; Zu, F.Q.; Li, Q.; Xi, Y.; Li, X.F.; Ding, G.H.; Liu, H.M. Study on temperature dependence of resistivity in liquid In-Sn alloy. *Phys. Scr.* **2006**, *73*, 184–187.
49. Wang, L.; Bian, X.F.; Liu, J.T. Discontinuous structural phase transition of liquid metal and alloys. *Phys. Lett. A* **2004**, *326*, 429–435.
50. Sun, C.; Geng, H.R. Viscous and structural behaviors of molten In–Sn alloys. *Mater. Charact.* **2005**, *55*, 383–387.
51. Plevachuk, Y.; Sklyarchuk, V.; Yakymovych, A.; Willers, B.; Eckert, S. Electronic properties and viscosity of liquid Pb–Sn alloys. *J. Alloys Compd.* **2005**, *394*, 63–68.
52. Nasch, P.; Manghnani, M.M.; Secco, R.A. Anomalous behavior of sound velocity and attenuation in liquid Fe-Ni-S. *Science* **1997**, *227*, 219–221.
53. Ji, M.; Gong, X.G. *Ab Initio* molecular dynamics simulation on temperature-dependent properties of Al–Si liquid alloy. *J. Phys. Condens. Matter* **2004**, *16*, 2507–2514.
54. Liu, C.S.; Li, G.X.; Liang, Y.F.; Wu, A.Q. Quantitative analysis based on the pair distribution function for understanding the anomalous liquid-structure change in In20Sn80. *Phys. Rev. B* **2005**, *71*, 4204–4210.
55. Greenberg, Y.; Yahel, E.; Caspi, E.N.; Benmore, C.; Beuneu, B.; Dariel, M.P.; Makov, G. Evidence for a temperature-driven structural transformation in liquid bismuth. *Europhys. Lett.* **2009**, *86*, 36004.
56. Greenberg, Y.; Yahel, E.; Ganor, M.; Hevroni, R.; Korover, I.; Dariel, M.P.; Makov, G. High precision measurements of the temperature dependence of the sound velocity in selected liquid metals. *J. Non-Crystal. Solids* **2008**, *354*, 4094–4100.
57. Dahlborg, U.; Besser, M.; Calvo-Dahlborg, M.; Cuello, G.; Dewhurst, C.D.; Kramer, M.J.; Morris, J.R.; Sordelet, D.J. Structure of molten Al-Si alloys. *J. Non-Crystal. Solids* **2007**, *353*, 3005–3010.
58. Khairulin, R.A.; Stankus, S.V.; Sorokin, A.L. Determination of the two-melt phase boundary and study of the binary diffusion in liquid Bi–Ga system with a miscibility gap. *J. Non-Crystal. Solids* **2007**, *297*, 120–130.
59. Sivkov, G.; Yagodin, D.; Kofanov, S.; Gornov, O.; Volodin, S.; Popel, P.; Sidorov, V.; Bao, C.; Calvo-Dahlborg, M.; Dahlborg, U.; *et al.* Physical properties of the liquid Pd-18 at.% Si alloy. *J. Non-Crystal. Solids* **2007**, *353*, 3274–3728.
60. Wu, A.Q.; Guo, L.J.; Liu, C.S. Structural characteristics of liquid Sn. *Chin. Phys. Lett.* **2005**, *22*, 1991–1993.
61. Wang, Y.B.; Zhao, G.; Liu, C.S.; Zhu, Z.G. *Ab initio* molecular dynamics simulations on the structural change of liquid eutectic alloy Si15Te85 from 673 to 1373 K. *Phys. B Condens. Matter* **2010**, *405*, 785–792.

62. Zhou, C.; Hu, L.; Sun, Q.; Qin, J.; Bian, X.; Yue, Y. Indication of liquid-liquid phase transition in CuZr-based melts. *Appl. Phys. Lett.* **2013**, *103*, 171904.
63. Wang, Y.Q.; Wu, Y.Q.; Liu, J.T.; Bian, X.F. Discontinuous structural phase transition behaviour in multiple component alloy melts. *Chin. Phys. Lett.* **2006**, *23*, 2513–2515.
64. Wang, Z.M.; Geng, H.R.; Zhou, G.R.; Guo, Z.Q.; Teng, X.Y. Metastable microheterogeneity in liquid monotectic Bi-Ga alloys. *Int. J. Cast Metals Res.* **2011**, *24*, 65–69.
65. Li, X.F.; Zu, F.Q.; Yu, J.; Zhou, B. Effect of liquid-liquid transition on solidification of Bi-Sb10 wt% alloy. *Phase Transit.* **2008**, *81*, 43–50.
66. Stefanescu, D.M.; Upadhya, G.; Bandyopadhyay, D. Heat Transfer-Solidification Kinetics Modeling of Solidification of Castings. *Metall. Mater. Trans. A* **1990**, *21*, 997–1005.
67. Kurz, W.; Fisher, D.J. *Fundamentals of Solidification*; Trans Tech Publications: Zurich, Switzerland, 1998; p. 28.
68. Li, X.Y. Research on Effects of Melt State of SnAgBi Lead-Free Solders on Solidification Microstructure and Welding Joint Reliability. Ph.D. Thesis, Hefei University of Technology, Hefei, China, June 2013.

4

Effect of Strain Localization on Pitting Corrosion of an AlMgSi0.5 Alloy

Daniela Nickel †, Dagmar Dietrich †,*, Thomas Mehner, Philipp Frint, Dagobert Spieler and Thomas Lampke

Technische Universität Chemnitz, Professur Oberflächentechnik/Funktionswerkstoffe, 09107 Chemnitz, Germany; E-Mails: daniela.nickel@mb.tu-chemnitz.de (D.N.); thomas.mehner@mb.tu-chemnitz.de (T.M.); philipp.frint@mb.tu-chemnitz.de (P.F.); dagobert.spieler@mb.tu-chemnitz.de (D.S.); thomas.lampke@mb.tu-chemnitz.de (T.L.)

† These authors contributed equally to this work.

* Author to whom correspondence should be addressed;
E-Mail: dagmar.dietrich@mb.tu-chemnitz.de

Academic Editor: Heinz Werner Höppel

Abstract: The corrosion susceptibility of an age-hardened aluminum alloy in different processing conditions, especially after a single pass of equal-channel angular pressing (ECAP), is examined. The main question addressed is how corrosive attack is changed by strain localization. For that purpose, an AlMgSi0.5 alloy with a strain localized microstructure containing alternating shear bands was subjected to potentiodynamic polarization on a macro-scale and micro-scale using the micro-capillary technique. Pitting potentials and the corrosion appearance (pit depth, corroded area fractions and volumes) are discussed with respect to microstructural evolution due to casting, extrusion and ECAP. Size, shape and orientation of grains, constituent particle fragmentation, cell size and microstrain were analyzed. Stable pitting of shear bands results in less positive potentials compared to adjacent microstructure. More pits emerge in the shear bands, but the pit depth is reduced significantly. This is attributed to higher microstrains influencing the stability of the passivation layer and the reduced size of grains and constituent particles. The size of the crystallographic pits is associated with the deformation-induced cell size of the aluminum alloy.

Keywords: AlMgSi0.5 alloy; ECAP; pitting corrosion; micro-capillary cell; deformation localization

1. Introduction

Aluminum alloys are used in a wide range of technical applications, especially when considering the lightweight aspect. Mechanical properties are influenced by microstructural heterogeneities formed by adding alloying elements mostly combined with thermal or thermo-mechanical treatments. Solution heat treatment and aging lead to the formation of solid solutions and intermetallic compounds (ICs) stimulating mechanisms of solid solution and precipitation strengthening. Mechanical-based treatments cause strain hardening by multiplication of dislocations or grain refinement in the deformed metal. An effective deformation technique is equal-channel angular pressing (ECAP), also known as equal-channel angular extrusion, originally invented by Segal in 1972, first described in his Sc.D. thesis in 1974 and patented in 1977 (Patent of the USSR, No. 575892) [1,2]. A massive metallic billet is pressed through two intersecting channels whereby the material is severely deformed by exceptional, intensive and oriented simple shear, which is localized inside thin, so-called shear bands [1]. Accordingly, after one ECAP pass, one family of shear bands is activated. Repetitive pressing induces new shear bands depending on the rotation between the successive passes. Grain refinement down to ultrafine grains (UFG), re-distribution and fragmentation of second phases or impurities if existing in the material and texture evolution differ according to the applied processing route. In addition, the bulk material is characterized by high densities of dislocations and grain boundaries, internal elastic strains as well as residual stress thus delivering improved mechanical properties [3,4].

The effect of any microstructural heterogeneity results in an electrochemical heterogeneous behavior influencing the susceptibility to corrosion. In recent years, significant progress has been achieved in the understanding of both the microstructure of high strength alloys and its influence on corrosion [5]. Dominant features like the surface fraction of grain boundaries and dislocations as well as the distribution of second phases including constituent and impurity particles, dispersoids and precipitates are studied with respect to their corrosion susceptibility. The improved physico-chemical characterization of the composition and spatial distribution of second phases together with advanced electrochemical techniques expanded the understanding of the influence of intermetallic particles on the corrosion of conventionally grained (CG) alloys. A comprehensive review is given by Hughes *et al.* [5]. Recently, electrochemical properties of UFG aluminum and its alloys in different ECAP conditions were studied in a variety of publications (Table 1). Pitting corrosion starts with the local destruction of the natural oxide layer by critical anions (e.g., chloride ions) and is a dominant corrosion type of passivating materials. In electrochemical studies used to investigate the susceptibility to pitting, the pitting potential marks the breakthrough of the passive layer. In high-purity aluminum and single-phase aluminum model alloys, the pitting potential is lowered [6,7] or unaffected [8] compared to the CG material and is ascribed to the effect of grain refinement. Ralston *et al.* [9] correlate the corrosion rate in terms of oxide formation and film ion conduction with reactive surfaces provided by enhanced grain boundary density. Brunner *et al.* [7] found a lowered pitting potential due to the transition from CG to UFG material but no significant influence with the number of ECAP passes. The pit morphology after polarization occurred less crystallographically, deeper and more localized in the UFG material compared to the CG material in which laterally spreading crystallographic filiform corrosion dominated. The dislocation density and the deformation state were supposed to be the dominating factors for the different propagation mode of pitting corrosion: propagation in depth for UFG and laterally spreading for CG material [7]. The variation in passive film stability and the existence of local galvanic cells stimulate the anodic dissolution of the surrounding matrix material. Coherent solute enriched clusters can act as local cathodes in precipitation hardening alloys. Ralston *et al.* [10] found a critical particle size (of some nm) below which they do not act as distinct galvanic cells, but promote susceptibility to pitting if they increase.

Table 1. Review of pitting corrosion results published for Al and its alloys.

Material and processing	Effect of pass number on E_{pit}	Dominant pit initiation	Pit morphology with increasing pass number	Analytical techniques	Reference
high purity Al 2, 6, 10 ECAP passes (route Bc)	lowered	dislocations, grain boundaries		PP	[6]
Al-2.5 wt% Cu-1.5 wt% Mg alloy aging, no ECAP	unaffected/lowered	critical size of GP zones (solute enriched regions)		PP, CT	[10]
Al-Mg model alloys (varying Mg content) 4, 8, 12 ECAP passes (route Bc)	lowered/unaffected	deformation state more dominant than grain size	after PP: pits deeper and more localized, hindered repassivation with increasing pass number, filiform, decreasing crystallographic character	PP, EIS	[7]
high purity Al 1, 4, 8 ECAP passes (route Bc)	unaffected	more reactive surfaces for passivation (grain size, grain boundaries)		PP	[8]
AA 1050 1, 2, 3, 5 passes	raised	fragmentation of Si-containing constituent particles	after PT: decreased pit size, less severe attack	PP, PT, EIS	[11]
AA 1100 8 ECAP passes (route Bc); AA5052 4 ECAP passes (route Bc)	raised	fragmentation of Fe-Al impurities, increased formation rate of passivation film	after PT: decreased pit size, crystallographic pits	PP, EIS	[12]
Al-5.4 wt% Ni and Al-5 wt% Cu 2, 4, 6 ECAP passes (route Bc)	raised	fragmentation of constituent particles, microstructural homogenization	after PP: fewer pits in the shear bands	PP	[6]
EN AW-6082 1, 2, 8 ECAP passes (route E), 1 ECAP pass + post ageing	raised	fragmentation and redistribution of constituent particles	after PP: decreased pit depth	PP, EIS, CV	[13,14]
low purity Al (Al-0.5 wt% Si-0.15 wt% Fe) 16 ECAP passes	raised	more stable passivation film due to strain induced crystalline defects and residual (micro)stress	after PT: decreased pit size uniform distribution	PP, PT, IT, EIS	[15]
low purity Al (99.1 wt%) with α-AlFeSi phases 2, 4 ECAP passes (route A)	lowered	fragmentation of constituent particles (α-phase), grain refinement and higher dislocation density	after IT: crystallographic pitting	IT, PP, EIS	[16]
low purity Al (99.1 wt%) with Al_8Fe_2Si phase 1, 2, 3, 4, 5 ECAP passes (route A)	unaffected	fragmentation of constituent particles, increased volume fraction of grain boundaries	after IT and PP: reduced intergranular corrosion increased pitting increased density and decreased pit depths	IT, PP, EIS, 3D optical profilometry	[17]

Note: PP, potentiodynamic polarization test; PT, potentiostatic polarization test; EIS, electrochemical impedance spectroscopy; CV, cyclovoltammetry; CT, current transient experiments; IT, immersion test; Epit, pitting potential.

Constituent particles in multi-phase materials mostly act as cathodic inhomogeneity performing a key role in pit initiation and propagation. With an increasing number of ECAP passes, the pitting potential of a variety of multi-phase aluminum alloys, which were reported to increase [11,12,15], decrease [16] or stay unaffected [17]. Correspondingly, for AA 5052, after four ECAP passes, the pitting potential was distinctly shifted to more positive potentials [12]. The crystallographic pits were smaller and more uniformly distributed in the UFG material than in the as-cast condition. The authors suggested a higher oxide formation rate based on the enhanced grain boundary fraction and dislocation density in the UFG materials. Similarly, Song *et al.* [15] stated that strain-induced crystalline defects provide more nucleation sites for the formation of a denser and thicker oxide film. Akiyama *et al.* [6] found an increasing pitting potential of Ni and Cu containing aluminum alloys due to ECAP processing. Fewer pits in the shear band areas were attributed to the microstructural homogenization and refinement of primary particles [6]. The amount, fragmentation and redistribution of constituent particles, containing mostly Si and Fe along with Al, were considered as the main factors for the modified pitting susceptibility and morphology of multiphase alloys [6,11,12,16].

Korchef *et al.* [16] found decreasing pitting potentials explained by the increasing fragmentation of α-AlFeSi particles acting as cathodes and driving anodic dissolution of the surrounding matrix. Jilani *et al.* [17] found no influence on the pitting potential with an increasing number of ECAP passes, but an increasing pitting density and decreasing pit depth. The change in the pitting morphology was assumed because of the fragmentation of Al_8Fe_2Si particles and/or the related increase of the volume fraction of grain boundaries acting as potential sites for pitting. Finally, the corrosion resistance of UFG materials processed by accumulative roll bonding shows similarly contrasting results [18,19].

Table 1 presents the reviewed literature together with own results [13,14] demonstrating little consensus in the influence of ECAP processing on the corrosion behavior of aluminum alloys even within the same alloy. Certainly, the main challenge of corrosion studies in this regard is the lack of a standardized material condition suitable as benchmark for evaluating the effect of ECAP processing on the corrosion behavior.

In order to overcome this obstacle, the present work adapts the testing and evaluating area to the strain localizations generated by a single ECAP pass processing. The selected ECAP condition presents concurrently ECAP-affected (shear bands) and non-affected (matrix) areas, which can be locally tested using a micro-capillary cell technique [20] and independently evaluated because only one family of shear bands is activated. Additional impacts by further processing are excluded. A commercial alloy AlMgSi0.5 with a well-known microstructure [21–23] was selected. The results of the micro-scale electrochemical studies are compared to standard macro-scale experiments including the material conditions of the foregoing processing steps casting and extrusion. Finally, the induced pitting damage is separately quantified for shear bands and matrix zones. The results are correlated with the microstructure of the material.

2. Experimental Section

2.1. Sample Preparation

A commercial AlMgSi0.5 alloy (EN AW-6060, chemical composition given in Table 2) was used in different material conditions. Samples were taken from a cast billet (referred to as cast) and from bars after appropriate processing steps. Homogenization and solution annealing at 560 °C for 10 h was followed by water quenching. The ingot was extruded at room temperature (RT) reducing the initial diameter of 110 mm to a cross-section of 50 × 50 mm² (samples referred to as RT-extruded). Severe plastic deformation was achieved using the combination of RT-extrusion with a subsequent large-scale ECAP single pass at RT [22,23] (samples referred to as ECAP). The internal angle of the die was 90°, the pressing speed 20 mm min^{-1} and the backpressure 200 MPa. The resulting effective strain was approximately 1.1 [24].

Table 2. Chemical composition of the investigated EN AW-6060 (wt%).

	Si	Fe	Cu	Mn	Mg	Zn	Al
EN AW-6060	0.45	0.16	0.003	0.026	0.49	0.003	bal.

Samples with longitudinal planes were cut of each billet. After mechanical grinding with SiC, the final polishing step of the metallographic preparation was dependent on the intended investigation method. For optical microscopy (OM), polishing in a silicon oxide suspension (OP-S, Struers GmbH, Willich, Germany) and grain boundary etching for microstructural contrasting (2% NaOH) was applied. For electrochemical measurements, polishing with 1 μm diamond suspension was used to achieve a fairly chemically unaffected surface. After corrosion experiments, the surface was carefully rinsed with de-ionized water and ethanol and subjected to vibrational polishing using OP-S to provide a clean and deformation-free surface for backscattered electron (BSE, Zeiss, Jena, Germany) imaging and electron backscatter diffraction (EBSD, AMETEK GmbH, Wiesbaden, Germanyin the scanning electron microscope (SEM, Zeiss, Jena, Germany [25].

2.2. Electrochemical Measurements

Deformation localizations (shear bands) were detected by OM on the etched ECAP sample surface and marked by indentations. The electrochemical experiments were accomplished in an aerated and unbuffered 0.1 M NaCl solution (pH 7) at (22.0 ± 0.5) °C. The samples were exposed to the solution for 15 min to stabilize the open circuit potential (OCP). Polarization curves were measured using a high-resolution potentiostat IMP83 PCT-BC (Jaissle Elektronik, Waiblingen, Germany). After determining the starting potential, potentiodynamic polarization experiments were done covering the potential range from 200 mV below the corrosion potential (between −1300 mV and −1100 mV depending on the material condition) to 1000 mV with a scan rate of 1 mVs^{-1}. For experiments with an exposed area of 10 mm in diameter (referred to as macro-scale), a three-electrode corrosion cell with a platinum plate as counter electrode and an Ag/AgCl (3 M KCl) reference electrode was used according to the standard specification DIN 50918. Macro-scale measurements were repeated three times.

For local micro-electrochemical measurements (referred to as micro-scale), a glass capillary with a ground tip of 200 µm in diameter was used. A layer of silicone rubber worked as a seal to prevent leakage. The capillary size was selected in order to cover either the shear band or the matrix of the material in ECAP condition. The three-electrode set-up consisting of the glass capillary, a 0.5 mm platinum wire counter electrode and a Ag/AgCl (3 M KCl) reference electrode was mounted in an objective nosepiece. With the samples fixed on the stage of the OM, exact positioning on the microstructural localization (shear band or matrix) was obtained. The set-up of the micro-cell system is described in detail in [20,26]. Micro-scale measurements were repeated nine up to 13 times depending on the material condition.

2.3. Microstructural Characterization

Characterization of the microstructure and the pit morphology were performed in a field-emission scanning electron microscope (FE-SEM, NEON40EsB, Carl Zeiss MicroImaging GmbH, Jena, Germany) equipped with an EBSD system (EDAX TSL OIM 5.2, AMETEK GmbH, Wiesbaden, Germany. EBSD patterns were acquired at an acceleration voltage of 15 kV with a 60 µm aperture in high-current mode under a sample tilt angle of 70°. After data acquisition, a slight cleanup procedure for the nearest neighbor pixels was conducted comprising neighbor confidence index (CI) correlation and grain CI standardization with a minimum CI of 0.1.

Material removal by corrosion was quantitatively evaluated by using a 3D optical profilometer (Mikro-CAD compact, GFMesstechnik GmbH, Teltow, Germany). Corroded surface areas were determined by means of the software a4i analysis (GFMesstechnik GmbH, Teltow, Germanycity). Depth and volume of the resulting corrosion pits were measured using replicas of the tested regions. Pits with a depth larger than 5 µm were considered.

X-ray diffraction (XRD) analysis (sin2ψ method) was applied for the determination of local residual stresses and microstrains. The diffractometer D8 Discover (Bruker AXS, Karlsruhe, Germany) was equipped with a Cu anode, a Ni filter for monochromatization, polycap optics and a 2D detector (Vantec-500, Karlsruhe, Germany. The {311} lattice planes of Al were considered for the residual stress measurements using several measurement directions (0°, 45°, 90°–rotation around the sample normal) each with tilt angles between −50° and 50° (in 10° steps) and a measurement time of 900 s for each angle. A small aperture (100 µm) in the primary beam, adequate to the dimensions of the strain localizations, and a point detector (SOL-XE) with a 0.6 mm detector slit were used for the determination of microstrains and coherence lengths in a line scan. The step size of the line scan was 100 µm. The program LEPTOS was used for stress evaluation. The line widths of the strong textured {220} lattice planes were evaluated by means of the program TOPAS under consideration of the device-related line-broadening effects to estimate the coherence length.

3. Results and Discussion

3.1. Evolution of Microstructure

The cast material (Figure 1a) is characterized by equiaxed grains with a mean diameter of 100 µm and numerous constituent particles decorating the grain boundaries. The IC particles in the cast condition are a few micrometers in size; homogenization and solution annealing resulted in a slight enlargement. The extrusion process transformed the alloy microstructure to elongated grains with an aspect ratio around 0.5 associated with the redistribution of the IC particles located at the grain boundaries. The single-pass ECAP process introduced a series of deformation localizations (shear bands) intercalated in fairly unchanged matrix regions. The grains in the shear bands become more elongated to an aspect ratio below 0.3. The IC particles are re-aligned with the aluminium grains and slightly fragmented (Figure 1b). Additionally, the deformation processing introduced a sub-grain structure with a size around 1 μm (Figure 2c,d). This is in agreement with transmission electron microscopic results achieved at the same alloy [23].

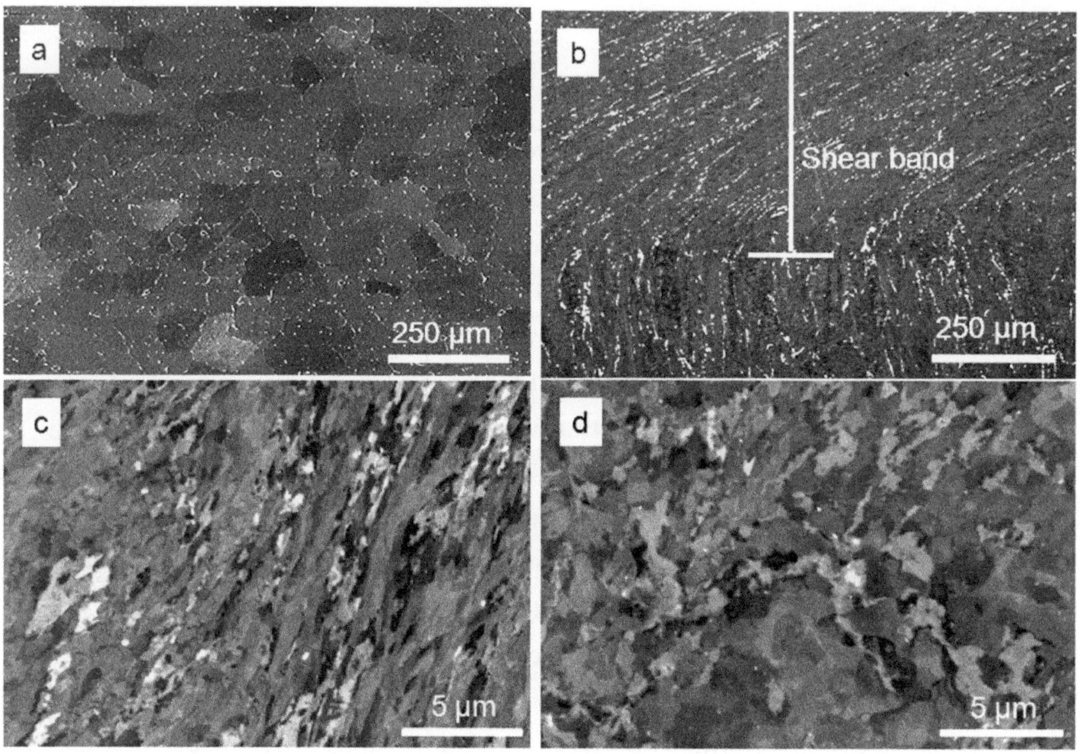

Figure 1. Microstructure of the AlMgSi0.5 alloy in (**a**) cast condition showing constituent particles in the grain boundaries; (**b**) ECAP (equal-channel angular pressing) condition showing shear bands (a part of it is marked by a white bar) introduced into the matrix (below); (**c**) elongated grains in the shear band and (**d**) fairly unchanged grains in the matrix, which is similar to the extruded condition.

According to Birbilis *et al.* [27], the IC particles of particular interest with regard to localized corrosion are those with the greatest quantity in size or frequency—these are AlFeSi phases in the AlMgSi0.5 alloy. As shown for the extruded condition of the alloy [28], the iron and silicon containing constituent particles include compositions with different stoichiometry. Apart from the AlFeSi phases, the solid solution matrix of the as-cast billet contains dissolved Mg and Si, which precipitate to fine β and β′ particles during quenching [29,30]. Certain observations show that the hardening precipitates in the studied AlMgSi0.5 alloy undergo fragmentation and change in coherency by ECAP [31,32].

Apart from grain sizes and aspect ratios, orientation maps (Figure 2a–c) were derived from EBSD measurements which illustrate the microstructural evolution of the material with respect to grain orientation or texture. The random orientation of the cast material (Figure 2a) is transformed by RT extrusion to a fiber texture with the main component <110> in extrusion direction (Figure 2b). The grain orientation map of the ECAP condition (Figure 2c) shows a shear band in the center with a texture fairly similar to the matrix. Deformation localizations are clearly reflected by dark regions in EBSD pattern quality maps as shown later in Figure 9.

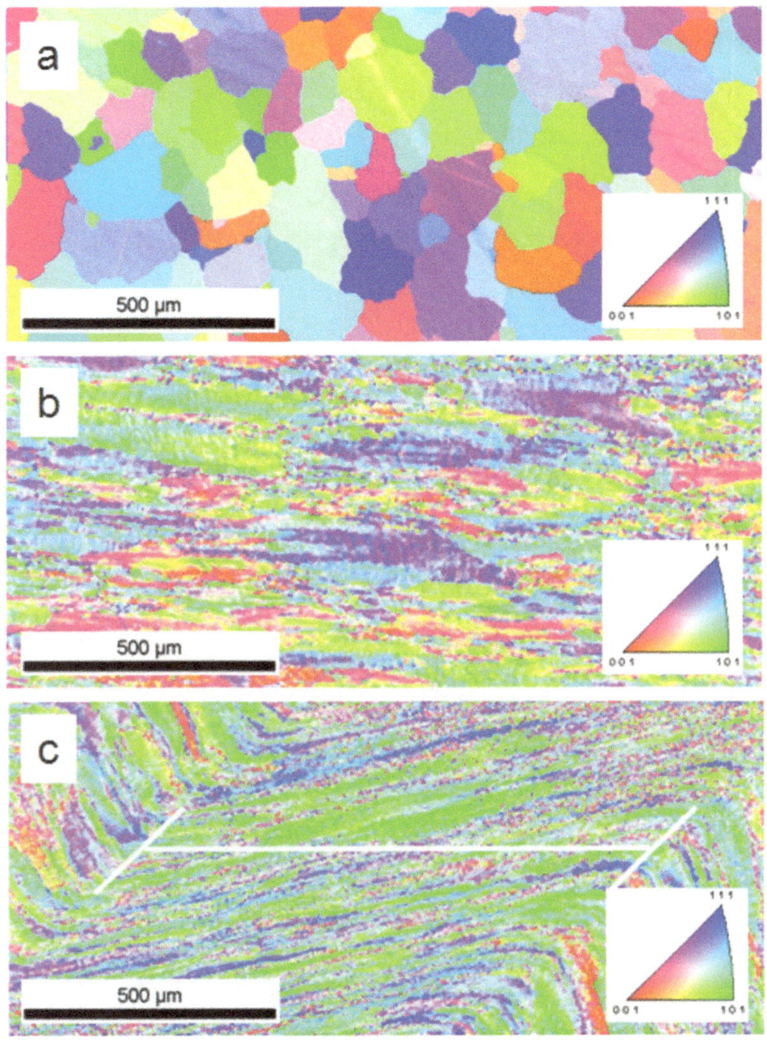

Figure 2. Orientation distribution maps of the material in the (**a**) cast; (**b**) the RT-extruded and (**c**) the ECAP condition showing the grain shape transformation to elongated grains of reduced size and texture evolution by extrusion and ECAP; the shear band in (**c**) is marked by a white bar.

The measurement of residual stresses and microstrains was focused on the deformation localizations of the material in ECAP condition. Almost no differences could be detected for the principal stress components σ1 and σ2 inside shear bands and matrix regions. The measured components of the residual stress are σ1 ≈ 60 MPa and σ2 ≈ −20 MPa.

Microstrains have been specified referring to the full width at half maximum of the Gaussian part of the line profile. Measurements of several points on the material in the RT-extruded condition were taken for comparison and reveal a fairly constant microstrain of 0.21. The microstrain in the ECAP condition varies between 0 and 0.34. A line scan was measured with an increment of 100 µm across three shear bands intercalated in the matrix. Figure 3 shows the local distribution of microstrain in correlation to the shear bands revealed by grain boundary etching. The shear band-matrix transitions are marked by white lines. It can be seen that the widths of deformation localizations comprise a wide range. For reasons of representation, the values of the microstrain are grouped in bins. Color-coded according to the legend, the microstrain scan is inserted in the optical micrograph at the position of the measurement. As expected, higher values of microstrain are found in the shear bands and lower values in the matrix (Figure 3). The broad matrix band on the right shows an additional subdivision—possibly because of the actual three-dimensional distribution of shear bands, which is not observable in the plain cross-section. Apparently, the microstrains are a result of the deformation localization introduced by the severe deformation during the ECAP pass.

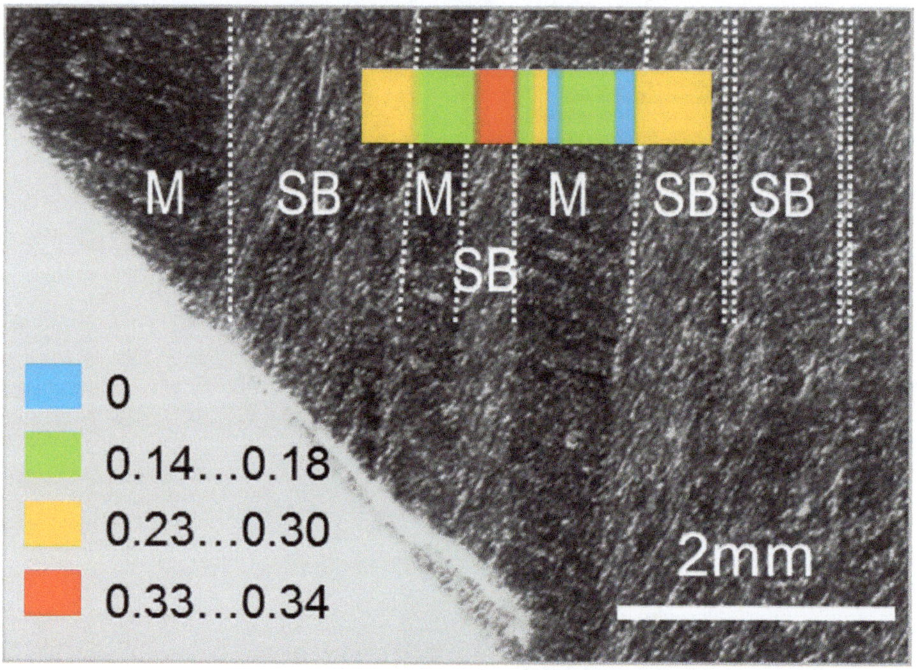

Figure 3. Local distribution of the microstrain in a line scan across three shear bands of the material in the ECAP condition; the region of interest is shown with respect to the deformation localizations shown by grain boundary etching (for color-coding, see legend).

Crystallite size was determined by the Lorentzian part of the line profile. The measured characteristic is the distance that scatters the incident X-ray beam coherently. Apart from grains and sub-grains defined by high- and low-angle boundaries, coherent regions are bounded by accumulated dislocations and are commonly assigned as cells. Cell sizes are averaged out at 180 nm in the extruded condition and typically range from 80 nm–170 nm for both matrix and shear bands of the ECAP condition, which is consistent with the published cell size data determined by STEM [32]. Interestingly, the mean cell size in the matrix region is smaller (≈100 nm) compared to the shear band with the largest cells in the region with the highest microstrain.

3.2. Pitting Corrosion Potential Derived from Macro- and Micro-Electrochemical Tests

Pitting corrosion is the most common corrosion type for aluminum alloys. The local destruction of the passivation film is caused by anions, e.g., chloride ions. Potentiodynamic polarization curves were measured to evaluate pitting susceptibility. The most negative potential above which pits nucleate and grow (pitting potential) was determined as the decisive corrosion characteristic. Figure 4 gives an example for the curves measured on the material in ECAP condition. A grouping of the pitting potentials depending on the position of the micro-capillary either on the shear bands or on the matrix is observed.

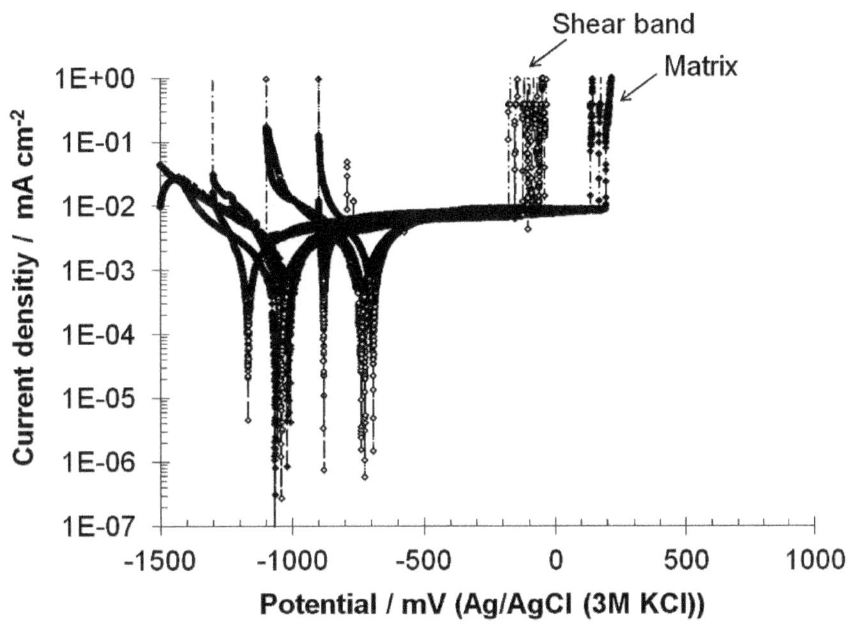

Figure 4. Potentiodynamic polarization curves measured on the shear bands and the matrix.

Pitting potentials and the corresponding corrosion appearance are shown in relation to the material condition and to the exposed area of 10 mm (macro-scale) and 200 μm (micro-scale) in diameter in Figure 5. The pitting potentials determined by macro-scale electrochemical measurements are very similar regardless of the material condition (Figure 5). Averaged out −430 mV referring to Ag/AgCl (3 M KCl), a variation of ±30 mV for the cast condition, of ±90 mV for the RT-extruded condition and of ±40 mV for the ECAP condition is ascertained. This is in agreement with the results of [8] and [17], which stated that the pitting potential of low purity aluminum does not vary with grain size or with the size of the Al_8Fe_2Si phase. The macro-scale pitting potentials are shifted to at least 700 mV lower values compared to micro-scale pitting potentials. This potential shift was already evidenced for a 2024 alloy in T3 temper state [33] and for Al-Zn-Mg-Cu alloys [34]. Evidently, this is because of the number of pit initiating constituent particles present in the testing area. More constituent particles exist in the area exposed in macro-scale experiments compared to the lower number in the area exposed in micro-scale experiments. For the same reason, the pitting potentials of the micro-scale measurements are spread over a wider potential range, also mentioned in [27].

Micro-scale experiments result in pitting potentials from −80 mV to +340 mV for the cast condition, more negative potentials from −160 mV to +300 mV for the RT extrusion condition and similarly from −180 mV to +270 mV for the ECAP condition. The scatter of the pitting potentials observed for all material conditions is suspected to be mainly the effect of various surface fractions of second phases in the exposed area according to Birbilis *et al.* [27]. Furthermore, the micro-scale experiments show a clear grouping of the higher pitting potentials related to the matrix (between +100 mV and +270 mV) and lower pitting potentials related to the shear bands (between −180 mV and +50 mV), whereas no influence of material conditions can be seen in the macro-scale experiments. Thus, micro-scale experiments allow for the differentiation of the pitting potentials in ECAP induced deformation localizations. The selected shear bands exceed a width of 600 μm thus providing a test region surely larger than the used opening of the micro-capillary (200 μm in diameter). Compared to the matrix, the significantly lower pitting potential in the shear band indicates more weak points with respect to corrosion in this region.

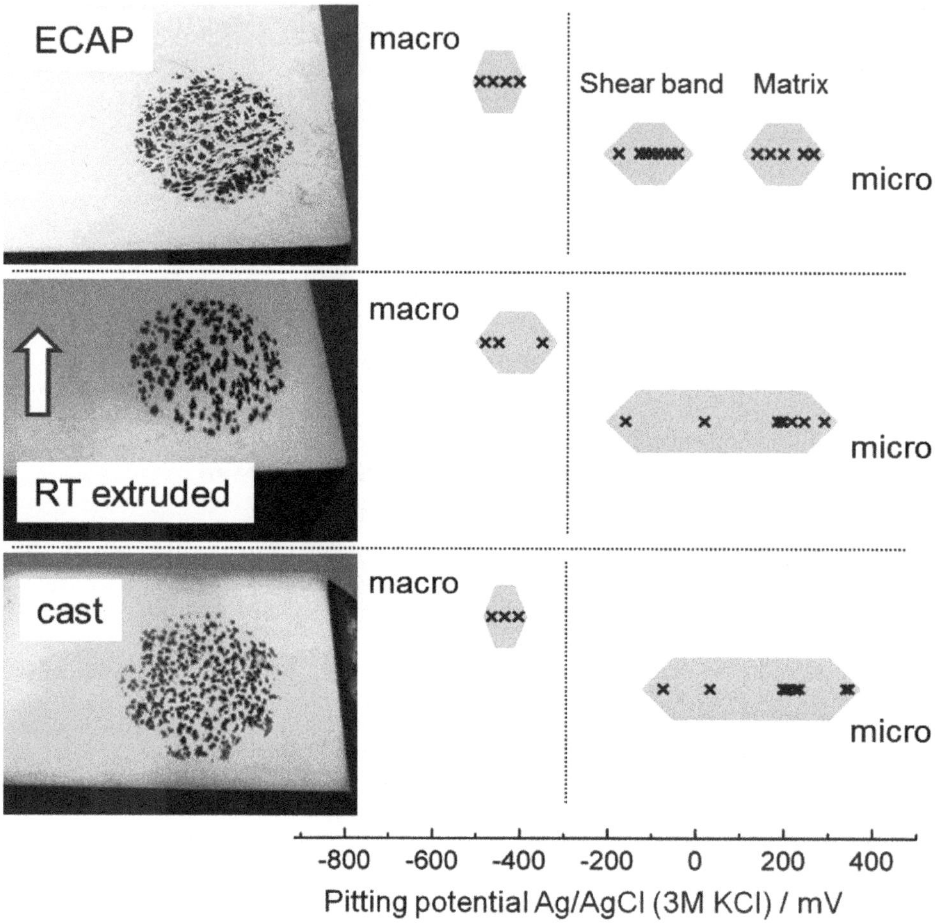

Figure 5. Documentation of the corrosion appearance after macro-scale potentiodynamic polarization of the ECAP condition, the extruded condition (extrusion direction marked by arrow) and the cast condition showing the modification from uniform to oriented (left from top to bottom) with corresponding pitting potentials in 0.1 M NaCl determined by macro-scale tests (middle) and micro-scale tests (right) at 1 mA cm^{-2}.

3.3. Pit Morphology

In Figure 6a–c, the secondary electron images (top) and the corresponding orientation maps derived from EBSD (below) show typical examples of the corrosion appearance as a consequence of different processing history. Pitting seems to start as well as to be terminated at the precipitation decorated grain boundaries thus evolving different corrosion appearance with respect to the grain shape.

The dominant feature of the alloy microstructure along with the modification of grain size and grain shape is the distribution and fragmentation of second-phase intermetallic particles (constituent particles) with respect to electrochemical characteristics. The constituent particles in an AlMgSi alloy are electrochemically nobler compared to the surrounding aluminum solid solution [16,17,27]. Due to the galvanic coupling between the AlSiFe phases and the surrounding Al matrix in the Cl^{-}-containing test environment, the naturally occurring passivation layer will be reduced and the surrounding aluminum dissolved. Starting at the precipitations mostly localized at the grain boundaries, the corrosion paths regularly start at the grain boundaries, spread into the grain and are terminated at the opposite grain boundary when the constituent particles are undermined and lose electrical contact. The large equiaxed grains of the cast condition are mirrored by a similar corrosion appearance. Similarly, the aligned and elongated grains of the material in the RT extrusion and ECAP condition are associated with an aligned and elongated corrosion appearance. The fragmentation of constituent particles by ECAP processing provides numerous starting sites for corrosion, but also faster removal of the smaller particles by undermining the surrounding aluminum. Thus, the propagation of the corrosion attack in vertical direction is decreased.

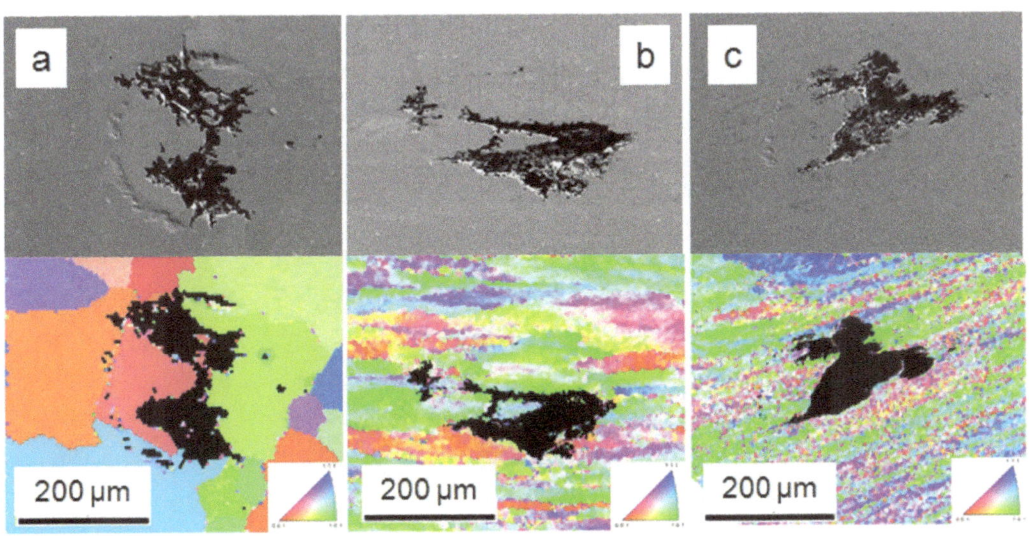

Figure 6. Corrosion appearance after micro-scale potentiodynamic polarization of the (**a**) cast condition; (**b**) extruded condition (extrusion direction is horizontal) and (**c**) in a shear band of the ECAP condition.

The pitting appearance produced during macro-scale testing shows a striking correlation to the microstructure developed by the processing history (Figure 5). The uniform corrosion in the cast condition is transformed to a non-uniform appearance corresponding to the directed and elongated grains after extrusion and ECAP. This behavior is accompanied by an obviously shallower corrosion attack in the ECAP condition which will be quantitatively evaluated and discussed later on. Crystallographic pitting is evidenced regardless of the material condition but with a remarkable influence on the size of the individual {100} facetted pits. Large crystallographic pits around 1 µm in length of the edge can be seen in the cast condition (Figure 7a), around half as long are the edges of crystallographic pits in the in RT-extruded condition (Figure 7b) and even more reduced in the ECAP condition (Figure 7c,d). Particularly noteworthy is the fact that the reduced size of individual crystallographic pits in the matrix zones in the ECAP condition is more similar to those in the shear bands. This was unexpected, since the mean grain size and the primary precipitate size are fairly equal in the RT-extruded condition and in the matrix of the single pass ECAP condition.

ECAP processing enhances the fraction of low- and high-angle grain boundaries, the dislocation density and the fragmentation of strengthening precipitates of the studied aluminum alloy [23,32]. A bimodal grain size distribution and an increased dislocation density appear already after one ECAP pass in both matrix and shear bands [32]. The dislocations accumulate to a complex network of deformation induced cell boundaries [23]. A schematic representation of the cell distribution and corresponding STEM images are shown in Figure 8 (adopted from [23]). The size of the cells as the smallest elements which subdivide grains is evaluated by XRD in this work and ranges between 80 nm and 180 nm.

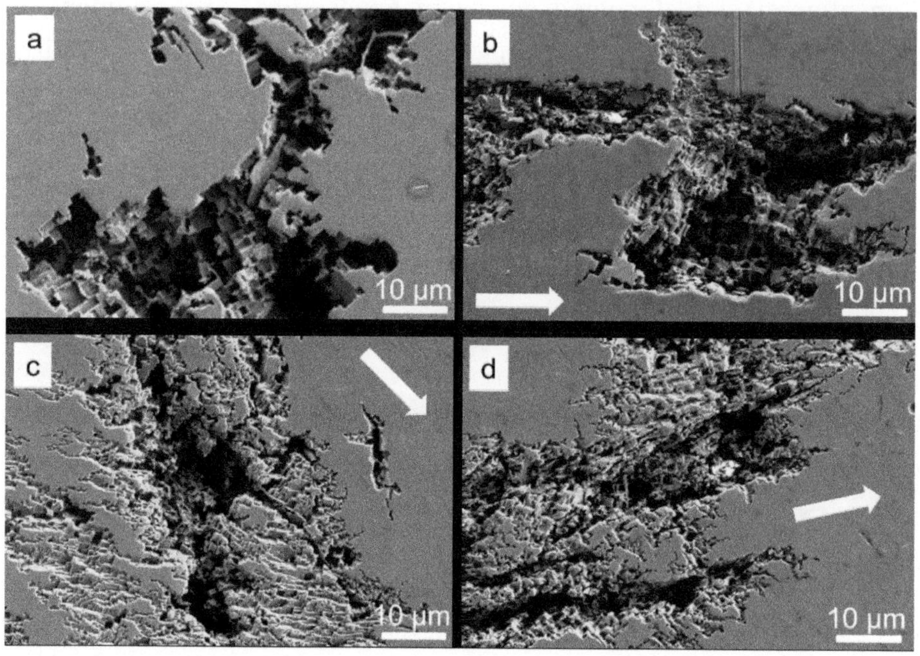

Figure 7. Formation of crystallographic {100} facetted pits due to micro-scale potentiodynamic polarization in the (**a**) cast condition; (**b**) in the extruded condition (extrusion direction and grain alignment marked by arrow); (**c**) in ECAP condition in the matrix and (**d**) in a shear band (deformation direction marked by arrows).

Apart from the decreasing pitting potential in shear bands, we observe a significant reduction of the single crystallographic pit size in both the matrix and shear bands in the ECAP condition. This observation correlates with the reduced dimensions of the cells evaluated by XRD and STEM. We conclude that the cell size and the microstrains are the distinguishing features to control the pit growth in the ECAP condition.

The reduced corrosion depth after ECAP was already indicated by several authors [11–15,17]. Contrary results were reported for Al-Ni and Al-Cu alloys containing other second phases [6]. In an Al-Mg model alloy [7], more localized and deeper pits were found in the UFG condition. Furthermore, the pitting potential evidenced influence of the number of ECAP passes. However, tests after one ECAP pass were not considered.

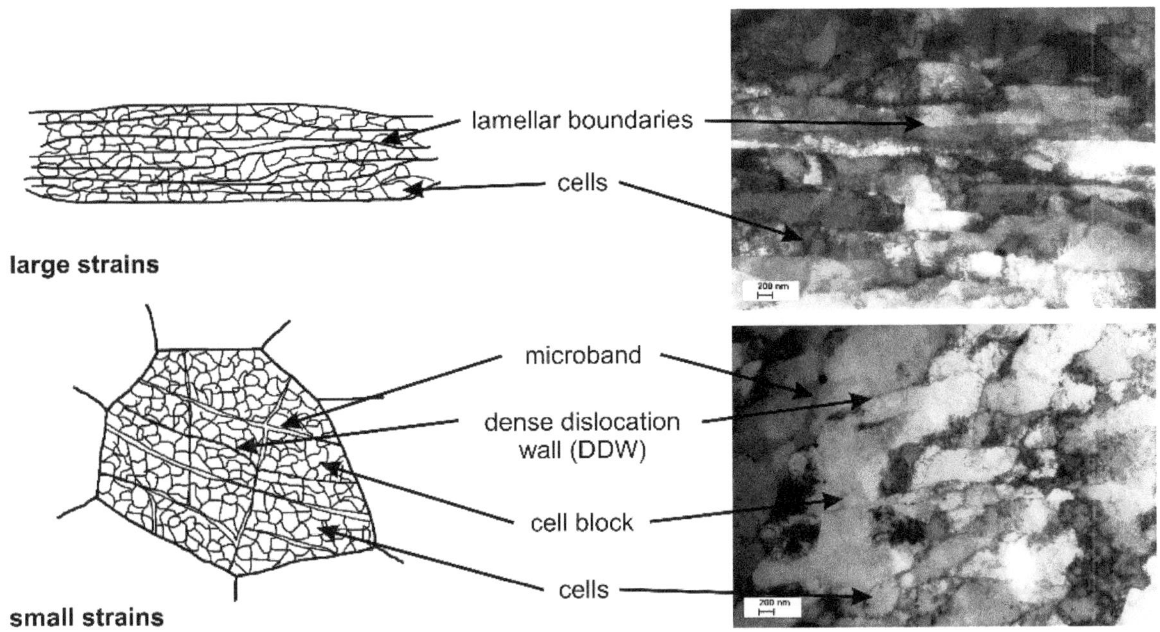

Figure 8. Schematic representation of the deformation microstructure and grain subdivision at different strains (large/small) and corresponding STEM-micrographs of the investigated aluminum alloy [23].

Jilani *et al.* [17] qualitatively discussed the change in the corrosion appearance from a low density of deep pits to a high density of less deep pits with an increasing ECAP pass number. This is in agreement with our observation. At first sight, the removal of material in the ECAP condition seems to be more concentrated in the regions without deformation localizations. The simple comparison of corrosion areas leads to this conclusion, but the volume of the eroded material has to be considered. To substantiate our findings by quantifying the material removal, several metrological characteristics were determined using a 3D optical profilometer. A depth mapping of the complete area exposed to macro-scale testing after ECAP is shown in Figure 9 with a magnified detail in the inset. Pit depths between 5 μm (red) and 75 μm (blue) were recorded and are shown by a color code. The contiguous EBSD pattern quality map corresponds to the position of the detailed depth map and illustrates the deformation localizations by darker gray tones (the corroded areas are shown in black). Several metric features were analyzed to reveal differences in the matrix and shear bands and to compare the other material conditions: the fraction of the corroded area, the corroded volume, the average corrosion depth, the maximal pit depth, the average pit depth and the density of pits (Table 3). The average corrosion damage refers to a uniform removal of the material across the tested area. The average pit depth refers to the total corroded volume divided by the total corroded area. The pit density is related to the total area used for macro-scale testing.

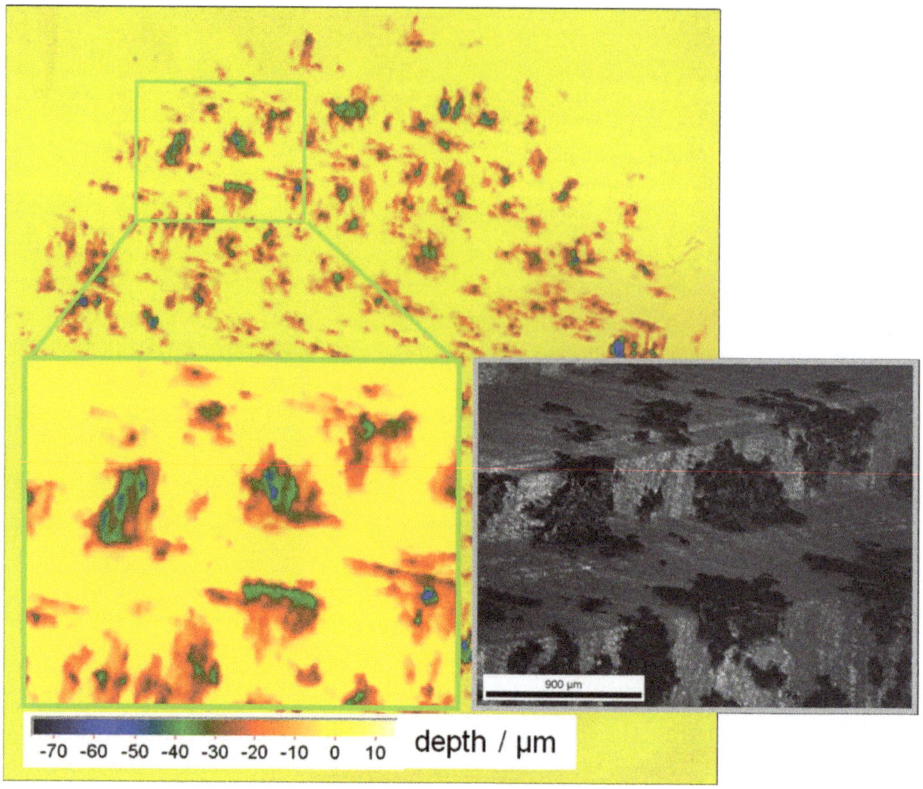

Figure 9. Depth mapping of pitting corrosion after macro-scale potentiodynamic polarization of the ECAP material (depth scale color-coded); the left inset shows the details; the right inset shows the corresponding EBSD pattern quality map indicating deformation localization (position of the insets is marked by the box in the overview map).

Table 3. Characteristics of the corrosion appearance for the studied conditions after macro-scale potentiodynamic polarization (Ø 10 mm).

State	Corroded area fraction (%)	Corroded volume (mm^3)	Average corrosion depth (µm)	Maximal pit depth (µm)	Average pit depth (µm)	Density of pits (pits/mm^2)
cast	34	0.57	7.3	100	19.6	4.6
RT-extruded	37	0.75	9.6	120	28.1	3.9
ECAP						
sum	38	0.43	5.1	75	13.5	8.7
matrix	45	0.24	6.0	75	13.3	8.0
shear band	32	0.19	4.4	75	13.6	9.3

The fraction of the corroded surface area of the cast condition is about 10% smaller than that of the RT-extruded and ECAP condition, but the corroded volume follow the order ECAP < cast < RT-extruded. The same applies to the average corrosion depth, the maximal pit depth and the average pit depth. Hence, the corrosion of the RT-extruded condition is the most local one of the considered conditions and thus most susceptible to corrosion damage. In contrast, the material in the ECAP condition corrodes in a more uniform way with remarkably lower maximal and average pit depths compared to the other conditions. The pit densities verify the statement of the more uniform corrosion of material in the ECAP condition. *Vice versa*, lower pit densities in association with enhanced corroded volumes for the cast and the RT-extruded conditions correspond to more local corrosion.

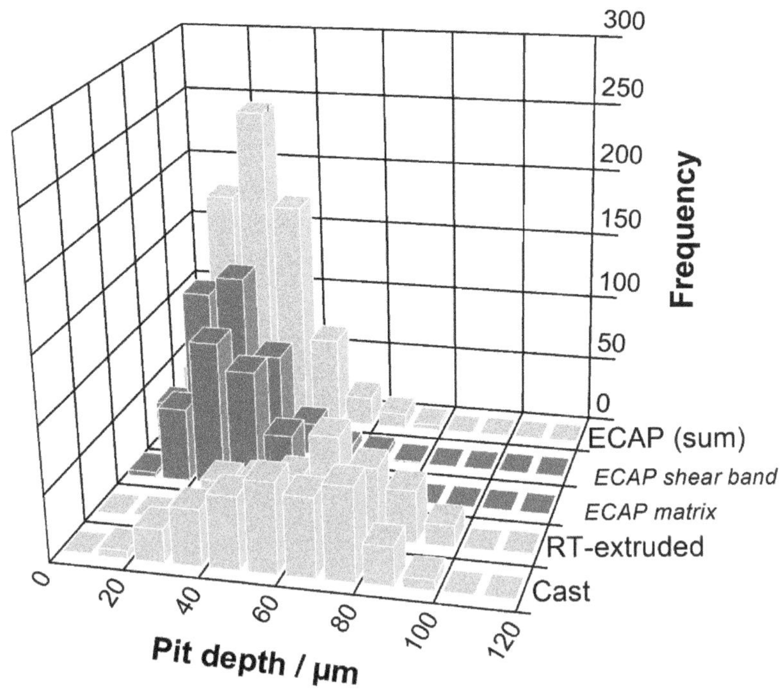

Figure 10. Frequency distribution of pit depths for different material conditions, additionally matrix and shear band are considered separately.

In order to substantiate the result, pit depth frequencies are shown for the cast, RT-extruded and ECAP condition with a detailed breakdown of the matrix and shear band (Figure 10). The pit depths are grouped in 10 µm bins. A fairly similar broad distribution of pit depths between 10 µm and 100 µm was found for the cast and the RT-extruded conditions. The maximum of the frequency distribution is about 65 µm. A tendency to lower pit depths in the cast condition can be noticed. The ECAP condition shows a strikingly enhanced number of pits. More than twice as many pits are evidenced in the material after one ECAP pass, but the depth distribution is quite sharp with its maximum being only about 30 µm. In order to differentiate the corrosion attack between the shear bands and the matrix, distinct frequency distributions of pit depths are shown (Figure 10). A slightly higher number of pits in the shear band contribute to the total pit frequency distribution. However, the pit depths in the shear band tend to be lower compared to the matrix which is attributable to the fragmentation and re-distribution of the constituent particles and moreover, to the enhanced microstrains in the shear bands. Thus, offering additional sites for the pit initiation and growth, pitting corrosion results in a uniform attack and prevents local deep damage.

4. Conclusions

With the focus on deformation localizations, our study demonstrates the effect of extrusion and ECAP on pit initiation and pit growth in an aluminum alloy EN AW-6060. The refined microstructure together with a heterogeneous distribution of microstrains and an increased dislocation density are the controlling microstructural features.

The influence of deformation localizations on pit initiation is proven by local electrochemical measurements restricted to single shear bands of the alloy in the single pass ECAP condition. Deformation localizations show an enhanced pitting susceptibility in relation to the cast and extruded condition. There was a strong indication that the enhanced pitting susceptibility is related to local microstrains.

However, easier pit initiation is associated with more homogeneous and shallow pit propagation in the AlMgSi0.5 alloy after a single ECAP pass. Regardless of the material condition, dissolution of complete grains was observed starting and stopping at primary precipitates predominantly located at the grain boundaries. The origin of corrosion is the cathodic behavior of the primary precipitates with respect to the AlMgSi0.5 alloy. When the dissolution front approaches the next grain boundary, the decorating precipitates are undermined and the local electrochemical cell disappears. Hence, grain fining and precipitate fragmentation increase pitting density but reduce pitting depth.

The pitting morphology shows characteristic channels formed by {100} facetted crystallographic pits. In contrast to the uniform microstrain in the material in the extruded condition, significant differences in microstrains in the shear bands and the alloy matrix are shown and supposed to be the reason for the particular corrosion appearance. The crystallographic pits appear modified in size with regard to the material condition. The reduced size of the pits in the material in ECAP condition is attributed to the cells as the smallest elements subdividing the grains by accumulation of dislocations (cell boundaries) as a response to the microstrains generated by the single ECAP step. The cell size corresponds with the size of the crystallographic pits.

Acknowledgments

Financial support was granted by the Deutsche Forschungsgemeinschaft (German Research Foundation) via the collaborative research centre "High strength aluminium-based lightweight materials for safety components" (DFG SFB 692) as well as via the project LA 1274/27-1. The technical assistance of A. Schulze, E. Benedix, C. Gläser, K. Muhr, G. Röllig, and A. Rauscher is gratefully acknowledged.

Author Contributions

D.N. is the primary author of the paper, and performed analysis of the experimental data. D.D., T.M., P.F. and D.S. contributed to the experimental research work (SEM and EBSD, XRD, electrochemical measurements and analysis). T.L. supervised the work and discussed the results and analysis with the other authors.

Conflicts of Interest

The authors declare no conflict of interest.

References

1. Segal, V.M. Equal channel angular extrusion: from macromechanics to structure formation. *Mater. Sci. Eng. A* **1999**, *271*, 322–333.

2. Segal, V.M. The Method of Material Preparation for Subsequent Working. Patent of the USSR 1977, No. 575892.
3. Vinogradov, A.; Mimaki, T.; Hashimoto, S.; Valiev, R. On the corrosion behaviour of ultra-fine grain copper. *Scripta Mater.* **1999**, *41*, 319–326.
4. Valiev, R.Z.; Langdon, T.G. Principles of equal-channel angular pressing as a processing tool for grain refinement. *Prog. Mater. Sci.* **2006**, *51*, 881–981.
5. Hughes, A.E.; Birbilis, N.; Mol, J.M.C.; Garcia, S.J.; Zhou, X.; Thompson, G.E. High Strength Al-Alloys: Microstructure, Corrosion and Principles of Protection. *Recent Trends in Processing and Degradation of Aluminium Alloys*; InTech: Rijeka, Croatia, 2011.
6. Akiyama, E.; Zhang, Z.; Watanabe, Y.; Tsuzaki, K. Effects of severe plastic deformation on the corrosion behavior of aluminum alloys. *J. Solid State Electrochem.* **2009**, *13*, 277–282.
7. Brunner, J.G.; May, J.; Höppel, H.W.; Göken, M.; Virtanen, S. Localized corrosion of ultrafine-grained Al–Mg model alloys. *Electrochim. Acta* **2010**, *55*, 1966–1970.
8. Ralston, K.D.; Fabianic, D.; Birbilis, N. Effect of grain size on corrosion of high purity aluminium. *Electrochim. Acta* **2011**, *56*, 1729–1736.
9. Ralston, K.D.; Birbilis, N. Effect of grain size on corrosion: A review. *Corrosion* **2010**, doi:10.5006/1.3462912.
10. Ralston, K.D.; Birbilis, N.; Weyand, M.; Hutchinson, C.R. The effect of precipitate size on the yield strength-pitting corrosion correlation in Al-Cu-Mg alloys. *Acta Mater.* **2010**, *58*, 5941–5948.
11. Chung, M.K.; Choi, Y.S.; Kim, J.G.; Kim, Y.M.; Lee, J.C. Effect of the number of ECAP pass time on the electrochemical properties of 1050 Al alloys. *Mater. Sci. Eng. A* **2004**, *366*, 282–291.
12. Son, I.J.; Nakano, H.; Oue, S.; Kobayashi, S.; Fukushima, H.; Horita, Z. Pitting corrosion resistance of ultrafine-grained aluminum processed by severe plastic deformation. *Mater. Trans.* **2006**, *47*, 1163–1169.
13. Hockauf, M.; Meyer, L.W.; Nickel, D.; Alisch, G.; Lampke, T.; Wielage, B.; Krüger, L. Mechanical properties and corrosion behaviour of ultrafine-grained AA6082 produced by equal channel angular pressing. *J. Mater. Sci.* **2008**, *43*, 7409–7417.
14. Wielage, B.; Nickel, D.; Lampke, T.; Alisch, G.; Podlesak, H.; Darwich, S.; Hockauf M. Corrosion characteristics of an ultrafine-grained Al-Mg-Si alloy (AA6082). *Mater. Sci. Forum* **2008**, *584–586*, 988–993.
15. Song, D.; Ma, A.B.; Jiang, J.H.; Lin, P.H.; Yang, D.H. Corrosion behavior of ultra-fine grained industrial pure Al fabricated by ECAP. *Trans. Nonferrous Metals Soc. China* **2009**, *19*, 1065–1070.
16. Korchef, A.; Kahoul, A. Corrosion Behavior of Commercial Aluminum Alloy Processed by Equal Channel Angular Pressing. *Inter. J. Corros.* **2013**, doi:10.1155/2013/983261.
17. Jilani, O.; Njah, N.; Ponthiaux, P. Transition from intergranular to pitting corrosion in fine grained aluminum processed by equal channel angular pressing. *Corros. Sci.* **2014**, *87*, 259–264.
18. Naeini, M.F.; Shariat, M.H.; Eizadjou, M. On the chloride induced pitting of ultra fine grains 5052 aluminum alloy produced by accumulative roll bonding process. *J. Alloys Compd.* **2011**, *509*, 4696–4700.
19. Wei, W.; Wei, K.X.; Du, Q.B. Corrosion and tensile behaviors of ultra-fine grained Al–Mn alloy produced by accumulative roll bonding. *Mater. Sci. Eng. A* **2007**, *454*, 536–541.

20. Suter, T.; Peter, T.; Böhni, H. Microelectrochemical investigations of MnS inclusions. *Mater. Sci. Forum* **1995**, *192–194*, 25–39.
21. Yulinova, A.; Nickel, D.; Frint, P.; Lampke, T. Electrochemical properties of AL-6060 alloy after industrial scale ECAP. *Mater. Sci.* **2012**, *48*, 191–196.
22. Frint, P.; Hockauf, M.; Halle, T.; Strehl, G.; Lampke, T.; Wagner, M.F.X. Microstructural features and mechanical properties after industrial scale ECAP of an Al-6060 alloy. *Mater. Sci. Forum* **2011**, *667–669*, 1153–1158.
23. Frint, P.; Hockauf, M.; Dietrich, D.; Halle, T.; Wagner, M.F.X.; Lampke, T. Influence of strain gradients on the grain refinement during industrial scale ECAP. *Mater. Sci. Eng. Technol.* **2011**, *42*, 680–685.
24. Iwahashi, Y.; Wang, J.T.; Horita, Z.; Nemoto, M.; Langdon, T.G. Principle of equal-channel angular pressing for the processing of ultra-fine grained materials. *Scripta Mater.* **1996**, *35*, 143–146.
25. Dietrich, D.; Berek, H.; Schulze, A.; Scharf, I.; Lampke, T. EBSD and STEM on aluminium alloys subjected to severe plastic deformation. *Prac. Metall.* **2011**, *43*, 136–150.
26. Böhni, H.; Suter, T.; Schreyer, A. Micro- and nanotechniques to study localized corrosion. *Electrochim. Acta* **1995**, *40*, 1361–1368.
27. Birbilis, N.; Buchheit, R.G. Electrochemical characteristics of intermetallic phases in aluminum alloys—An experimental survey and discussion. *J. Electrochem. Soc.* **2005**, *152*, B140–B151.
28. Couper, M.J.; Rinderer, B.; Yao, J.Y. Characterisation of AlFeSi intermetallics in 6000 series aluminium alloy extrusions. *Mater. Sci. Forum* **2006**, *519–521*, 303–308.
29. Milkereit, B.; Wanderka, N.; Schick, C.; Kessler, O. Continuous cooling precipitation diagrams of Al–Mg–Si alloys. *Mater. Sci. Eng.* **2012**, *A550*, 87–96.
30. Birol, Y. Precipitation during homogenization cooling in AlMgSi alloys. *Trans. Nonferrous Metals Soc. China* **2013**, *23*, 1875–1881.
31. Bournane, M.; Berezina, A.; Davydenko, O.; Monastyrska, T.; Molebny, O.; Spuskanyuk, V.; Kotko, A. Effect of severe plastic deformation on structure and properties of Al-Mg-Si alloy of 6060 type. *Mater. Sci. Metall. Eng.* **2013**, *1*, 13–21.
32. Hockauf, K.; Wagner, M.F.X.; Halle, T.; Niendorf, T.; Hockauf, M.; Lampke, T. Influence of precipitates on low-cycle fatigue and crack growth behavior in an ultrafine-grained aluminum alloy. *Acta Mater.* **2014**, *80*, 250–263.
33. Suter, T.; Alkire, R.C. Microelectrochemical studies of pit initiation at single inclusions in Al 2024-T3. *J. Electrochem. Soc.* **2001**, *148*, B36–B42.
34. Wloka, J.; Hack, T.; Virtanen, S. Local electrochemical properties of laser beam-welded high-strength Al–Zn–Mg–Cu alloys. *Mater. Corros.* **2008**, *59*, 5–13.

5

Selective Laser Melting of Ti-45Nb Alloy

Holger Schwab [1,*], Konda Gokuldoss Prashanth [1,†], Lukas Löber [1,‡], Uta Kühn [1] and Jürgen Eckert [1,2]

[1] IFW Dresden, Institut für Komplexe Materialien, Postfach 270116, D-01171 Dresden, Germany; E-Mails: l.loeber@ifw-dresden.de (L.L.); u.kuehn@ifw-dresden.de (U.K.); j.eckert@ifw-dresden.de (J.E.)

[2] TU Dresden, Institut für Werkstoffwissenschaft, D-01062 Dresden, Germany

† Present Address: R & D Engineer, Additive manufacturing Center, Sandvik AB, 81181 Sandviken, Sweden; E-Mail: kgprashanth@gmail.com or prashanth.konda_gokuldoss@sandvik.com.

‡ Present Address: R & D Engineer, Sauer GmbH–DMG MORI SEIKI, D-87459 Pfronten, Germany.

* Author to whom correspondence should be addressed; E-Mail: h.schwab@ifw-dresden.de

Academic Editor: Hugo F. Lopez

Abstract: Ti-45Nb is one of the potential alloys that can be applied for biomedical applications as implants due to its low Young's modulus. Ti-45Nb (wt.%) gas atomized powders were used to produce bulk samples by selective laser melting with three different parameter sets (energy inputs). A β-phase microstructure consisting of elliptical grains with an enriched edge of titanium was observed by scanning electron microscopy and X-ray diffraction studies. The mechanical properties of these samples were evaluated using hardness and compression tests, which suggested that the strength of the samples increases with increasing energy input within the range considered.

Keywords: selective laser melting; Ti-45Nb; additive manufacturing

---

1. Introduction

Ti-based alloys, Co-based alloys and stainless steels are alloy systems that are widely used as biomedical implants [1]. However, most of these alloys show higher Young's modulus (>70 GPa) compared to the Young's modulus of the cortical bone (10–30 GPa) [2]. These differences cause significant stress shielding [3] due to a mechanical stress mismatch between the implant and the bone, results in loosening of the implant or potential mismatch of the implant in the human body [4,5].

β-titanium alloys like Ti-45Nb are one of the candidates for such applications because of its low Young's modulus (~62 GPa) due to the presence of the high-temperature bcc β-phase [6], stabilized by the high niobium content. Also, a high cooling rate will favor the solidification of the bcc β-phase [7]. This relatively low Young's modulus reduces the risk of the undesired stress shielding effect rendering Ti-45Nb a potential material for biomedical implants [8,9]. In addition, Ti-45Nb possesses excellent ductility [10] and good corrosion resistance [8] apart from the excellent biomedical properties rendered by this alloy. The high affinity of titanium towards oxygen yields a self-generating passive layer consisting of TiO_2 [7,11] resulting in superior chemical resistance making it a prominent competitor in the biomedical arena.

Selective Laser Melting (SLM), an additive manufacturing technique developed in the 1980s [12,13], offers the possibility to produce parts with intricate shapes and near net design [14,15] along with high cooling rates resulting in parts with superior properties [15]. Hence, SLM has attracted the attention of both scientific community and industries, where there is a strong demand for the production of novel materials, alloy development and production of near-net shaped materials. SLM involves the production of parts by the interaction of a laser beam selectively with a powder bed [16]. The selective interaction process can be directed from a 3D CAD (Computer Aided Design) model, which gives the processing parameters like the laser power, scanning speed, hatch style etc. layer by layer [16]. The present work focuses on the production of a Ti-45Nb biomedical alloy using the SLM process, followed by structural and microstructural characterization. The mechanical properties of the specimens were evaluated for microhardness and compression tests.

2. Experimental Section

2.1. Powder Fabrication

Ti-Nb rods (4 kg), used to produce powder were commercially acquired from ATI (Albany, NY, USA). Spherical gas atomized powder (GAP) was produced adopting the EIGA method (Electrode Induction-melting Gas Atomization) in an argon (purity 99.999%) atmosphere, by the company TLS Technik (Bitterfeld-Wolfen, Germany). Wet sieving (fluid: Isopropyl alcohol) was employed to select the powder particles ranging between $20 < d < 100$ μm. Bulk samples were produced from the GAP using SLM Solutions SLM 250HL device, equipped with a continuous Nd-YAG laser (P_{max}: 400 W, continuous mode, λ: 1054 nm and build chamber size: 250 mm × 250 mm × 350 mm) with a beam spot size of ~80 μm. The samples were produced on a Ti-45Nb substrate plate and the building chamber was flooded with high purity argon gas (purity 99.9%) to avoid oxidation of the samples during the SLM process. To prevent the influx of oxygen from the atmosphere the building chamber was kept under an overpressure of 10 mbar.

2.2. Parameter Selection

Selected combinations of laser powder P_L, laser speed v, hatch distance along with the energy E_V input employed and the relative density of the samples obtained are shown in Table 1. A strip hatch style with an angle of 74° was set between the layers to ensure the production of defect free samples. This so called strip or cross hatching avoids the formation of interconnected porosity [17]. A layer thickness of 100 μm between layers was used for all experiments.

2.3. Sample Characterization

Structural characterization was carried out by X-ray diffraction (XRD) using D3290 PANalytical X'pert PRO (Almelo, The Netherlands) with Co-Kα radiation (λ = 0.17889 nm) in Bragg-Brentano configuration. Microstructural investigations were carried out by scanning electron microscopy (SEM) in both back scattered and secondary electron modes using Gemini 1530 microscope (Carl Zeiss AG, Oberkochen, Germany) fitted with an energy dispersive X-ray spectrometer (EDX). The density of the samples was evaluated by the Archimedes principle using a Sartorius YDK-01 (0D) device (Sartorius, Göttingen, Germany). Cylindrical samples of 5 mm diameter and 10 mm height (DIN 50106) were produced for compression tests and were tested an INSTRON (model 5869) device at a cross head speed of 0.005 mm/s. The measurements were stopped at a strain of 35%. The hardness was measured with a HMV Shimadzu micro hardness tester with a load of 100 g for 10 s on polished samples. Ten indents were performed under each condition and the average value is reported.

Table 1. Selective Laser Melting (SLM) parameters used for building bulk samples and their related densities measured by Archimedes method.

v [mm/s]	P_L [W]	hatching [%]	E_V [J/mm^3]	rel. density [%]
35	150	25	137	97.7 ± 1.3
35	250	25	172	99.4 ± 0.5
35	250	50	257	99.5 ± 0.2

3. Results and Discussion

3.1. Ti-45Nb GAP Starting Material

The GAP used for the present study consists of spherical particles in the range of 20 < d < 100 μm (Figure 1). The powder show the presence of satellites, which may influence the flow ability of the powder (Figure 1) [18]. The XRD pattern (Figure 2) reveals the presence of a single β-phase (Im-3m), whereas the rod used for gas atomization process consists of two phases (hexagonal α-phase and β-phase).

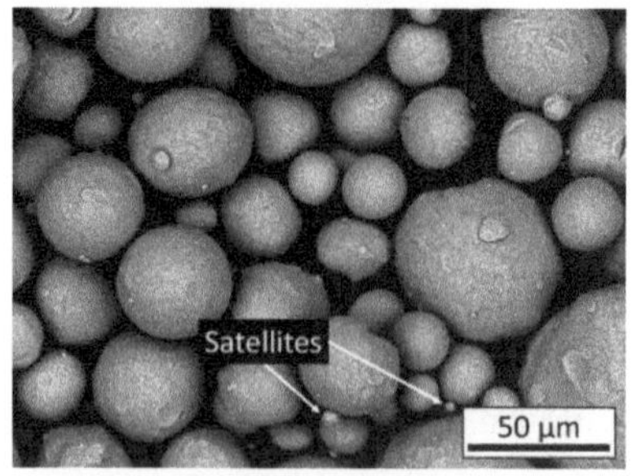

Figure 1. SEM image (Secondary electron mode, 15 kV) showing the spherical shaped Ti-45Nb gas atomized powder (GAP) particles with the presence of satellites.

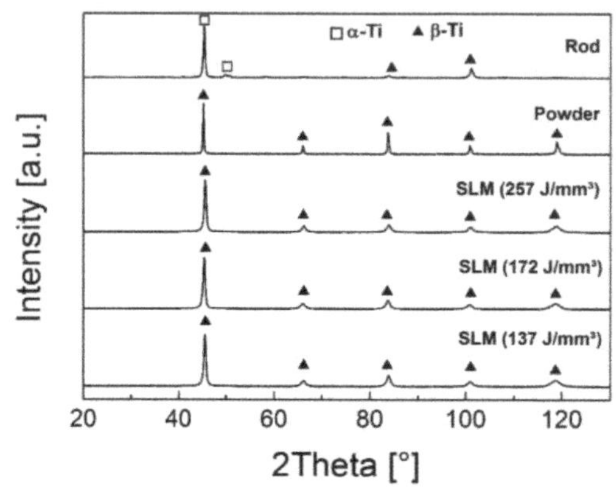

Figure 2. X-ray diffraction patterns of the initial Ti-45Nb rod, the gas atomized powder and of bulk samples produced by SLM with different energy input.

3.2. Bulk SLM Samples

The XRD patterns show that all the samples produced by SLM process employ three different parameter sets (see Table 1) that exhibit the presence of single phased β structure (space group Im-3m) (Figure 2), suggesting that the variation in the processing parameters does not have a significant influence on the phase formation in the Ti-45Nb alloy. Compared to the XRD patterns of the initial cast rod and the GAP, the SLM samples exhibit broad diffraction peaks. The peak broadening can be a result of the presence of internal stresses in the material and/or a fine grained microstructure [12,13,15].

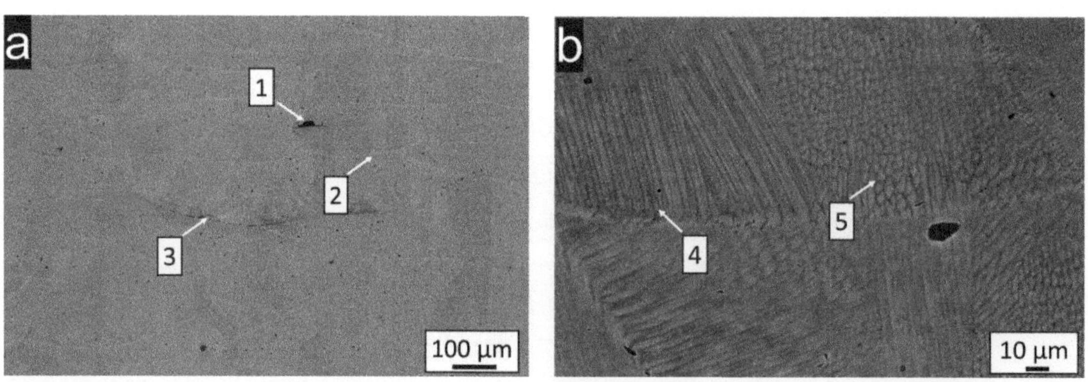

Figure 3. Scanning electron microscopy images of the SLM sample produced with the following parameter set: laser speed, 35 mm/s; power, 250 W and hatching overlap, 50% (Back scattered electron mode, (**a**) 10 kV (**b**) 20 kV, etchant: Kroll's agent).

3.2.1. Microstructure

The SEM microstructure investigations (Figure 3a) show the presence of pores (<5 μm in diameter) in the material, suggesting that the sample is not completely dense (Table 1). Gas inclusions generated during the melting and re-melting process may result in the formation of spherical porosity in the samples [19]. It can be observed from the longitudinal section of the sample that these pores are distributed all over the samples and they do not have a preferential site. Some irregular pores were also observed, which is characteristic of the additive manufacturing process. Some typical SLM features like the scan track boundaries (2) can also be observed clearly along with the presence of partially molten powder particles (3). The SEM images show that the width of the scan tracks was observed not to be uniform, which might be due to the rotation of the hatches to 74° between the layers. The solidification process is dominated by an epitaxial growth mechanism (Figure 3) [20]. The solidified layer acts as substrate for the solidifying melt leading to the formation of elliptical-shaped grains orientated perpendicular to the solidification front [20]. The elliptical-shaped grains grow along the same orientation as their neighbors and are separated by thin grain boundaries. Several of these elliptical-shaped grains with same orientation form one cell separated by primary grain boundaries. A similar behavior can also be seen in other materials like AlSi10Mg, Inconel 939 or 316 L [21–24]. The grain growth occurs in a wavelike fashion and the boundary of the elliptical grains are rich in titanium (Figure 4), which is similar to the cellular kind of microstructure observed in Al-12Si alloys, where the cellular boundaries are rich in Si [15]. The elliptical grains are ~5 μm wide and ranges up to 100 μm in length.

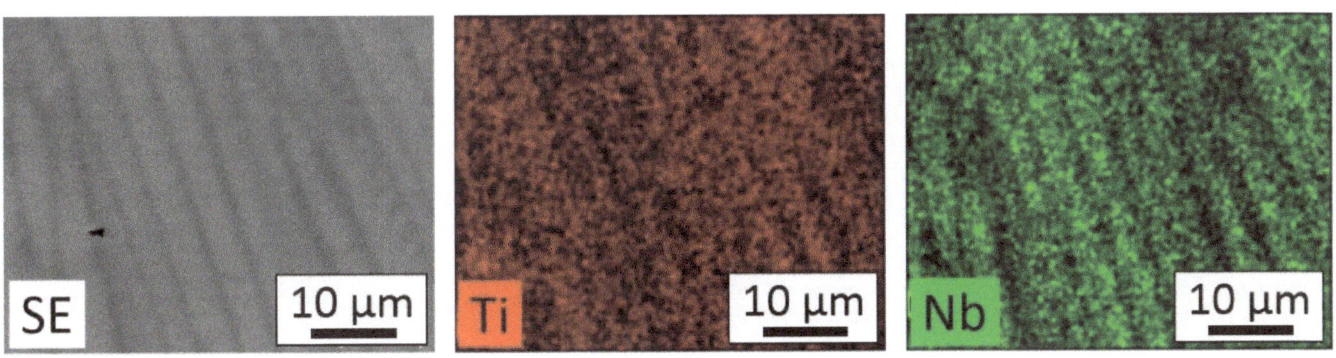

Figure 4. Energy dispersive X-ray spectrometer (EDX) mapping of the elliptical-shaped grains and their segregation edges for samples obtained with the SLM parameters: laser speed, 35 mm/s and laser power, 250 W (etchant: Kroll's agent, 11 kV).

3.2.2. Micro Hardness

The micro hardness data of the SLM samples produced with different parameters is displayed in Figure 5. A micro hardness of ~211 $HV_{0.1}$ is observed for an energy input of 257 J/mm^3 and the value drops to ~192 $HV_{0.1}$ as the energy input is reduced to 137 J/mm^3. The results reveal that the micro hardness is directly correlated with the energy input supplied to the material. With increasing energy input, the thermal gradients change, thereby changing the local heat flux and in turn the cooling rates observed. We expect this to have an influence on the microstructure formation, grain sizes and in turn the hardness (as observed in Figure 5) [20,25–27].

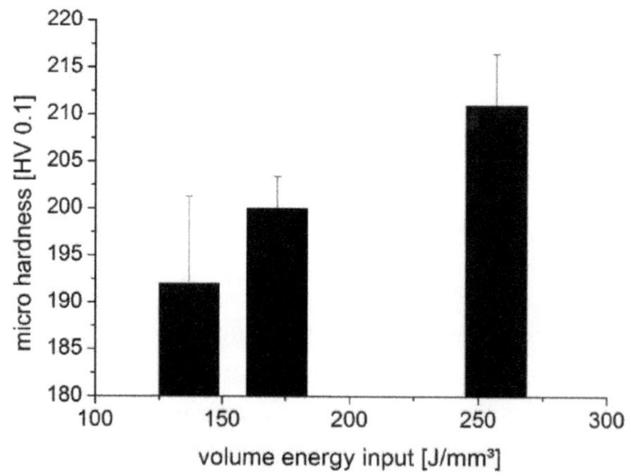

Figure 5. Vickers microhardness data of the samples produced by SLM with different energy input.

3.2.3. Compression Tests

Typical room temperature compression test curves are shown in Figure 6. The compression test was stopped at 35% strain due to technical limitation that may arise from the strain measurement system. It can be observed that the yield strength as well as the compression strength at 35% increases with increasing energy input and the density of the samples. A highest compressive strength at 35% strain of ~723 MPa is observed for the sample with the highest energy input of 257 J/mm^3 (99.5% dense) corroborating the Vickers microhardness results with respect to the increase in strength owing to fine microstructural features observed in the SLM specimens. The strengths of the material under compression do not show a significant variation between the energy densities 137 J/mm^3 and 172 J/mm^3. The compression test results along with the hardness results and density data suggests that the energy input plays a distinct role in creating a fine defect free microstructure and in turn the mechanical properties.

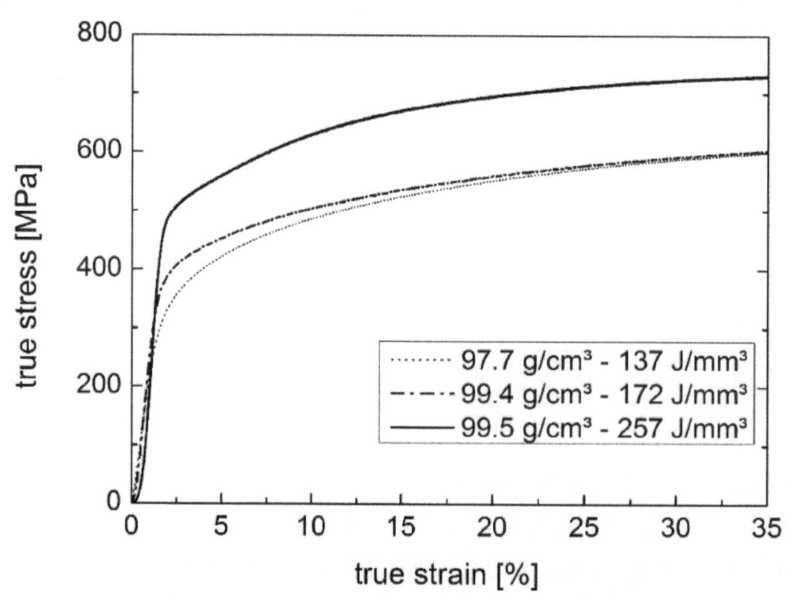

Figure 6. Compression test curves of the samples produced by SLM with different energy input.

However, the compressive modulus was observed to be ~30–50 GPa higher than the Ti-45Nb alloy prepared by other conventional routes [6,28,29], which might be attributed to the fine microstructure. Thanks to the advantage of such additive manufacturing processes, modification in the modulus can be incorporated in the form of introducing structures in the samples instead of the bulk samples. This suggests that additive manufacturing processes like selective laser melting are suitable for producing custom specific biomaterials.

4. Conclusions

Ti-45(wt.%)Nb samples were successfully produced using the SLM technique with three different parameter sets. The XRD patterns show the presence of β-phase with broadening of the peaks compared to the GAP, suggesting the presence of internal stresses and/or grain refinement. The SEM images reveal the presence of elliptical grains with Ti along the boundaries. Both the Vickers micro-hardness as well as the compression test results prove that the strength of the SLM samples increases with increasing energy input.

Author Contributions

Holger Schwab performed the all experiments and characterization studies and created the initial draft. Konda Gokuldoss Prashanth aided in the preparation of the manuscript and conceived the final manuscript. Lukas Löber helped Holger Schwab with the design of the experiments. Uta Kühn supervised the research and Jürgen Eckert contributed to the overall development of the main concepts of this study.

Conflicts of Interest

The authors declare no conflict of interest.

References

1. Epple, M. *Biomaterialien und Biomineralisation. Eine Einführung für Naturwissenschaftler, Mediziner und Ingenieure*; Springer: Wiesbaden, Germany, 2003.
2. Calin, M.; Helth, A.; Gutierrez Moreno, J.J.; Bönisch, M.; Brackmann, V.; Giebeler, L.; Gemming, T.; Lekka, C.E.; Gebert, A.; Schnettler, R.; *et al.* Elastic softening of β-type Ti–Nb alloys by indium (In) additions. *J. Mechan. Behav. Biomed. Mater.* **2014**, *39*, 162–174.
3. Niinomi, M.; Nakai, M.; Hieda, J. Development of new metallic alloys for biomedical applications. *Acta Biomater.* **2012**, *8*, 3888–3903.
4. Haase, K.; Rouhi, G. Prediction of stress shielding around an orthopedic screw: Using stress and strain energy density as mechanical stimuli. *Computers Biol. Med.* **2013**, *43*, 1748–1757.
5. Zhuravleva, K.; Bönisch, M.; Scudino, S.; Calin, M.; Schultz, L.; Eckert, J.; Gebert, A. Phase transformations in ball-milled Ti–40Nb and Ti–45Nb powders upon quenching from the β-phase region. *Powder Technol.* **2014**, *253*, 166–171.
6. Hanada, S.; Matsumoto, H.; Watanabe, S. Mechanical compatibility of titanium implants in hard tissues. *Int. Congress Ser.* **2005**, *1284*, 239–247.

7. Peters, M.; Leyens, C. *Titan und Titanlegierungen*; Wiley-VCH Verlag GmbH: Weinheim, Germany, 2002.
8. ATI 45Nb™ Alloy, Technical Data Sheet. Available online: https://www.atimetals.com/Documents/ti-45nb_tds_en_v1.pdf (accessed on 24 April 2015).
9. Gostin, P.F.; Helth, A.; Voss, A.; Sueptitz, R.; Calin, M.; Eckert, J.; Gebert, A. Surface treatment, corrosion behavior, and apatite-forming ability of Ti-45Nb implant alloy. *J. Biomed. Mater. Res.* **2013**, *101B*, 269–278.
10. Chen, Y.; Li, J.; Tang, B.; Kou, H.; Zhang, F.; Chang, H.; Zhou, L. Grain boundary character distribution and texture evolution in cold-drawn Ti–45Nb wires. *Mater. Lett.* **2013**, *98*, 254–257.
11. Handtrack, D. *Herstellung, Gefüge und Eigenschaften Nanokristalliner und Dispersionsverfestigter Titanwerkstoff*; Shaker: Aachen, Germany, 2008.
12. Prashanth, K.G.; Shakur Shahabi, H.; Attar, H.; Srivastava, V.C.; Ellendt, N.; Uhlenwinkel, V.; Eckert, J.; Scudino, S. Production of high strength Al85Nd8Ni5Co2 alloy by selective laser melting. *Addit. Manuf.* **2015**, *6*, 1–5.
13. Kruth, J.-P.; Froyen, L.; van Vaerenbergh, J.; Mercelis, P.; Rombouts, M.; Lauwers, B. Selective laser melting of iron-based powder. *J. Mater. Process. Technol.* **2004**, *149*, 616–622.
14. Zhang, B.; Dembinski, L.; Coddet, C. The study of the laser parameters and environment variables effect on mechanical properties of high compact parts elaborated by selective laser melting 316l powder. *Mater. Sci. Eng. A* **2013**, *584*, 21–31.
15. Prashanth, K.G.; Scudino, S.; Klauss, H.J.; Surreddi, K.B.; Löber, L.; Wang, Z.; Chaubey, A.K.; Kühn, U.; Eckert, J. Microstructure and mechanical properties of Al–12Si produced by selective laser melting: Effect of heat treatment. *Mater. Sci. Eng. A* **2014**, *590*, 153–160.
16. Löber, L.; Schimansky, F.P.; Kühn, U.; Pyczak, F.; Eckert, J. Selective laser melting of a beta-solidifying TNM-B1 titanium aluminide alloy. *J. Mater. Process. Technol.* **2014**, *214*, 1852–1860.
17. Yadroitsev, I.; Smurov, I. Selective laser melting technology: From the single laser melted track stability to 3D parts of complex shape. *Phys. Procedia* **2010**, *5*, 551–560.
18. Biamino, S.; Penna, A.; Ackelid, U.; Sabbadini, S.; Tassa, O.; Fino, P.; Pavese, M.; Gennaro, P.; Badini, C. Electron beam melting of Ti–48Tl–2Tr–2Tb alloy: Microstructure and mechanical properties investigation. *Intermetallics* **2011**, *19*, 776–781.
19. Ardila, L.C.; Garciandia, F.; González-Díaz, J.B.; Álvarez, P.; Echeverria, A.; Petite, M.M.; Deffley, R.; Ochoa, J. Effect of IN718 recycled powder reuse on properties of parts manufactured by means of selective laser melting. *Phys. Procedia* **2014**, *56*, 99–107.
20. Yadroitsev, I.; Krakhmalev, P.; Yadroitsava, I.; Johansson, S.; Smurov, I. Energy input effect on morphology and microstructure of selective laser melting single track from metallic powder. *J. Mater. Process. Technol.* **2013**, *213*, 606–613.
21. Thijs, L.; Kempen, K.; Kruth, J.-P.; van Humbeeck, J. Fine-structured aluminium products with controllable texture by selective laser melting of pre-alloyed AlSi10Mg powder. *Acta Mater.* **2013**, *61*, 1809–1819.
22. Su, X.; Yang, Y. Research on track overlapping during selective laser melting of powders. *J. Mater. Process. Technol.* **2012**, *212*, 2074–2079.

23. Kanagarajah, P.; Brenne, F.; Niendorf, T.; Maier, H.J. Inconel 939 processed by selective laser melting: Effect of microstructure and temperature on the mechanical properties under static and cyclic loading. *Mater. Sci. Eng. A* **2013**, *588*, 188–195.
24. Ma, P.; Prashanth, K.; Scudino, S.; Jia, Y.; Wang, H.; Zou, C.; Wei, Z.; Eckert, J. Influence of annealing on mechanical properties of Al-20Si processed by selective laser melting. *Metals* **2014**, *4*, 28–36.
25. Kruth, J.-P.; Badrossamay, M.; Yasa, E.; Deckers, J.; Thijs, L.; van Humbeeck, J. Part and material properties in selective laser melting of metals. In Proceedings of the 16th International Symposium on Electromachining, Shanghai, China, 19–23 April 2010.
26. Attar, H.; Prashanth, K.G.; Chaubey, A.K.; Calin, M.; Zhang, L.C.; Scudino, S.; Eckert, J. Comparison of wear properties of commercially pure titanium prepared by selective laser melting and casting processes. *Mater. Lett.* **2015**, *142*, 38–41.
27. Attar, H.; Calin, M.; Zhang, L.C.; Scudino, S.; Eckert, J. Manufacture by selective laser melting and mechanical behavior of commercially pure titanium. *Mater. Sci. Eng. A* **2014**, *593*, 170–177.
28. Ozaki, T.; Matsumoto, H.; Watanabe, S.; Hanada, S. Beta Ti alloys with low Young's modulus. *Mater. Trans.* **2004**, *45*, 2776–2779.
29. Hon, Y.-H.; Wang, J.-Y.; Pan, Y.-N. Composition/phase structure and properties of Titanium-Niobium alloys. *Mater. Trans.* **2003**, *44*, 2384–2390.

Influence of Zr Addition on Structure and Performance of Rare Earth Mg-Based Alloys as Anodes in Ni/MH Battery

Shujun Qiu [1,2], Jianling Huang [2], Hailiang Chu [1,2,*], Yongjin Zou [1,2], Cuili Xiang [1,2], Huanzhi Zhang [1,2], Fen Xu [1,2], Lixian Sun [1,2,*], Liuzhang Ouyang [3] and Huaiying Zhou [1,2]

1 Guangxi Key Laboratory of Information Materials, Guilin University of Electronic Technology, Guilin 541004, China; E-Mails: qiushujun@guet.edu.cn (S.Q.); zouy@guet.edu.cn (Y.Z.); xiangcuili@guet.edu.cn (C.X.); zhanghuanzhi@dicp.ac.cn (H.Z.); xufen@lnnu.edu.cn (F.X.); zhy@guet.edu.cn (H.Z.)

2 School of Materials Science and Engineering, Guilin University of Electronic Technology, Guilin 541004, China; E-Mail: jianlinghuang@126.com

3 School of Materials Science and Engineering, South China University of Technology, Guangzhou 510641, China; E-Mail: meouyang@scut.edu.cn

* Authors to whom correspondence should be addressed; E-Mails: chuhailiang@guet.edu.cn (H.C.); sunlx@guet.edu.cn (L.S.)

Academic Editor: Hugo F. Lopez

Abstract: In this study, the substitution of Mg with Zr in $La_{0.7}Mg_{0.3}(Ni_{0.85}Co_{0.15})_{3.5}$ was carried out with the purpose of improving the electrochemical performances. The structural and hydrogen storage properties in both gas-solid reaction and the electrochemical system were systematically studied on $La_{0.7}(Mg_{0.3-x}Zr_x)(Ni_{0.85}Co_{0.15})_{3.5}$ (x = 0.05, 0.1, 0.2, 0.3) alloys. Each tested alloy is composed of $LaNi_3$ phase, $LaNi_5$ phase and $ZrNi_3$ phase with different phase abundances. The electrochemical studies indicated that all Zr-substituted anodes possessed a much higher cycling capacity retention than pristine $La_{0.7}Mg_{0.3}(Ni_{0.85}Co_{0.15})_{3.5}$. However, the maximum discharge capacity was reduced with the increase of Zr content. The potential-step tests showed that the diffusion of hydrogen atoms inside the anodes was decelerated after the introduction of Zr.

Keywords: hydrogen storage; rare earth Mg-based alloy; Ni/MH battery; elemental substitution; electrochemical performance

---

1. Introduction

Nowadays Ni/MH rechargeable batteries are employed diffusely in all kinds of portable electronic products because of longer working life and less environmental pollution [1,2]. Many classes of alloys have been used as anodes in Ni/MH batteries in past decades. At the very beginning, the first case was AB_5-type alloys contained rare earth elements [3]. However, the lower theoretical discharge capacity impedes their practical application. Subsequently, Zr-based or Ti-based alloys [4], V-based solid solution alloys [5] and Mg-based amorphous alloys [6,7] were extensively studied. However, Ti-based or Zr-based anodes need too many charge/discharge cycles to achieve their maximum discharge capacities, while Mg-based anodes quickly lose their discharge capacities because of the corrosion of Mg in KOH electrolyte [8–11].

In 2000, a newly developed alloys with the chemical composition of RMg_2Ni_9 (R represents Ca or rare earth elements) were reported by Kadir *et al.* [10–13], among which $(La_{0.65}Ca_{0.35})(Mg_{1.32}Ca_{0.68})Ni_9$ had a relatively high capacity of 1.87 wt% H_2 in gas-solid system at 283 K under 3.30 MPa H_2 pressure. Chen and coworkers demonstrated that a much higher discharge capacity of 356 mAh/g was achieved for $LaCaMgNi_9$ alloy [8]. For the case of nonstoichiometric La–Mg–Ni–Co-type alloys, the discharge capacity could reach to an extremely high value of 410 mAh/g for $La_{0.7}Mg_{0.3}Ni_{2.8}Co_{0.5}$ alloy with the comparatively good cycle durability during the initial 30 cycles [14]. Furthermore, the partial substitution of Ni with Mn or Al could give rise to an enhancement of the electrochemical properties [15–17]. Although the overall electrochemical properties are improved to some extent in the aforementioned literatures, the charge/discharge cyclic stability and the high-rate dischargeability (*HRD*) of anodes are urgently requiring further improvement. In general, the elemental substitution to form multi-component alloy is perceived as a very accessible method to enhance the electrochemical performances [18–20]. Metallic Zr, as an element for absorbing hydrogen, tends to form an oxide passive film insoluble in 6 M KOH electrolyte, which prevents further corrosion of anodes [21]. This was evidenced in Mg-based alloys, in which the introduction of Zr could result in a remarkable amelioration of the cyclic life [21] and maximum discharge capacity [22].

In this paper, Zr was selected as a partially substituting element for Mg with the purpose of the performance enhancement of La–Mg–Ni–Co-type alloys as anodes in Ni/MH battery. The structural and hydrogen storage properties in both the electrochemical system and gas-solid reaction of $La_{0.7}(Mg_{0.3-x}Zr_x)(Ni_{0.85}Co_{0.15})_{3.5}$ ($x = 0.05$–0.3) alloys were systematically studied.

2. Experimental Section

2.1. Materials Preparation

The purity of all raw materials is above 99%. Under argon atmosphere, $La_{0.7}(Mg_{0.3-x}Zr_x)(Ni_{0.85}Co_{0.15})_{3.5}$ (x = 0.05, 0.1, 0.2, 0.3) hydrogen storage alloys were prepared by magnetic levitation melting in a water-cooled copper crucible. To obtain homogeneous samples, the ingots were turned over and melted again for three times. Then, the ingots were mechanically crushed and ground into the powder using a mortar and pestle. The global chemical composition of each alloy was measured on an ICP–OES (Optima 8000, Perkin Elmer Inc., Waltham, MA, USA).

2.2. Materials Characterization

The phase structure was characterized on a PANalytical X'pert diffractometer with Cu K_α radiation. The diffraction patterns were analyzed using Topas software [23]. P–C isotherms were determined by a conventional Sievert-type apparatus. The activation was performed at 303 K under 3 MPa H_2 pressure. The following dehydrogenation was carried out at 573 K under dynamic vacuum for 3 h. After repeated three cycles, P–C isotherms were carried out at 303 K.

2.3. Electrochemical Tests

Each tested anode was fabricated through homogenously mixing alloy powder with Ni powder. The mixture was then pressed into a pellet with a diameter of 10 mm under a pressure of 30 MPa. Each side of the anode pellet was coated with a rounded foam nickel sheet of about 25 mm in a diameter, then pressed at 6 MPa and tightly spot-welded at the edge of foam nickel. A nickel lead wire was attached to this pressed foam nickel sheet by spot welding. The electrochemical measurements were performed in a standard open trielectrode electrolysis cell comprising a working anode of alloy pellet studied, a sintered $Ni(OH)_2$/NiOOH counter electrode, and a Hg/HgO reference electrode immersed in 6 M KOH electrolyte. High-rate dischargeability (*HRD*) and charge/discharge cycles were tested on an automatic instrument (LAND). The anodes were charged for 3 h at a current density of 300 mAh/g, rested for 5 min and then discharged to the cut-off potential of −0.6 V *versus* Hg/HgO reference electrode at a current density of 100 mAh/g. To determine the high rate dischargeability of the alloy anodes, the discharge capacities at different discharge current density were measured. The other electrochemical measurements including linear polarization (LP), electrochemical impedance spectroscopy (EIS), anodic polarization (AP) and potential-step were conducted on a Zahner Elektrik IM6e electrochemical workstation. AP and LP were measured by scanning the anode potential at a rate of 0.1 mV/s from −5 to 5 mV (*versus* open circuit potential) and 5 mV/s from 0 to 600 mV (*versus* open circuit potential) at 50% depth of discharge (DOD), respectively. EIS was conducted from 10 kHz to 5 mHz with an amplitude of 5 mV *versus* open circuit potential at 50% DOD. Before EIS tests, the anodes were fully activated by charge-discharge for three cycles.

3. Results and Discussion

XRD patterns for $La_{0.7}(Mg_{0.3-x}Zr_x)(Ni_{0.85}Co_{0.15})_{3.5}$ alloys were shown in Figure 1. Each alloy is composed of $LaNi_3$ phase and $LaNi_5$ phase except for a residual phase of $ZrNi_3$. The detailed information for crystal structure including phase composition, phase abundance, cell constants, and cell volume are summarized in Table 1. The solubility of Zr in $LaNi_5$ phase is very limited [24] and can be seen from the negligible change in the cell volume of $LaNi_5$ phase (Table 1). However, Zr must participate in $LaNi_3$ phase which is obviously from the enlarged unit cell of $LaNi_3$ phase due to Zr partially replacement of the smaller Mg and/or entrance into the Ni-site from x = 0.05 to 0.3. With Zr content increasing from 0.05 to 0.3, the phase abundance of $LaNi_3$ increases, while the phase abundance of $LaNi_5$ decreases. This indicates that the introduction of Zr into alloys facilitates the formation of $LaNi_3$ phase, which is different from the gradual substitution of La with Zr in La–Mg–Ni-type alloys [25]. It is noted that the phase abundance of non-hydrogen-absorbing phase

($ZrNi_3$) increases with x increasing. Therefore, such a variation of phase abundance will give rise to some effects on the electrochemical properties and gaseous hydrogen storage of $La_{0.7}(Mg_{0.3-x}Zr_x)(Ni_{0.85}Co_{0.15})_{3.5}$ alloys.

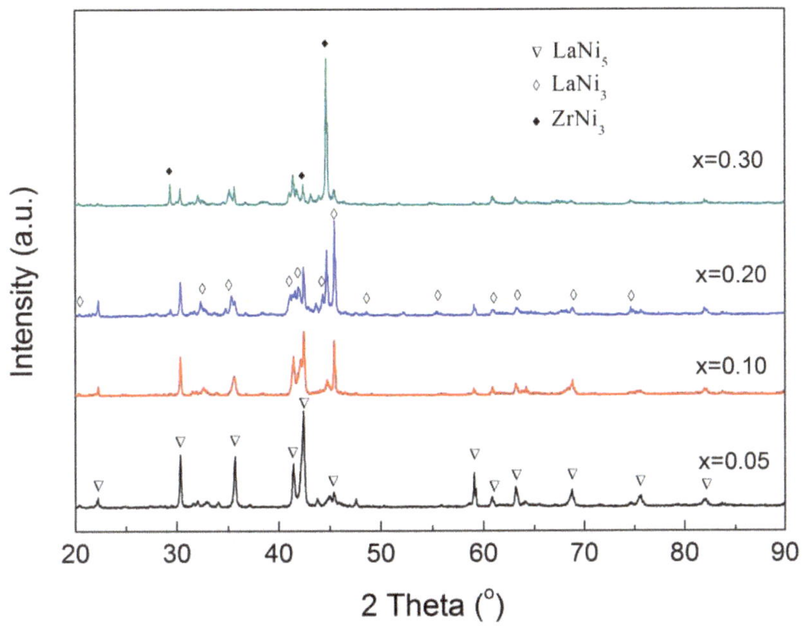

Figure 1. XRD patterns of $La_{0.7}(Mg_{0.3-x}Zr_x)(Ni_{0.85}Co_{0.15})_{3.5}$ alloys.

Table 1. The detailed structural information of $La_{0.7}(Mg_{0.3-x}Zr_x)(Ni_{0.85}Co_{0.15})_{3.5}$ alloys.

Samples	Phases	Space group (no.)	Phase abundance (wt%)	Lattice parameters (Å)		Cell volume (Å³)	Parameters of fit
				a	*c*		
	$LaNi_5$	P6/mmm (191)	65.5	5.0301(5)	3.9906(5)	87.44(2)	
$x = 0.05$	$LaNi_3$	R-3m (166)	24.1	5.0388(1)	24.391(1)	536.3(3)	$R_{wp} = 2.7$; $R_p = 2.9$
	$ZrNi_3$	$P6_3$/mmc (194)	10.4	5.2805(3)	4.2780(3)	103.3(1)	
	$LaNi_5$	P6/mmm (191)	49.1	5.0255(6)	3.9862(5)	87.19(2)	
$x = 0.1$	$LaNi_3$	R-3m (166)	35.3	5.0484(1)	24.601(1)	543.0(3)	$R_{wp} = 3.0$; $R_p = 3.4$
	$ZrNi_3$	$P6_3$/mmc (194)	15.6	5.3003(4)	4.2805(6)	104.1(2)	
	$LaNi_5$	P6/mmm (191)	33.2	5.0232(6)	3.9832(4)	87.04(2)	
$x = 0.2$	$LaNi_3$	R-3m (166)	40.6	5.0645(2)	24.824(1)	551.4(5)	$R_{wp} = 4.8$; $R_p = 5.0$
	$ZrNi_3$	$P6_3$/mmc (194)	26.2	5.3166(3)	4.2858(6)	104.9(20)	
	$LaNi_5$	P6/mmm (191)	23.5	5.0300(1)	3.9872(8)	87.36(4)	
$x = 0.3$	$LaNi_3$	R-3m (166)	47.8	5.0797(2)	25.087(1)	560.6(4)	$R_{wp} = 3.9$; $R_p = 4.3$
	$ZrNi_3$	$P6_3$/mmc (194)	28.7	5.4320(1)	3.9806(1)	101.7(5)	

To understand the hydrogen storage properties of $La_{0.7}(Mg_{0.3-x}Zr_x)(Ni_{0.85}Co_{0.15})_{3.5}$ alloys in gas-solid system, P–C isotherms were conducted through a Sieverts' method [26] at 303 K. As shown in Figure 2, the plateau pressure clearly increases while the absorbed hydrogen capacity decreases from 1.51 (x = 0.05) to 1.02 wt% (x = 0.3), which is related to the gradual increase of the amount for non-hydrogen-absorbing $ZrNi_3$ in the alloys.

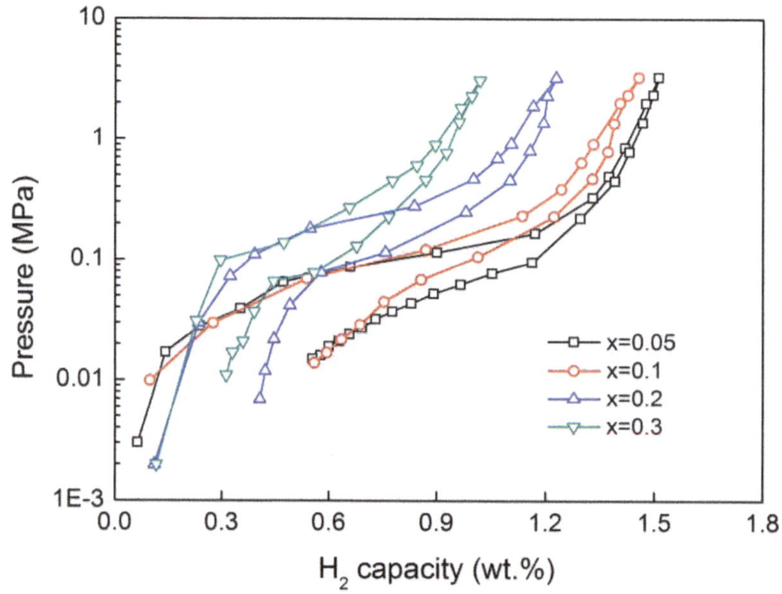

Figure 2. P–C isotherms of $La_{0.7}(Mg_{0.3-x}Zr_x)(Ni_{0.85}Co_{0.15})_{3.5}$ alloys.

Figure 3 shows charge/discharge cyclic life for $La_{0.7}(Mg_{0.3-x}Zr_x)(Ni_{0.85}Co_{0.15})_{3.5}$ anodes, and Table 2 lists their detailed electrochemical properties. All anodes could achieve their maximum discharge capacities (denoted as C_{max}) within three cycles, indicating that they can be easily activated. The C_{max} is reduced from 341.1 ($x = 0.05$) to 176.4 mAh/g ($x = 0.3$), which agrees well with the results obtained from P–C isotherms. Two factors may be responsible for this variation. First, LaMg-rich particles in the alloys serve as preferred sites for absorbing hydrogen, which is beneficial to the improvement of the discharge capacity [27]. As Mg content in the alloy is reduced by increasing substitution with Zr, the number of LaMg-rich particles is reduced, which gives rise to the gradual decrease of the discharge capacity. Second, the increasing amount of non-hydrogen-absorbing $ZrNi_3$ could also result in the reduction of the discharge capacity. Figure 4 indicates that the capacities in both gas-solid hydrogen storage and the electrochemical reaction have similar reduction with increasing abundance of $ZrNi_3$ phase. However, the cyclic stability is gradually improved with increasing content of Zr. For instance, the capacity retention after 60 cycles (denoted as C_{60}/C_{max}) increases from 53.9% ($x = 0.05$) to 60.8% ($x = 0.2$), then reduces to 58.4% ($x = 0.3$). Note that C_{60}/C_{max} of all anodes with Zr-substitution in our study is higher than that of Zr-free anode ($C_{60}/C_{max} = 45.9\%$) [28]. In addition, C_{100}/C_{max} is still kept at 40.3% for anode with $x = 0.1$. As we all know, the capacity fading is mainly resulting from the particle pulverization upon cycling and the surface oxidation and/or corrosion of Mg, Ni and La elements [29]. In the La–Mg–Ni system, Mg and Ni were involved in a mixed hydroxide of (Mg, Ni)$(OH)_2$ and La transformed into nanoporous $La(OH)_3$ needles [30]. The cycling improvement in this study could be related to the fact that the extent of oxidation/corrosion of Mg and La is reduced because ZrO_2 was formed as a passive film on anode surface. As Zr content increases, the thickness of formed ZrO_2 film gradually increases at $x = 0.05$–0.2, which prevents the corrosion/oxidation of La and Mg and thus gives rise to an improvement of cyclic stability [25]. Similar improvements are observed in Zr-substituted Mg-based [21] and Ti-based anodes [31].

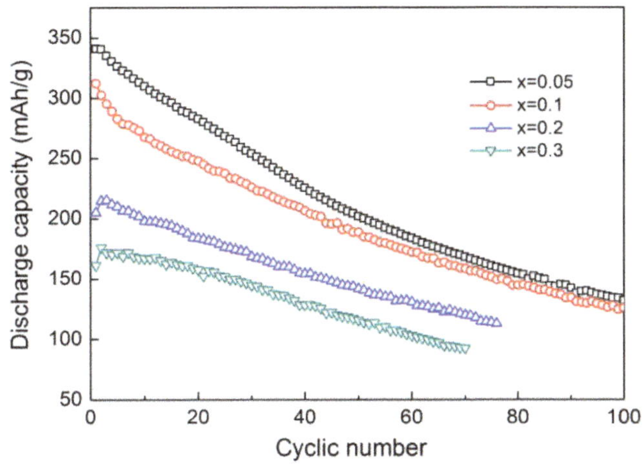

Figure 3. Cyclic life curves of $La_{0.7}(Mg_{0.3-x}Zr_x)(Ni_{0.85}Co_{0.15})_{3.5}$ anodes.

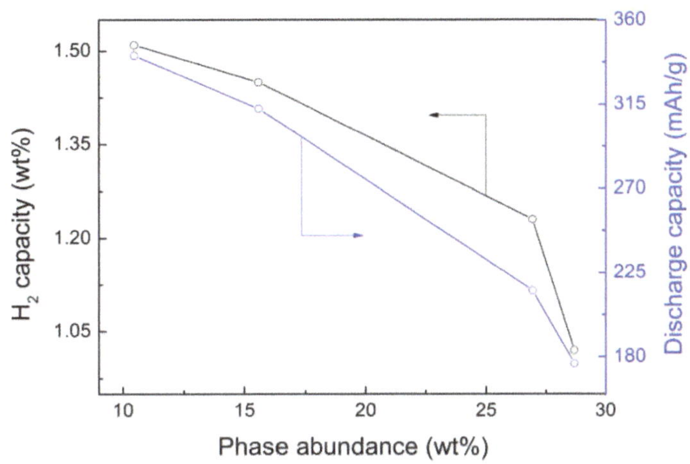

Figure 4. Relationship between hydrogen capacity (discharge capacity) and phase abundance of $ZrNi_3$.

Table 2. The electrochemical and gaseous hydrogen storage properties of $La_{0.7}(Mg_{0.3-x}Zr_x)(Ni_{0.85}Co_{0.15})_{3.5}$ alloys.

Samples	H_2 (wt.%)	C_{max} (mAh/g)	N_a	C_{60}/C_{max} (%)	HRD_{800} (%)
$x = 0.05$	1.51 ± 0.02	341.1 ± 0.5	1	53.9 ± 0.5	94.2 ± 0.8
$x = 0.1$	1.45 ± 0.01	312.8 ± 0.8	1	55.2 ± 0.3	90.6 ± 0.6
$x = 0.2$	1.23 ± 0.03	215.5 ± 0.6	3	60.8 ± 0.6	82.5 ± 0.3
$x = 0.3$	1.02 ± 0.03	176.4 ± 0.4	2	58.4 ± 0.7	78.9 ± 0.4

Figure 5 shows *HRD* of $La_{0.7}(Mg_{0.3-x}Zr_x)(Ni_{0.85}Co_{0.15})_{3.5}$ anodes at several higher discharge current densities. At 800 mA/g, *HRD* is reduced from 94.2% ($x = 0.05$) to 78.9% ($x = 0.3$). As a rule of thumb, *HRD* is an important parameter for Ni/MH batteries, which is principally determined by the anode reaction kinetics. The detailed kinetic properties can be jointly described using the exchange current density I_0, the limiting current density I_L, and the hydrogen diffusion coefficient D, which will be discussed in the forthcoming sections.

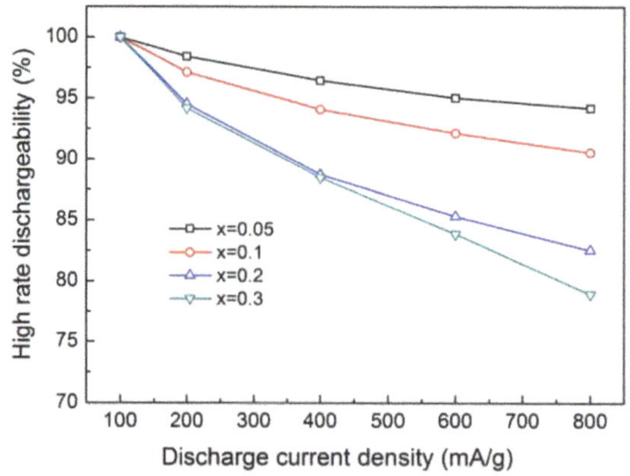

Figure 5. High-rate dischargeability (*HRD*) of $La_{0.7}(Mg_{0.3-x}Zr_x)(Ni_{0.85}Co_{0.15})_{3.5}$ anodes.

Figure 6 shows EIS of $La_{0.7}(Mg_{0.3-x}Zr_x)(Ni_{0.85}Co_{0.15})_{3.5}$ anodes. Each EIS is comprised of a tilted line and two semicircles. Based on Kuriyama's interpretation analyzed by an equivalent circuit (see the inset in Figure 6), the middle-frequency semicircle was ascribed to charge-transfer resistance [32]. Clearly shown in Figure 6, the radius of the middle-frequency semicircle is first increased to $x = 0.2$ and then remarkably reduced at $x = 0.3$, which demonstrates that the charge-transfer reaction resistance has the same variation trend. It was reported that a thicker passive film on anode surface gives rise to a higher charge-transfer reaction resistance [21]. In the present work, with the increase of Zr content, the passive film is gradually thickening, which blocks the charge-transfer reaction when x increases to $x = 0.2$. For the case of $x = 0.3$, Mg was completely substituted by Zr, the charge-transfer resistance is reduced to a certain extent, which can be related to the absence of MgO and/or $Mg(OH)_2$. As a consequence, the thickness of passive film is reduced at $x = 0.3$ and thus the charge-transfer reaction is facilitated.

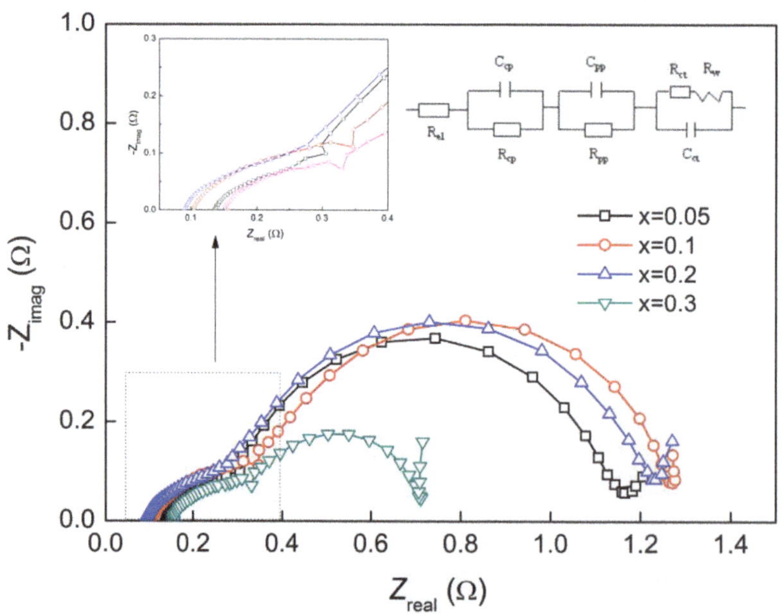

Figure 6. Electrochemical impedance spectroscopy (EIS) of $La_{0.7}(Mg_{0.3-x}Zr_x)(Ni_{0.85}Co_{0.15})_{3.5}$ anodes. The insets show the enlargement of EIS in high-frequency region and an equivalent circuit.

Table 3. Several parameters related to the anode reaction kinetics of $La_{0.7}(Mg_{0.3-x}Zr_x)(Ni_{0.85}Co_{0.15})_{3.5}$ alloys.

Samples	Polarization resistance, R_p (mΩ)	Exchange current density, I_0 (mA/g)	Limiting current density, I_L (mA/g)	Hydrogen diffusion coefficient, D (cm²/s)
$x = 0.05$	132.9 ± 1.0	196.5 ± 0.6	3432.1 ± 8.1	$(2.85 \pm 0.04) \times 10^{-10}$
$x = 0.1$	141.7 ± 0.8	184.3 ± 0.8	2871.1 ± 5.4	$(2.62 \pm 0.02) \times 10^{-10}$
$x = 0.2$	148.7 ± 0.4	175.6 ± 0.7	2802.4 ± 3.4	$(2.60 \pm 0.01) \times 10^{-10}$
$x = 0.3$	80.4 ± 0.7	324.7 ± 0.9	2489.4 ± 9.4	$(2.39 \pm 0.03) \times 10^{-10}$

For the anode reaction kinetics, I_0 is another important parameter [33,34], which can be determined from polarization resistance R_p. R_p could be determined from LP curve shown in Figure 7 and are summarized in Table 3. R_p increases first from 132.9 ($x = 0.05$) to 148.7 mΩ ($x = 0.2$) and then reduces to 80.4 mΩ ($x = 0.3$). Its variation is in line with charge-transfer resistance from EIS. Based on R_p, I_0 can be calculated through the Equation (1) [34]:

$$I_0 = \frac{RT}{FR_p} \tag{1}$$

in which R, F, and T are gas constant, Faraday constant and absolute temperature, respectively. Its variation with x increasing indicates that the anode reaction kinetics is first retarded and then improved due to the reduction of the active sites resulting from a gradually thickening of the passive film. For $x = 0.3$, I_0 increases to a certain extent, which may be due to the reduction of passive film resulting from the absence of MgO and/or $Mg(OH)_2$ on Mg-free anode surface.

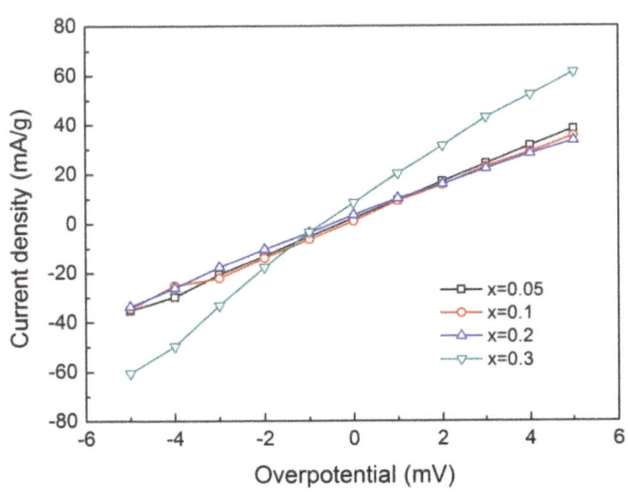

Figure 7. Linear polarization curves of $La_{0.7}(Mg_{0.3-x}Zr_x)(Ni_{0.85}Co_{0.15})_{3.5}$ anodes.

Figure 8 shows the anodic polarization (AP) curves. For each AP curve, with the increase of overpotential, there is a maximum for the anodic current density, denoted as limiting current density I_L, indicating an oxidation reaction occurring on anode surface and a resistance of further penetration for hydrogen atoms by generated oxidation products [35]. Therefore, I_L can be related to hydrogen diffusion inside anodes and the resistance on anode surface [33]. I_L decreases from 3432.1 ($x = 0.05$) to 2489.4 ($x = 0.3$), suggesting that the hydrogen diffusion is decelerated.

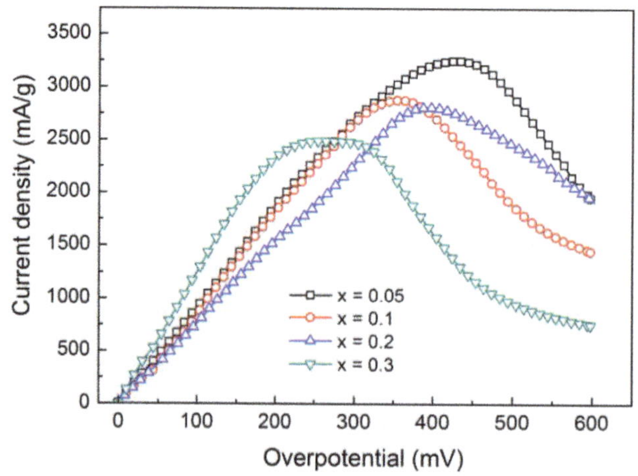

Figure 8. Anodic polarization curves of $La_{0.7}(Mg_{0.3-x}Zr_x)(Ni_{0.85}Co_{0.15})_{3.5}$ anodes.

Figure 9 shows the discharging curves at a constant potential. Each curve of semilograrithmic plot has two regions during the tested time [36]. In the beginning of each curve, the current is quickly reduced because of a rapid hydrogen depletion in electro-oxidation. For the case of the latter region, the current shows a slow and linear decrease due to hydrogen diffusion as a rate-determining step. Under this circumstance, hydrogen is provided inside the anodes based on the hydrogen concentration. Modeled as the finite hydrogen diffusion [34], *D* could be calculated using the fitted slope of the linear part through the Equation (2) [37]:

$$\log i = \log\left(\frac{6FD}{da^2}(C_0 - C_s)\right) - \frac{\pi^2}{2.303}\frac{D}{a^2}t \tag{2}$$

where i, D, d, a, C_0, C_s, and t represent the diffusion current density (A/g), the hydrogen diffusion coefficient (cm^2/s), the alloy density (g/cm^3), the alloy particle radius (cm), the initial hydrogen concentration (mol/cm^3), the surface hydrogen concentration (mol/cm^3), and the discharge time (s), respectively. Using the alloy particle radius of 25 μm roughly determined by sieving, *D* is calculated and also summarized in Table 3. *D* is monotonously reduced from 2.85×10^{-10} to 2.39×10^{-10} cm^2/s, in line with the results from AP curves.

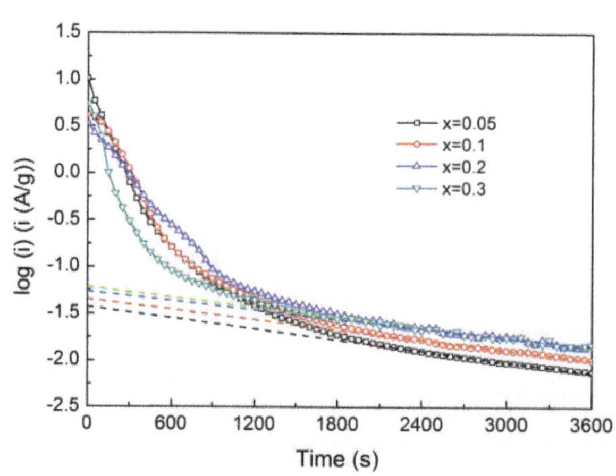

Figure 9. Semilogarithmic curves of discharge curves of $La_{0.7}(Mg_{0.3-x}Zr_x)(Ni_{0.85}Co_{0.15})_{3.5}$ anodes.

4. Conclusions

The structure and hydrogen storage properties in both gas-solid reaction and the electrochemical system of $La_{0.7}(Mg_{0.3-x}Zr_x)(Ni_{0.85}Co_{0.15})_{3.5}$ alloys were systemically investigated. XRD patterns show that all studied alloys are composed of $LaNi_3$ phase, $LaNi_5$ phase and $ZrNi_3$ residual phase. The electrochemical tests indicate that C_{60}/C_{max} first increases from 53.9% (x = 0.05) to 60.8% (x = 0.2), then reduces to 58.4% (x = 0.3) and that C_{max} is reduced from 341.1 (x = 0.05) to 176.4 mAh/g (x = 0.3), consistent with the capacities obtained from P–C isotherms in gas-solid system. The electrochemical tests such as EIS, LP and AP indicate that the Zr substitution for Mg retards the anode reaction kinetics. Furthermore, D is gradually reduced from 2.85×10^{-10} to 2.39×10^{-10} cm^2/s with the increase of Zr content.

Acknowledgments

This research was jointly sponsored by Guangxi Key Laboratory of Information Materials (Guilin University of Electronic Technology), P.R. China (Project No. 1210908–03–K), National Natural Science Foundation of China (51361006, 51401059, 51461010, 51361005, 51371060, 51201041, and 51201042), Guangxi University Research Project (YB2014132, 2013ZD023) and Guangxi Natural Science Foundation (2014GXNSFAA118043, 2013GXNSFBA019239, 2013GXNSFBA019034, 2014GXNSFAA118333). This work was partially supported by the Guangxi Collaborative Innovation Center of Structure and Property for New Energy Materials.

Conflicts of Interest

The authors declare no conflict of interest.

References

1. Liu, Y.F.; Pan, H.G.; Gao, M.X.; Wang, Q.D. Advanced hydrogen storage alloys for Ni/MH rechargeable batteries. *J. Mater. Chem.* **2011**, *21*, 4743–4755.
2. Zhao, X.Y.; Ma, L.Q. Recent progress in hydrogen storage alloys for nickel/metal hydride secondary batteries. *Int. J. Hydrogen Energy* **2009**, *34*, 4788–4796.
3. Li, W.; Du, Y.L. Effects of Y Substitution for La on the microstructure and electrochemical properties of $LaNi_{3.55}Mn_{0.4}Al_{0.3}Co_{0.75}$ hydrogen storage alloys. *Mater. Trans.* **2008**, *49*, 2229–2232.
4. Chu, H.L.; Zhang, Y.; Sun, L.X.; Qiu, S.J.; Qi, Y.N.; Xu, F.; Yuan, H.T. Structure and electrochemical properties of composite electrodes synthesized by mechanical milling Ni-free $TiMn_2$-based alloy with La-based alloys. *J. Alloys Compd.* **2007**, *446–447*, 614–619.
5. Tsukahara, M.; Kamiya, T.; Takahashi, K.; Kawabata, A.; Sakurai, S.; Shi, J.; Takeshita, H.T.; Kuriyama, N.; Sakai, T. Hydrogen storage and electrode properties of V-based solid solution type alloys prepared by a thermic process. *J. Electrochem. Soc.* **2000**, *147*, 2941–2944.
6. Zhang, Y.; Chen, L.X.; Lei, Y.Q.; Wang, Q.D. The effect of partial substitution of Ti with Zr, Cr or V in the $Mg_{35}Ti_{10}Ni_{55}$ electrode alloy on its electrochemical performance. *Electrochim. Acta* **2002**, *47*, 1739–1746.

7. Iwakura, C.; Inoue, H.; Furukawa, N.; Nohara, S. Effects of carbon crystallinity on hydriding-dehydriding and charge-discharge characteristics of MgNi alloy–carbon material composites. *Mater. Trans.* **2002**, *43*, 2706–2710.
8. Chen, J.; Kuriyama, N.; Takeshita, H.T.; Tanaka, H.; Sakai, T.; Haruta, M. Hydrogen storage alloys with $PuNi_3$-type structure as metal hydride electrodes. *Electrochem. Solid–State Lett.* **2000**, *3*, 249–252.
9. Qiu, S.J.; Chu, H.L.; Zhang, J.; Zhang, Y.; Sun, L.X.; Xu, F.; Sun, D.L.; Ouyang, L.Z.; Zhu, M.; Grolier, J.P.E.; Frenkel, M. Effect of La partial substitution for Zr on the structural and electrochemical properties of $Ti_{0.17}Zr_{0.08-x}La_xV_{0.35}Cr_{0.1}Ni_{0.3}$ (x = 0–0.04) electrode alloys. *Int. J. Hydrogen Energy* **2009**, *34*, 7246–7252.
10. Kadir, K.; Kuriyama, N.; Sakai, T.; Uehara, I.; Eriksson, L. Structural investigation and hydrogen capacity of $CaMg_2Ni_9$: A new phase in the AB_2C_9 system isostructural with $LaMg_2Ni_9$. *J. Alloys Compd.* **1999**, *284*, 145–154.
11. Kadir, K.; Sakai, T.; Uehara, I. Structural investigation and hydrogen storage capacity of $LaMg_2Ni_9$ and $(La_{0.65}Ca_{0.35})(Mg_{1.32}Ca_{0.68})Ni_9$ of the AB_2C_9 type structure. *J. Alloys Compd.* **2000**, *302*, 112–117.
12. Kadir, K.; Sakai, T.; Uehara, I. Synthesis and structure determination of a new series of hydrogen storage alloys; RMg_2Ni_9 (R=La, Ce, Pr, Nd, Sm and Gd) built from $MgNi_2$ Laves-type layers alternating with AB_5 layers. *J. Alloys Compd.* **1997**, *257*, 115–121.
13. Kadir, K.; Sakai, T.; Uehara, I. Structural investigation and hydrogen capacity of YMg_2Ni_9 and $(Y_{0.5}Ca_{0.5})(MgCa)Ni_9$: New phases in the AB_2C_9 system isostructural with $LaMg_2Ni_9$. *J. Alloys Compd.* **1999**, *287*, 264–270.
14. Kohno, T.; Yoshida, H.; Kawashima, F.; Inaba, T.; Sakai, I.; Yamamoto, M.; Kanda, M. Hydrogen storage properties of new ternary system alloys: La_2MgNi_9, $La_5Mg_2Ni_{23}$, La_3MgNi_{14}. *J. Alloys Compd.* **2000**, *311*, L5–L7.
15. Liu, Y.F.; Cao, Y.H.; Huang, L.; Gao, M.X.; Pan, H.G. Rare earth–Mg–Ni-based hydrogen storage alloys as negative electrode materials for Ni/MH batteries. *J. Alloys Compd.* **2011**, *509*, 675–686.
16. Liu, Y.F.; Pan, H.G.; Gao, M.X.; Zhu, Y.F.; Lei, Y.Q.; Wang, Q.D. The effect of Mn substitution for Ni on the structural and electrochemical properties of $La_{0.7}Mg_{0.3}Ni_{2.55-x}Co_{0.45}Mn_x$ hydrogen storage electrode alloys. *Int. J. Hydrogen Energy* **2004**, *29*, 297–305.
17. Pan, H.G.; Liu, Y.F.; Gao, M.X.; Zhu, Y.F.; Lei, Y.Q.; Wang, Q.D. Electrochemical properties of the $La_{0.7}Mg_{0.3}Ni_{2.65-x}Mn_{0.1}Co_{0.75}Al_x$ (x = 0–0.5) hydrogen storage alloy electrodes. *J. Electrochem. Soc.* **2005**, *152*, A326–A332.
18. Wu, J.M.; Li, J.; Zhang, W.P.; Muo, F.J.; Tai, L.C.; Xu, R.G. The study on the hydrogen storage properties of $MmNi_{3.55}Co_{0.75}Mn_{0.7-x}Al_x$ compounds. *J. Alloys Compd.* **1997**, *248*, 180–184.
19. Notten, P.H.L.; Latroche, M.; Percheron-Guegan, A. The Influence of Mn on the crystallography and electrochemistry of nonstoichiometric AB_5-type hydride forming compounds. *J. Electrochem. Soc.* **1999**, *146*, 3181–3189.
20. Chu, H.L.; Zhang, Y.; Qiu, S.J.; Qi, Y.N.; Sun, L.X.; Xu, F.; Wang, Q.; Dong, C. Electrochemical performances of cobalt-free $La_{0.7}Mg_{0.3}Ni_{3.5-x}(MnAl_2)_x$ (x = 0–0.20) hydrogen storage alloy electrodes. *J. Alloys Compd.* **2008**, *457*, 90–96.

21. Zhang, Y.; Lei, Y.Q.; Chen, L.X.; Yuan, J.; Zhang, Z.H.; Wang, Q.D. The effect of partial substitution of Zr for Ti on the electrochemical properties and surface passivation film of $Mg_{35}Ti_{10-x}Zr_xNi_{55}$ ($x = 1, 3, 5, 7, 9$) electrode alloys. *J. Alloys Compd.* **2002**, *337*, 296–302.
22. Han, S.S.; Lee, Y.L.; Goo, N.H.; Jeong, W.T.; Lee, K.S. Improvement of electrode performances of Mg_2Ni by mechanical alloying. *J. Alloys Compd.* **2002**, *330*, 841–845.
23. TOPAS V4: General Profile and Structure Analysis Software for Powder Diffraction Data. Bruker AXS: Karlsruhe, Germany, 2008.
24. Msika, E.; Latroche, M.; Cuevas, F.; Percheron-Guégan, A. Zr-substitution in $LaNi_5$-type hydride compound by room temperature ball milling. *Mater. Sci. Eng. B* **2004**, *108*, 91–95.
25. Pan, H.G.; Yue, Y.J.; Gao, M.X.; Wu, X.F.; Chen, N.; Lei, Y.Q.; Wang, Q.D. The effect of substitution of Zr for La on the electrochemical properties of $La_{0.7-x}Zr_xMg_{0.3}Ni_{2.45}Mn_{0.1}Co_{0.75}Al_{0.2}$ hydrogen storage electrode alloys. *J. Alloys Compd.* **2005**, *397*, 269–275.
26. Chu, H.L.; Zhang, Y.; Sun, L.X.; Tian, Q.F.; Xu, F.; Zhang, T.; Yuan, H.T. Structures and hydrogen storage properties of $Mg_{45}M_5Co_{50}$ (M = Zr, Ni, Al) ternary alloys by mechanical alloying. *Int. J. Electrochem. Sci.* **2006**, *1*, 47–54.
27. Seo, C.Y.; Choi, S.J.; Choi, J.; Park, C.N.; Lee, J.Y. Effect of V and Zr on the electrochemical properties of La-based AB_5-type metal hydride electrodes. *J. Alloys Compd.* **2003**, *351*, 255–263.
28. Pan, H.G.; Liu, Y.F.; Gao, M.X.; Zhu, Y.F.; Lei, Y.Q. The structural and electrochemical properties of $La_{0.7}Mg_{0.3}(Ni_{0.85}Co_{0.15})_x$ (x = 3.0–5.0) hydrogen storage alloys. *Int. J. Hydrogen Energy* **2003**, *28*, 1219–1228.
29. Chu, H.L.; Qiu, S.J.; Sun, L.X.; Zhang, Y.; Xu, F.; Jiang, T.; Li, W.X.; Zhu, M.; Hu, W.Y. The improved electrochemical properties of novel La–Mg–Ni-based hydrogen storage composites. *Electrochim. Acta* **2007**, *52*, 6700–6706.
30. Monnier, J.; Chen, H.; Joiret, S.; Bourgon, J.; Latroche, M. Identification of a new pseudo-binary hydroxide during calendar corrosion of $(La, Mg)_2Ni_7$-type hydrogen storage alloys for Nickel–Metal Hydride batteries. *J. Power Sources* **2014**, *266*, 162–169.
31. Zhu, Y.F.; Pan, H.G.; Gao, M.X.; Ma, J.X.; Li, S.Q.; Wang, Q.D. The effect of Zr substitution for Ti on the microstructures and electrochemical properties of electrode alloys $Ti_{1-x}Zr_xV_{1.6}Mn_{0.32}Cr_{0.48}Ni_{0.6}$. *Int. J. Hydrogen Energy* **2002**, *27*, 287–293.
32. Kuriyama, N.; Sakai, T.; Miyamura, H.; Uehara, I.; Ishikawa, H.; Iwasaki, T. Electrochemical impedance spectra and deterioration mechanism of metal hydride electrodes. *J. Electrochem. Soc.* **1992**, *139*, L72–L73.
33. Ratnakumar, B.V.; Witham, C.; Bowman, R.C., Jr.; Hightower, A.; Fultz, B. Electrochemical studies on $LaNi_{5-x}Sn_x$ metal hydride alloys. *J. Electrochem. Soc.* **1996**, *143*, 2578–2584.
34. Notten, P.H.L.; Hokkeling, P. Double phase hydride forming compounds: A new class of highly electrocatalytic materials. *J. Electrochem. Soc.* **1991**, *138*, 1877–1885.
35. Niu, H.; Northwood, D.O. Enhanced electrochemical properties of ball-milled Mg_2Ni electrodes. *Int. J. Hydrogen Energy* **2002**, *27*, 69–77.
36. Nishima, T.; Ura, H.; Uchida, I. Determination of chemical diffusion coefficients in metal hydride particles with a microelectrode technique. *J. Electrochem. Soc.* **1997**, *144*, 1273–1277.

37. Zheng, G.; Popov, B.N.; White, R.E. Electrochemical determination of the diffusion coefficient of hydrogen through an $LaNi_{4.25}Al_{0.75}$ electrode in alkaline aqueous solution. *J. Electrochem. Soc.* **1995**, *142*, 2695–2698.

Research on Semisolid Microstructural Evolution of 2024 Aluminum Alloy Prepared by Powder Thixoforming

Pubo Li [1], Tijun Chen [1,*], Suqing Zhang [1] and Renguo Guan [2]

[1] State Key Laboratory of Advanced Processing and Recycling of Nonferrous Metals, Lanzhou University of Technology, Lanzhou 730050, China; E-Mails: lipubogs@163.com (P.L.); zhangsuqing1985@163.com (S.Z.)

[2] School of Materials and Metallurgy, Northeastern University, Shenyang 110004, China; E-Mail: guanrg@smm.neu.edu.cn

* Author to whom correspondence should be addressed; E-Mail: chentj@lut.cn

Academic Editor: E. W. Jarfors

Abstract: A novel method, powder thixoforming, for net-shape forming of the particle-reinforced Aluminum matrix composites in semi-solid state has been proposed based on powder metallurgy combining with thixoforming technology. The microstructural evolution and phase transformations have been investigated during partial remelting of the 2024 bulk alloy, prepared by cold pressing of atomized alloy powders to clarify the mechanisms of how the consolidated powders evolve into small and spheroidal primary particles available for thixoforming. The effect of heating temperature on the resulting semisolid microstructure has also been discussed. The results indicate that the microstructural evolution includes three stages—the initial rapid coarsening of the fine grains within the powders, the formation of continuous liquid layer on the primary particle surface (the original powder), and the final coarsening—that result from the phase transformations of $\theta \rightarrow \alpha$, $\alpha \rightarrow L$, and $\alpha \rightarrow L$ and $L \rightarrow \alpha$, respectively. The coarsening rate of the primary particles is low, and one original powder always evolves into one spheroidal particle with a continuous liquid layer surface. Properly raising the heating temperature is beneficial for obtaining an ideal semisolid microstructure.

Keywords: powder thixoforming; cold pressing; coarsening rate; spheroidal particles

---

1. Introduction

SiC particle (SiC_p)-reinforced Aluminum matrix composites (SiC_p/Al) have been developed successfully over the past few decades, owing to their excellent properties such as light weight, high temperature strength, good wear resistance, and a low thermal expansion coefficient [1–5]. The existing fabrication techniques include powder metallurgy (PM), squeeze casting, pressure or pressureless infiltration, and spray forming [2]. It is known that the PM process is particularly popular due to the uniform distribution of reinforcements and flexible constituent design [6,7]. However, the PM-processed products always have a very high cost due to the complex processing procedures, and their microstructures are relatively loose due to difficultly eliminated pores, which seriously limit its wider application [6,8]. In addition, components with complex shape and/or large size are also very difficult to fabricate using this technology. However, thixoforming can significantly decrease or even eliminate porosities, and thus can produce components with compact microstructures. Furthermore, the mechanical properties of the produced components can be significantly improved compared to fully liquid casting [9]. In addition, this technology is also very suitable for producing large-sized components with complex shape [10]. The key procedure for the thixoforming technology is the production of a non-dendritic microstructure with fine globular grains, suspended in liquid phase [9]. To reach the desirable microstructure, the starting materials must be specially treated prior to partial remelting, *i.e.*, pre-deformed, refined, and stirred in the semi-solid state or during solidification [9,11,12]. If the bulk alloy prepared by pressing of the small and spheroidal alloy powders can be used as the feedstock of thixoforming, a new approach to producing a non-dendritic semisolid microstructure available for thixoforming will be developed. Therefore, combining the advantages of both PM and thixoforming technologies, a new technology that can prepare and form SiC_p/Al matrix composite, powder thixoforming, has been proposed. The mixing and cold pressing steps of PM are applied to prepare the start ingots with uniform distribution of reinforcing particles in the matrix. Then the ingots are partially remelted and thixoformed.

At present, the most similar technologies to powder thixoforming are semisolid powder processing [13] and pseudo-semisolid thixoforming [14,15]. The former has bad thixoformability due to the high solid fraction (>90%) and mainly focuses on the effects of SiC volume fraction and particle size, and forming pressure on the mechanical properties. The latter's forming temperature is the pseudo-semi-solid temperature at which the matrix (SiC_p or Al_2O_{3p}) is in solid state and the metal (Al) is in liquid state, and the solid fraction is also very high (60%–70%). Unfortunately, existing investigations on the proposed powder thixoforming have not been found.

It is well known that the microstructural evolution during partial remelting is another key topic for thixoforming because this process will have a significant influence on the resultant semisolid microstructure. However, existing references about how the powders evolve into primary particles in the semisolid microstructure have not addressed this problem. In addition, the microstructural evolution essentially results from phase transformations occurring during partial remelting, and research on the phase transformations is capable of providing some useful information toward clarify the microstructural evolution mechanisms. However, the existing investigations have not involved this aspect.

Therefore, the microstructural evolution of 2024 bulk alloy during partial remelting was first verified in order to preferably clarify that of the SiC_p/2024 Al-based composite. Simultaneously, the phase transformations occurring during partial remelting and the effect of heating temperature on the semisolid microstructure have also been discussed.

2. Experimental Section

2.1. Material

The material used is 2024 Aluminum alloy powder (Al_p) prepared by atomization (Zhejiang Bainianyin Industry & Trade Co., Ltd., Yongkang, Zhejiang, China), and has an average particle size of 20 μm and a spherical morphology (Figure 1). Its actual composition is Al-4.22Cu-1.35Mg-0.5Si-0.5Fe-0.5Mn (in wt%). To confirm the partial remelting temperature, this alloy was analyzed by a Pyris Diamond TG/DTA differential thermal analyzer (DTA; NETZSCH, Bavaria, Germany). As shown in Figure 2, its solidification range is 498–663.7 °C.

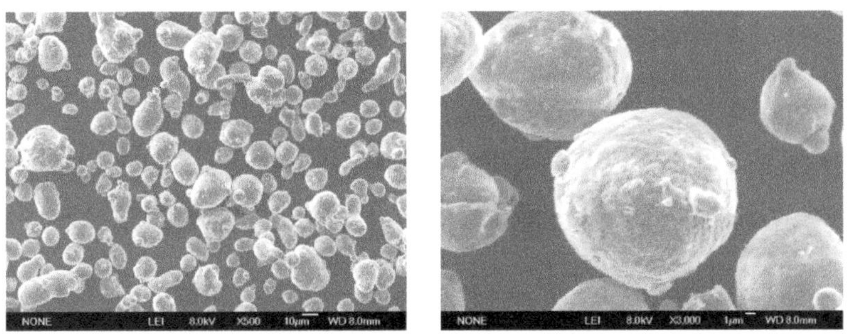

Figure 1. SEM micrographs of the as-received 2024 alloy powders.

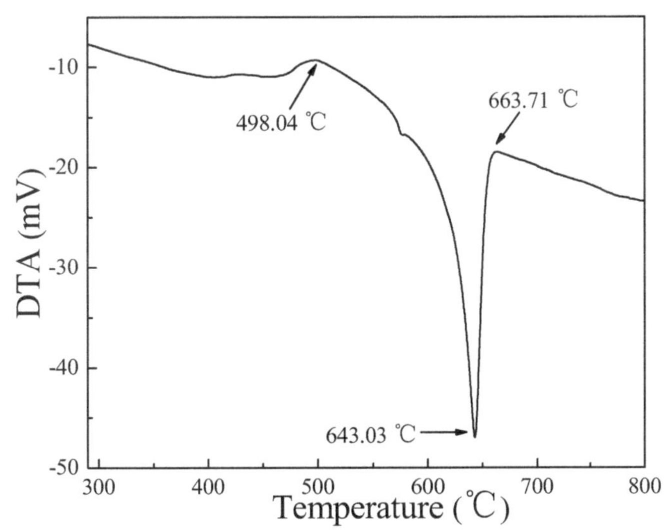

Figure 2. Differential thermal analyzer (DTA) curve of the 2024 alloy powders.

2.2. Experiments

Twenty grams of Al_p were uniaxially cold-compacted at 300 MPa pressure for 5 min to produce green billets with 30 mm in diameter and 7 mm in height. Repeating the experimental procedures, some billets were obtained. Then some of them were heated at a semisolid temperature of 635 °C for different times (0–60 min) to study the microstructural evolution of the bulk alloy during partial remelting. The other billets were heated for 30 min at different temperatures (625, 630, and 640 °C) to investigate the influence of heating temperature on semisolid microstructure. To examine the billets' temperature variation during partial remelting, one thermocouple was inserted into a hole in the center of one billet. The billets are composed of alloy powders, so it can be expected that clarifying the microstructural evolution of a single Al_p is very helpful to study that of the billet. For this purpose, some powders were heated for different durations (0–30 min) at a temperature of 635 °C.

2.3. Metallographic Characterization

All of the heated billets were quickly quenched in a Bi-based alloy melted at about 45 °C. Each of the resultant billets was cut along the radial direction into two specimens. Then a cross section of one specimen was finished and polished by standard metallographic techniques, followed by etching using Keller's etchant (Yantai Shuangshuang Chemical Co., Ltd., Haiyang, Shandong, China). The heated powders were water-quenched and then prepared into small specimens together with epoxy resin, and finally they were also prepared into standard metallographic specimens using the same procedures as the quenched billet specimens. All of the specimens were observed on a scanning electron microscope (SEM; FEI, Hillsboro, OR, USA) or/and an optical microscope (OM; Nikon Instruments, Shanghai, China), and some of the billet specimens were examined by energy dispersive spectrometer (EDS; FEI, Hillsboro, OR, USA) on the SEM. To quantitatively examine the liquid fraction and the primary particle size, the related OM images were analyzed by Image-Pro Plus software (Media Cybernetics Company, Silver Spring, MD, USA). At least three typical OM images with magnification of 500 times were examined. In order to illustrate the microstructural evolution, an X-ray diffractometer (XRD; Rigaku, Tokyo, Japan) was used to verify the phase constituents of the 2024 bulk alloy heated at 635 °C for different durations in order to deduce the phase transformations occurring during partial remelting. To verify the interactions between the powders or particles during partial remelting, such as if mergence occurs, and then further clarify the coarsening mechanisms of the primary particles in semisolid microstructure, the other billet specimens were machined into tensile bars. Tensile tests were carried out on a universal material testing machine (Jinan HengXu Testing Machine Technology Co., Ltd., Jinan, Shandong, China) at a nominal strain rate of 1 mm/min. The fracture surfaces were observed on the SEM.

3. Results and Discussion

3.1. Microstructural Evolution of the Alloy Powders during Partial Remelting

Figure 3 presents the microstructures of the alloy powders heated at 635 °C for different durations. It shows that the original microstructure of a powder consisted of fine α-Al grains and intergranular eutectics between the grains (shown by arrows in Figure 3a). The difference between grain boundaries and grain interior within powder is also found in 7075 Al powder [16,17]. The amount of the eutectics obviously decreases during the heating from 0 min to 1 min (comparing Figure 3a,b). The original powders (including many fine α grains) become compact particles without grain boundaries (one powder only contains one grain). Simultaneously, some white particles and globular holes appear (Figure 3b).

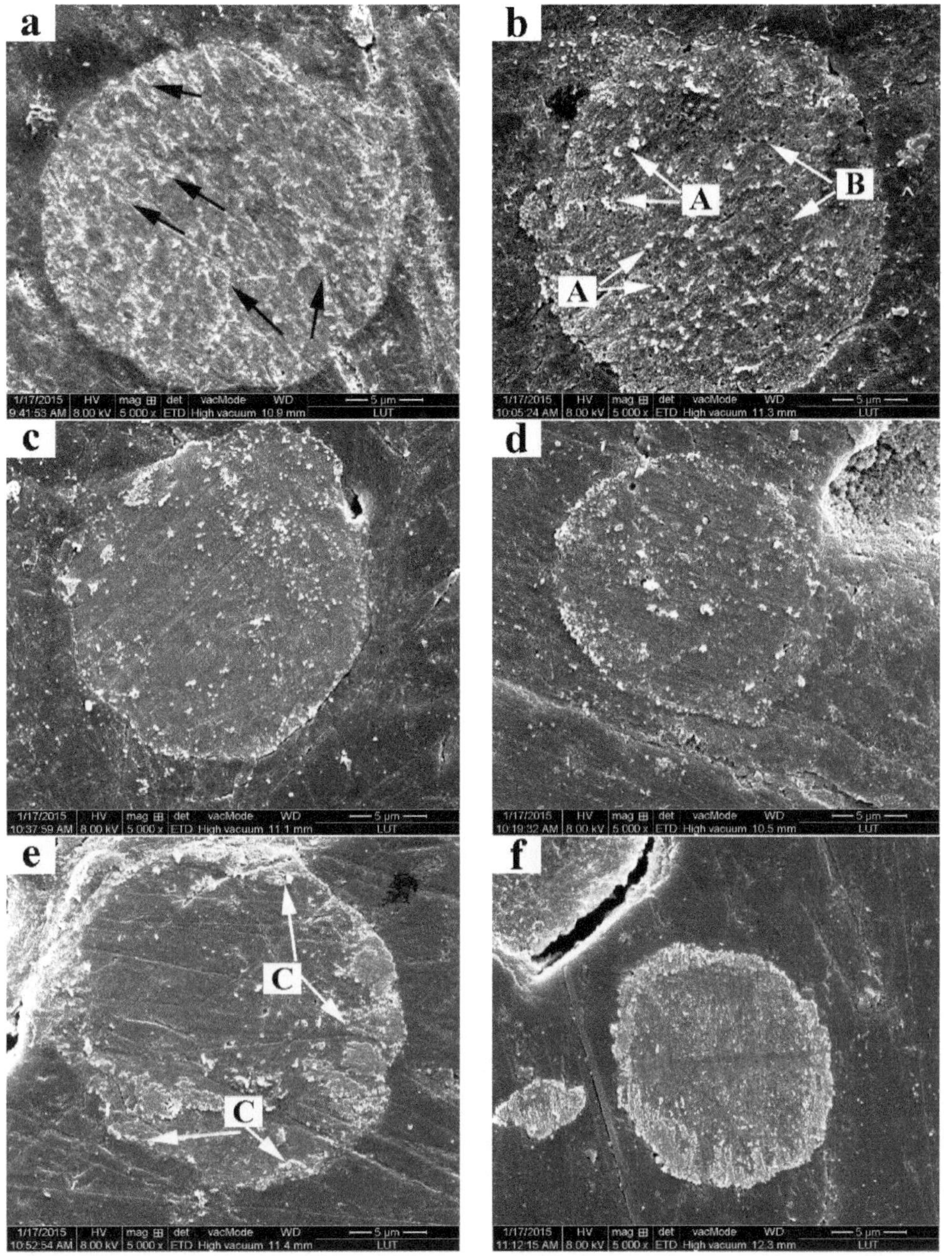

Figure 3. Micrographs of the alloy powders heated at 635 °C for different durations and then water-quenched. (**a**) 0 min; (**b**) 1 min; (**c**) 2 min; (**d**) 3 min; (**e**) 10 min; (**f**) 30 min.

The Al-4Cu-1.3Mg (wt%) alloy is consisted three phases—α-Al, θ-$CuAl_2$, and S-Al_2CuMg [18]. However, because the S phase within this Al_p is very small, the XRD can barely detect it (Figure 4). Therefore, it can be deemed that the eutectics contain an θ phase and α-Al phase in this paper. According to the binary Al-Cu phase diagram [19], the solubility of Cu in the primary α-Al phase increases and the binary Al-4Cu will experience a single α-Al phase interval (only considering the two main elements, Al and Cu) during reheating. So the eutectics should dissolve towards the primary α phase as the temperature rises, resulting in the decrease of the eutectic phases. This phenomenon is also found in other as-casting materials [20,21]. Most of the eutectics between the grains have dissolved or even disappeared because the eutectic size in the interdendritic zones is smaller. Based on this standpoint, it can be supposed that the disappearance of the interdendritic eutectics must result in the disappearance of grain boundaries, and then result in the coarsening of the fine α-Al grains through mergence. The original powders change into compact particles without grain boundaries. However, some small particles also appear when heated for 1 min (marked by the A arrows in Figure 3b). It can be expected that the temperature rise in the powders should be very fast and the temperature will reach and exceed the eutectic reaction temperature in a short time. That is, there is not enough time for the complete dissolution of eutectics. Then the residual interdendritic eutectics melt and become a liquid phase. From the spherical morphology of the white particles in Figure 3b, they should be liquid pools. The spherical liquid pools within the primary solid phase are also found in AZ91HP magnesium composites produced by the rotating cylinder method [22]. However, some holes inside the powders are also liquid phase due to their etching during the stage of preparing metallographic samples (marked by B in Figure 3b). Although the eutectics have not completely dissolved when heated for 1 min, the grain boundaries have already disappeared. Thus, grain coarsening is the main phenomenon during the period 0–1 min.

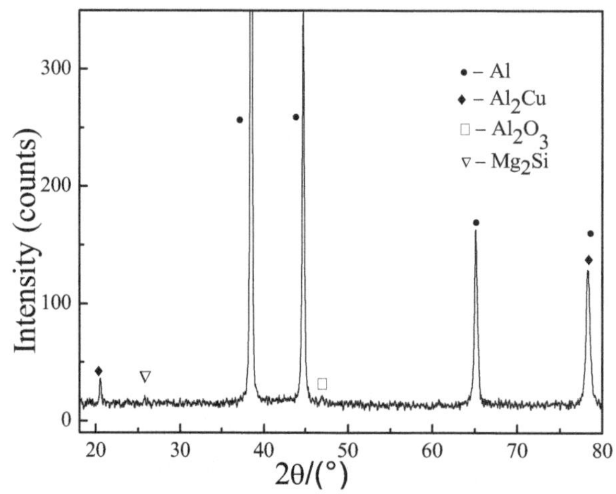

Figure 4. XRD spectrum of the 2024 Aluminum alloy.

As the reheating increases from 1 min to 10 min, it can be found that the liquid phase amount within the particles decreases and that the size increases (comparing Figure 3b,e). It can be expected that the solid/liquid interfacial energy within the powders should be very high because of the large solid/liquid interfacial area (Figure 3b). To decrease the interfacial energy, the liquid pools migrate to the particles' edges, as seen in Figure 3d,e. The liquid pools also coarsen through mergence (comparing Figure 3c,d). Simultaneously, the liquid phase should be rich in solutes (such as Cu) and some liquid phase dissolves towards the surrounding α Al phase due to the increase of the Cu solubility in the α phase with the rise in heating temperature. The above reasons result in the reduction of the liquid pool's amount and an increase of the liquid pool's size (Figure 3e). In addition, the liquid amount around the particles obviously increases as the liquid pools within the particles decrease, and a continuous liquid layer forms around the particles when heated for 10 min (marked by C in Figure 3e). The reason is that the primary particles (the original powders) partially melt, driven by the requirement for increasing liquid amount due to the temperature rise. Thus the main event occurring during the period of 1–10 min is the variations of the liquid pools and the formation of a continuous liquid layer on the primary particle surface.

When the heating time is further prolonged, the primary particles further melt due to the further rise of the temperature, which leads the liquid amount surrounding the primary particles to continuously increase (comparing Figure 3e,f). More importantly, one powder evolves into one primary particle with a continuous liquid layer surface. Because the original alloy powders are quite spheroidal, the primary particles in the semisolid microstructure always remain spheroidal (Figure 3). This phenomenon is different from the microstructural evolution of the as-cast materials, which generally needs to experience the spheroidization stage to obtain spheroidal primary particles [20,21]. Therefore, the main event occurring during the period of 10–30 min is the further increase of the liquid amount surrounding the primary particles.

Based on the above discussion, it can be concluded that the microstructural evolution of Al_p can be divided into two stages: the initial rapid coarsening of the fine grains due to the dissolution of the eutectics (0–1 min) and the formation of a continuous liquid layer on the primary particle surface, ascribed to the partial melting of the primary particles (1–30 min). One spherical powder evolves into one spheroidal primary particle with a continuous liquid layer surface.

3.2. Microstructural Evolution of the Bulk Alloy during Partial Remelting

Figure 5 indicates the microstructures of the 2024 bulk alloy heated at the semisolid temperature of 635 °C for different durations. It shows that the eutectics (shown by arrows in Figure 5a) gradually decrease and the grain boundaries within powders disappear during the heating period 0–3 min (Figure 5a–d). As the reheating increases from 3 min to 10 min, the liquid amount around the particles obviously increases, accompanied by the decrease of the liquid pools within the particles (Figure 5d–f). The above phenomena are basically consistent with the processes that the powders have experienced in this heating stage of 0–10 min.

For the same reasons as the powders, a continuous decrease in the eutectic amount and coarsening of the grains have occurred during the early heating stage 0–3 min (Figure 5a–d). The liquid phase within the particles appears after being heated for 3 min (marked by A in Figure 5d). Figure 6 shows that the heating rate of the bulk alloy during the 0–3 min stage is rapid, so the eutectics have no enough time to completely dissolve into the primary α phase when the specimen's temperature reaches the eutectic reaction temperature. The residual eutectic phases then melt to form liquid phase. However, some holes inside the particles also appear (marked by B in Figure 5d). As discussed above, the holes are also the liquid phase. It is just for this reason that the arrows B in Figure 5c and e represent the residual θ phase and liquid phase, respectively. The microstructure of the bulk alloy heated for 3 min—compact particles without grain boundaries—is similar to that of the powders heated for 1 min, but the liquid phase amount in the bulk alloy microstructure is obviously less than that of the powders (comparing Figures 3b and 5d). The interdendritic eutectics only decrease when heated for 1 min (comparing Figure 5a,b). The above phenomena should be attributed to the reason that the heating rate of the bulk alloy is relatively slower than that of the powders. Namely, the microstructural evolution rate of the bulk alloy lags behind that of the powders during the initial period of 0–3 min. Thus grain coarsening is the main phenomenon occurring in this period.

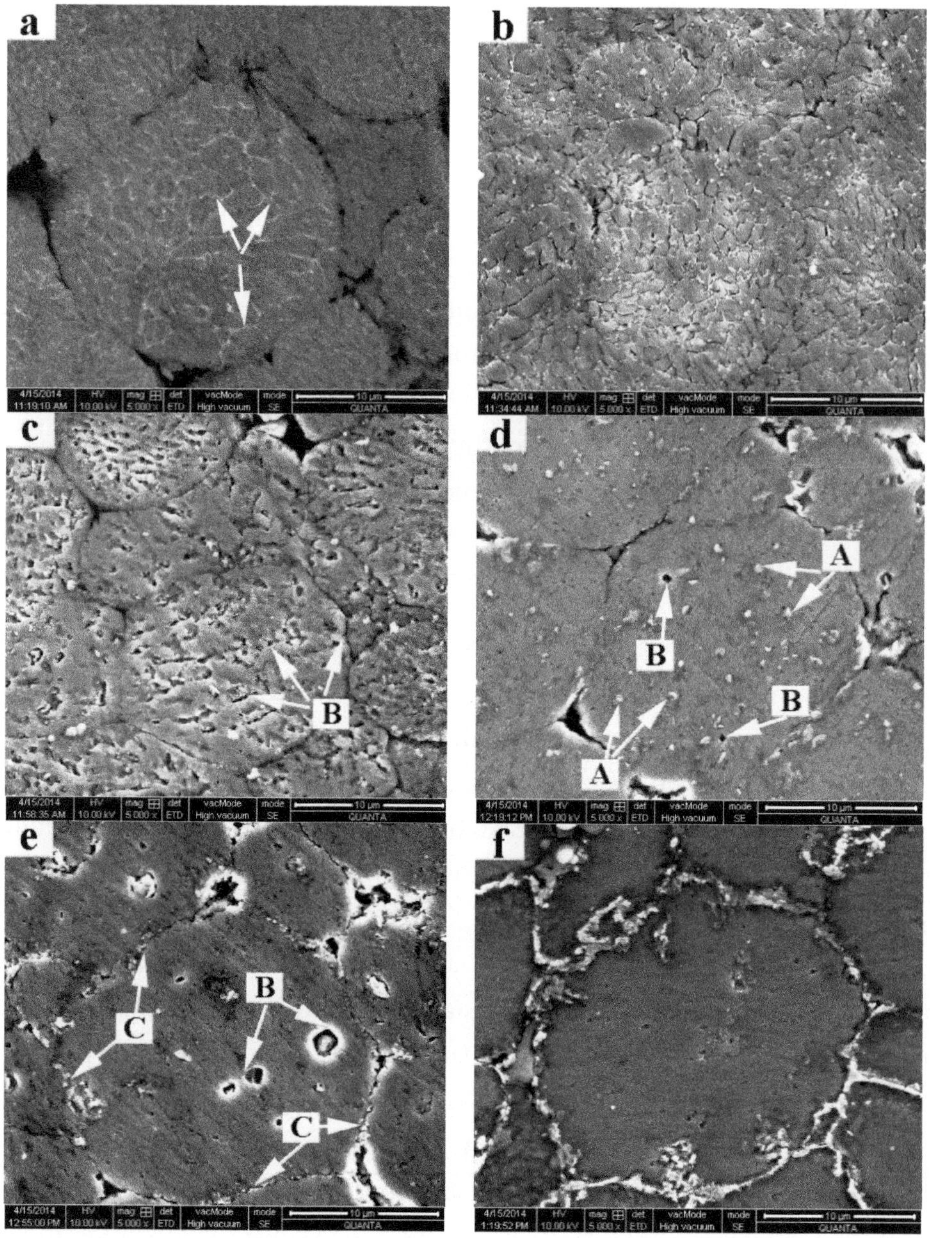

Figure 5. Micrographs of the 2024 bulk alloy heated at 635 °C for different durations and then quenched. (**a**) 0 min; (**b**) 1 min; (**c**) 2 min; (**d**) 3 min; (**e**) 5 min; (**f**) 10 min.

As the heating time is further prolonged, the liquid phase gradually distributes between the neighboring particles as that within the particles decreases (Figure 5d–f). As discussed above, the powders partially melt, driven by increasing liquid amount due to the temperature rise. Figure 5e shows that some sites melt to form discontinuous liquid layers or small-sized liquid pools between the primary particles (marked by C in Figure 5e). A continuous liquid layer forms around the primary particles when the particles further melt (Figure 5f), which is similar to the phenomenon that the structure separation has completed in the as-casting materials during partial remelting [20,21]. The liquid pools within particles also generate dissolution towards the α phase, migration to the particles' boundaries, and coalescence to decrease the solid/liquid interfacial energy (comparing Figure 5d,f). Therefore, the main phenomenon occurring during the 3–10 min period is basically similar to that of the powders when heated for 1 min.

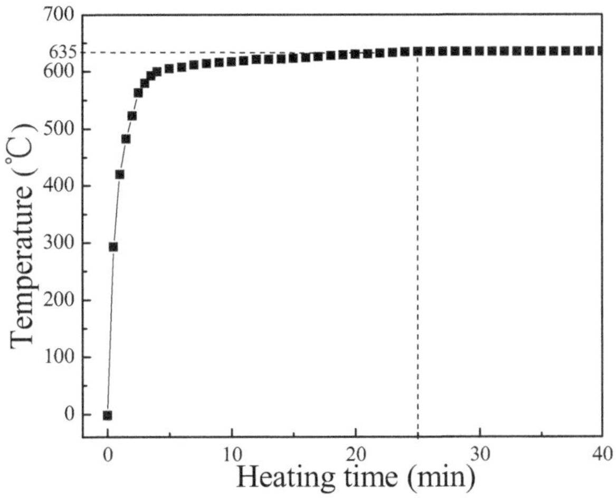

Figure 6. Temperature variation in the specimen with heating time.

Figure 7 shows the microstructures of the 2024 bulk alloy heated at 635 °C for over 10 min. It can be seen that the microstructures always consist of large and small particles during the heating period 10–60 min. The neighboring particles are separated by black continuous eutectics. However, the amount of the small particles increases as that of the large particles decreases when heated for 20 min (comparing Figure 7a,b). As the heating time is prolonged, it can be supposed that the liquid amount increases due to the temperature rise. The liquid phase solidifies into secondarily primary α-Al phase during quenching. However, the solidification rate of the liquid phase is very rapid during quenching, so the secondarily solidified particles are smaller than the primary α-Al particles anyhow. Thus it can be concluded that the relatively small particles in the quenched microstructures originate from the solidification of the liquid phase. Figure 8 shows the full-liquid microstructure when heated at 710 °C for 30 min. It shows that this microstructure is all small particles and intergranular eutectics. The result of quantitative examination of Figure 8 shows that the particle size is close to 5 μm. Therefore, the particles which are less than 5 μm are regarded as the liquid phase in this paper. Based on this standpoint, Figure 7a clearly shows that the liquid phase (marked by circles) distributes between the particles.

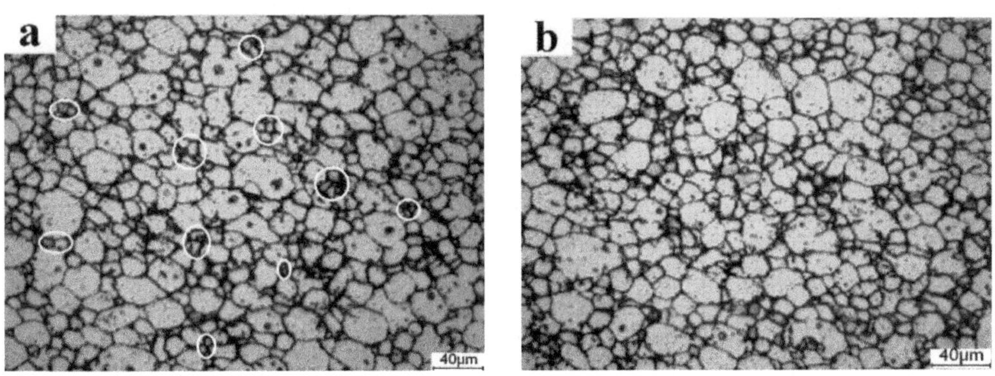

Figure 7. *Cont.*

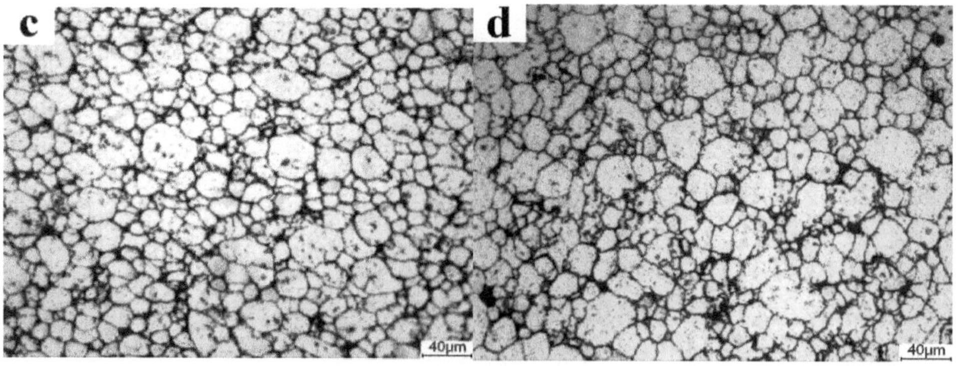

Figure 7. Micrographs of the 2024 bulk alloy heated at 635 °C for different durations and then quenched. (**a**) 10 min; (**b**) 20 min; (**c**) 30 min; (**d**) 60 min.

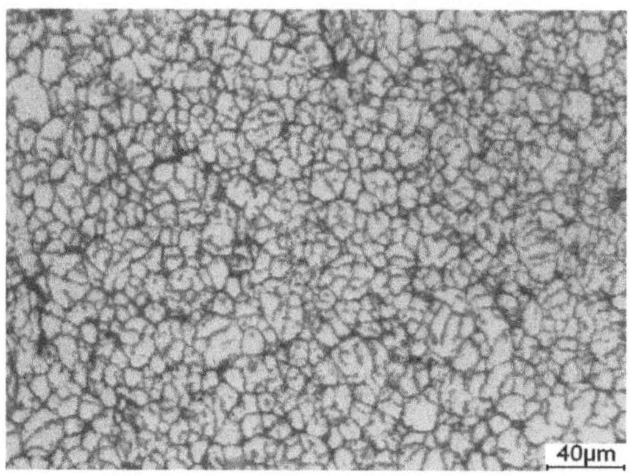

Figure 8. Micrograph of the bulk alloy heated at 710 °C for 30 min and then quenched.

As shown in Figure 7a–c, the liquid amount gradually increases through partial melting of the particles due to the temperature rise, and the primary particle size remains almost unchanged. Moreover, the shapes of the particles in the present microstructure are always spherical. The distance between the primary particles is small, and some of the neighboring particles are only separated by thin eutectics (Figure 7a). Under this condition, the solid/liquid interfacial energy should be very high due to the large solid/liquid interfacial area. Thus, the particles should coarsen during the subsequent heating driving by decreasing the interfacial energy. However, the coarsening does not operate during the 10–30 min period. Alumina layers have already formed on the surfaces of the alloy powders [23]. The great difference in the coefficient of thermal expansion between Aluminum and alumina will generate stress on the oxide shell during heating, leading to a dynamic oxide breakage and repair process if there is any oxygen in the atmosphere [24]. Namely, the alumina on the surfaces of the alloy powders cannot disappear as the heating proceeds, and thus the healed or original oxide can hinder the merging of the neighboring particles (Figure 7a–c). So it can be concluded that the low coarsening rate of the particles is due to the less active coalescence. In addition, because the as-received Al_p is spherical in shape with a particle size of 20 μm, and the coalescence of the particles which evolve from the original powders is impossible, the microstructure does not need to experience an additional spheroidization stage to obtain the spheroidal primary particles. Thus, it is not difficult to understand the present experimental results; the primary particles are always small and spheroidal. The main event occurring during the 10–30 min period is the further increase of the liquid amount.

Coarsening through mergence cannot be found during partial remelting (Figures 5 and 7). This can be more clearly demonstrated by the fracture surfaces of the bulk alloy heated at 635 °C for different durations (Figure 9). It shows that the plastic deformation of the powders occurs during cold pressing (shown by arrows in Figure 9a). Figure 9b shows that the powders are individual, which means that coalescence between neighboring powders is impossible during the initial heating stage of 0–3 min (Figure 9a,b). The appearance of sintering necks demonstrates that transformation from mechanical bonding to good metallurgical bonding among the powders occurs (marked by arrows in Figure 9b). This phenomenon is common in PM [25]. However, the mergence between particles does not accompany microstructural evolution during the 3–30 min period (Figure 9b–e). As discussed above, a continuous liquid layer on the primary particle surface gradually forms after being heated for 3 min, which can be better demonstrated by the fracture surfaces (marked by A in Figure 9c–e). In short, coarsening through mergence is impossible during partial remelting.

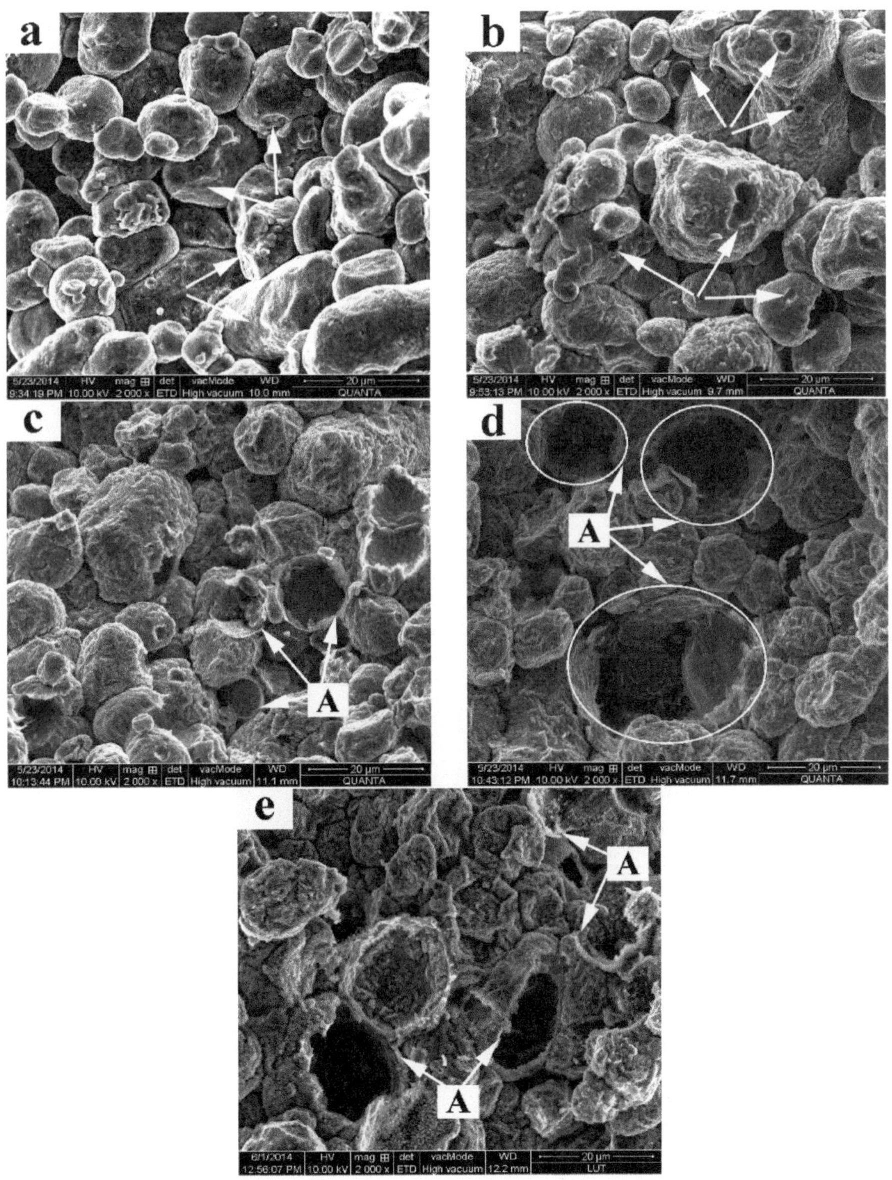

Figure 9. Fractographs of the bulk alloy heated at 635 °C for different durations and then quenched. (**a**) 0 min; (**b**) 3 min; (**c**) 5 min; (**d**) 10 min; (**e**) 30 min.

The liquid amount remains almost unchanged after being heated for 30 min (comparing Figure 7c,d), which can be more obviously seen by the quantitative examinations shown in Figure 10. Figure 6 indicates that the specimen's temperature reaches the final equilibrium temperature of 635 °C after being heated for 25 min, which implies that the system reaches its solid/liquid equilibrium state and the liquid amount is basically invariable after that. This phenomenon is consistent with metallographic analysis. The particle size slightly increases as the heating time is further prolonged (Figure 10). Thus the main event occurring after heating for 30 min is the slight coarsening of the primary particles.

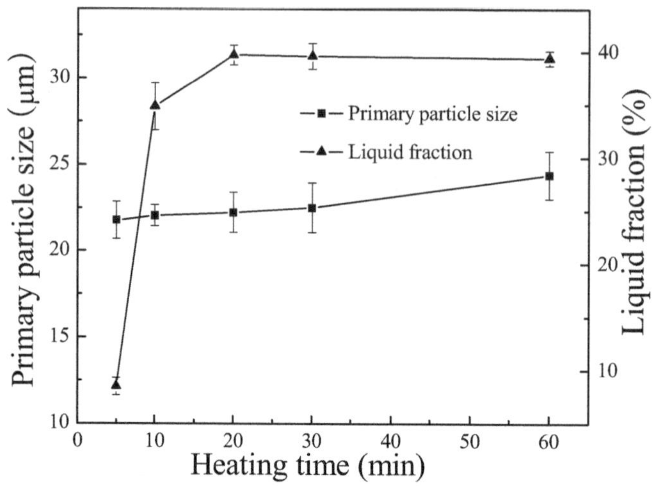

Figure 10. Variations in the primary particle size and liquid fraction of the bulk alloy with heating time after being heated for 5 min at 635 °C.

The coarsening mechanisms after being heated for 30 min should not be attributed to mergence. The particle size should sharply increase if coarsening through mergence operates. Thus the coarsening must be caused by another mechanism, Ostwald ripening. Ostwald ripening operates through the dissolution of small particles and the growth of large particles [26]. That is to say, the particle coarsening is achieved through atom diffusion and the resulting coarsening rate should be milder than that from mergence. The variation of particle size with heating time should obey the formula $D_{(t)}^3 - D_{(0)}^3 = Kt$, where $D_{(t)}$ is the average particle size at time t, $D_{(0)}$ is the initial particle size, and K is the coarsening rate constant [26], if the coarsening results from Ostwald ripening. However, Figure 11 indicates that the coarsening during the period from 25 to 60 min does not obey this regime. The alumina layer will hinder the diffusion ability of atoms, which should inhibit the Ostwald ripening. However, the oxide film can be effectively broken in some local zones due to the relative sliding and deformation of the powders during cold pressing, and local fresh contact surfaces between powders can be obtained [27]. As the temperature in the specimen rises, these contact zones melt to form liquid paths. So the atom diffusion among the particles may be accelerated from these sites. According to this standpoint, slight coarsening results from Ostwald ripening. However, compared with other as-casting materials [28], the coarsening rate of the primary particles after the semi-solid systems reach a solid/liquid equilibrium state is quite slow, even if the atoms' diffusion ability through the local fresh paths among particles is enhanced to a certain degree. More importantly, it can be supposed that one powder in the as-pressed microstructure will basically evolve into one spheroidal particle in the semisolid microstructure. Thus the final coarsening after 30 min results from the impeditive Ostwald ripening.

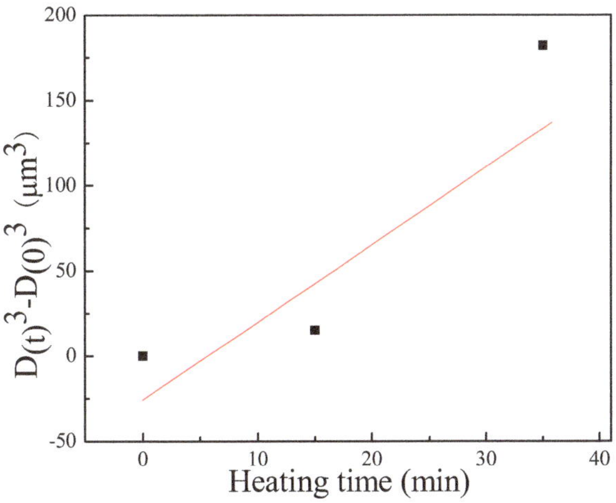

Figure 11. Cube of the primary particle size *versus* heating time, taking 25 min as starting time, $t = 0$.

Based on the above discussion, the microstructural evolution of the 2024 bulk alloy can be divided into three stages: the initial rapid coarsening of the fine grains due to the dissolution of eutectic θ phase (0–3 min), the formation of continuous liquid layer on the primary particle surface owing to the partial melting of the particles (3–30 min), and the final coarsening attributed to the influenced Ostwald ripening (30–60 min). The coarsening rate of the particles is so slow that one powder basically evolves into one spheroidal particle in the semisolid microstructure.

3.3. Phase Transformations during Partial Remelting

As discussed above, the main event occurring during the heating period of 0–3 min is the coarsening of the fine grains within the powders due to the dissolution of the Cu-rich θ phase towards the α-Al phase. In view of phase transformation, this process can be expressed as θ→α. Because of this reaction, the θ phase amount decreases (comparing the diffraction intensities of the θ phase shown in Figure 12), and the Cu content within the primary α phase increases (Table 1). Thus the initial coarsening is ascribed to the reaction of θ→α.

The residual θ phase melts to form liquid pools within the particles after being heated for 3 min (marked by arrows in Figure 5d). The essence of this process is owed to the reverse eutectic reaction, θ + α→L. The Cu content in the liquid phase decreases due to the melting of the Al-rich α phase (Table 1). Moreover, Figure 5d indicates that the liquid amount is small, so the solidified θ phase derived from the formed liquid phase after being quenched is negligible and the XRD can barely detect it (Figure 12). As the heating proceeds, a continuous liquid layer forms around the particles due to the partial melting of the particles (Figures 3e and 5f). The melting can be expressed by the reaction of α→L. Because the primary α-Al phase gradually melts, the Cu content in the liquid phase continuously decreases (Table 1). Moreover, the Cu solubility in the primary phase increases as the heating temperature rises [19], leading the Cu content in the liquid phase to further decrease and that in the primary α phase to increase during the period 3–10 min (Table 1). The liquid phase can solidify into the α phase and eutectics (θ + α) after being quenched, and the eutectic θ phase amount increases with the increase of liquid amount. So the diffraction peak of the θ phase gradually becomes intensive again as the heating time is further prolonged (comparing the diffraction intensities of the θ phase during the period of 3–10 min, shown in Figure 12). Therefore, the formation of the liquid pools within the particles and continuous liquid layer on the primary particle surface results from the reactions of θ + α→L and α→L respectively.

Table 1. Compositions of different structures of the bulk alloy heated at 635 °C for different durations.

Heating Time/Min	Structure	Al	Composition (wt%) Cu	Mg
0	Primary α	95.9	3.5	0.6
	Eutectic θ	70.0	28.5	1.5
1	Primary α	94.9	3.9	1.2
	Eutectic θ	75.7	23.0	1.3
2	Primary α	94.9	4.0	1.1
	Eutectic θ	88.4	10.8	0.8
3	Primary α	95.0	4.2	0.8
	Liquid phase	90.8	8.3	0.9
5	Primary α	93.9	4.8	1.3
	Liquid phase	91.7	7.8	0.5
10	Primary α	92.7	5.8	1.5
	Liquid phase	95.7	3.7	0.6
20	Primary α	94.4	4.6	1.0
	Liquid phase	95.2	3.9	0.9
30	Primary α	94.9	4.4	0.7
	Liquid phase	95.0	4.0	1.0

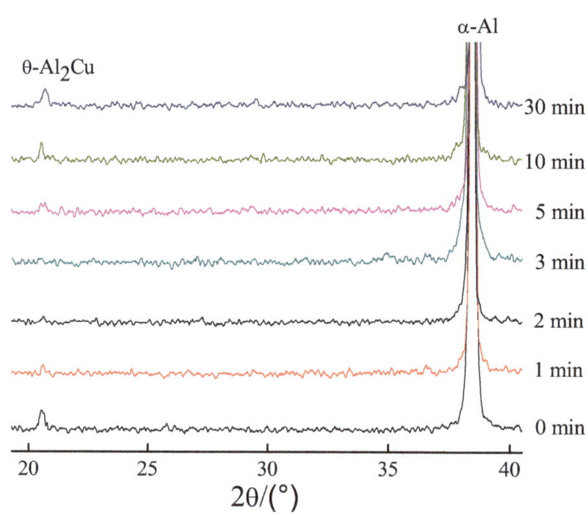

Figure 12. X-ray diffractograms of the bulk alloy heated at 635 °C for different durations and then quenched.

As the heating further proceeds, the liquid amount further increases (Figure 10). The essence of this process is still from the reaction of α→L. However, the Cu solubility in the primary α phase decreases when the heating temperature is in the solidification interval [19]. The Cu content in the primary phase decreases during the 10–30 min period, while that in the liquid phase increases (Table 1).

After being heated for 30 min, the final coarsening induced by the influenced Ostwald ripening operates. This process essentially refers to the dissolution and subsequent reprecipitation. Thus the final coarsening results from the two reverse reactions of α→L and L→α.

Therefore, it can be concluded that the microstructural evolution of the 2024 bulk alloy can be attributed to the phase transformations occurring during partial remelting. The initial coarsening is due to the reaction of $\theta \rightarrow \alpha$; the formation of the liquid pools within the particles and continuous liquid layer on the primary particle surface results from $\theta + \alpha \rightarrow L$ and $\alpha \rightarrow L$; and the final coarsening is attributed to the two reverse reactions of $\alpha \rightarrow L$ and $L \rightarrow \alpha$.

3.4. Effect of Heating Temperature on Semisolid Microstructure

Figure 13 shows the microstructures of the bulk alloy heated for 30 min at different temperatures. Combining with Figure 7c, it can be seen that the primary particle size decreases and the liquid amount increases as the temperature rises. The above changes can be more clearly seen in the quantitative examinations shown in Figure 14.

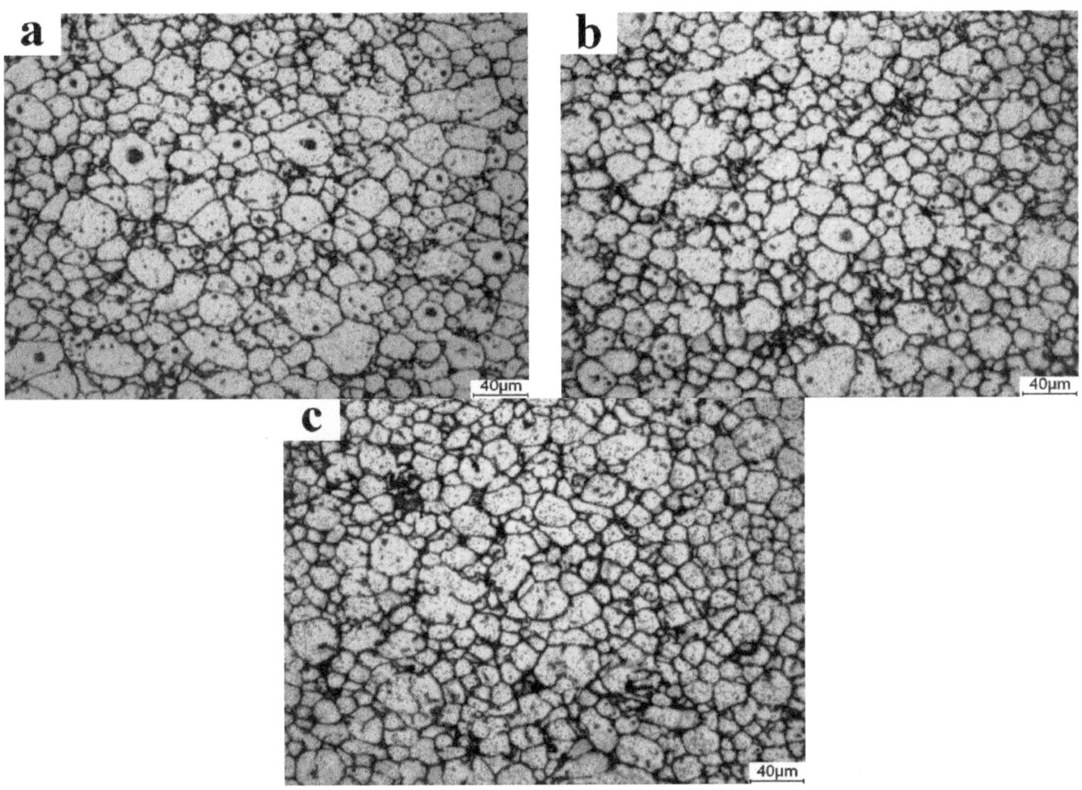

Figure 13. Micrographs of the bulk alloy heated for 30 min at (**a**) 625 °C; (**b**) 630 °C; (**c**) 640 °C.

As discussed above, the possibility of the merging of the primary particles is small when heated at 635 °C for different durations. For the same reason, mergence between particles is also not found at low temperature of 625 and 630 °C; then the microstructure consisted of individual primary particles uniformly suspended in some liquid phase (Figure 13a,b). As the temperature rises, the liquid amount further increases due to the further partial melting of the primary particles (comparing Figures 13b and 7c). More of the particles can melt than at low temperature, which thus decreases the primary particle size (comparing Figure 13a,c). In addition, it can be found that the primary particles in the semisolid microstructure are always spheroidal, which should be regarded as one of the overwhelming advantages to obtaining spheroidal particles by powder thixoforming.

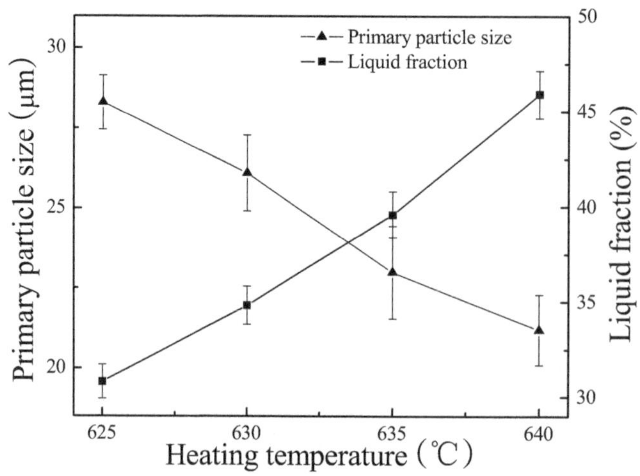

Figure 14. Variations in the primary particle size and liquid fraction with the heating temperature.

Therefore, the elevated temperature is beneficial for obtaining a semisolid ingot with small and spheroidal particles. However, it also generates a large amount of liquid phase if the temperature further rises (Figure 13c). Thus the potential of the semisolid cannot be developed until the temperature is properly controlled.

4. Conclusions

1. A semisolid microstructure with small and spheroidal primary particles (evolved from the original powders) can be obtained through powder thixoforming after the alloy powders are uniaxially cold-compacted and then partially remelted at 635 °C. Thus powder thixoforming is a feasible approach to producing semisolid microstructure available for thixoforming.

2. The microstructural evolution of the 2024 bulk alloy during partial remelting can be divided into three stages: the initial coarsening of the grains within powders, the formation of continuous liquid layer on the particle surface, and the final slight coarsening. The phase transformations correspond to the specific stages of the microstructural evolution. The above three stages are ascribed to the phase transformations of $\theta\rightarrow\alpha$, $\alpha\rightarrow L$, and $\alpha\rightarrow L$ and $L\rightarrow\alpha$, respectively.

3. The alumina on the surfaces of the alloy powders results in less active coalescence and low Ostwald ripening rate during partial remelting. The variation of the primary particle size is small. Thus it can be concluded that one spherical powder in the as-pressed microstructure evolves into one spheroidal particle with a continuous liquid layer surface in the semisolid microstructure.

4. Compared with traditional methods used to fabricate semisolid ingots, because the as-received Al_p is spherical and has a powder size of 20 μm, and the coarsening rate of the solid particles is low, it does not need to experience a spheroidization stage to obtain small, spheroidal particles.

5. Properly raising the heating temperature is the key to obtaining an ideal semisolid microstructure.

Acknowledgments

The authors wish to express thanks for the financial support from the Basci Scientific Fund of Gansu University (Grant No. G2014-07), the Program for New Century Excellent Talents in the University of

China (Grant No, NCET-10-0023), and Lanzhou University of Technology's Program for Hongliu Outstanding Youth.

Author Contributions

Pubo Li performed the experiments and wrote the paper under Tijun Chen's guidance, and contributed to all activities; Suqing Zhang contributed analysis tools; Pubo Li, Tijun Chen, and Renguo Guan contributed to the interpretation and discussion of results.

Conflicts of Interest

The authors declare no conflict of interest.

References

1. Fan, T.X.; Shi, Z.L.; Zhang, D.; Wu, R.J. The interfacial reaction characteristics in SiC/Al composite above liquidus during remelting. *Mater. Sci. Eng. A* **1998**, *257*, 281–286.
2. Cheng, N.P.; Zeng, S.M.; Liu, Z.Y. Preparation, microstructures and deformation behavior of SiCp/6066Al composites produced by PM route. *J. Mater. Process. Technol.* **2008**, *202*, 27–40.
3. Tham, L.M.; Gupta, M.; Cheng, L. Effect of limited matrix–reinforcement interfacial reaction on enhancing the mechanical properties of Aluminium–silicon carbide composites. *Acta Mater.* **2001**, *49*, 3243–3253.
4. Kim, Y.; Lee, J.-C. Processing and interfacial bonding strength of 2014 Al matrix composites reinforced with oxidized SiC particles. *Mater. Sci. Eng. A* **2006**, *420*, 8–12.
5. Guo, M.; Tsao, C.-Y. Tribological behavior of self-lubricating aluminium/SiC/graphite hybrid composites synthesized by the semi-solid powder-densification method. *Compos. Sci. Technol.* **2000**, *60*, 65–74.
6. Ogel, B.; Gurbuz, R. Microstructural characterization and tensile properties of hot pressed Al–SiC composites prepared from pure Al and Cu powders. *Mater. Sci. Eng. A* **2001**, *301*, 213–220.
7. Zhao, N.Q.; Nash, P.; Yang, X.J. The effect of mechanical alloying on SiC distribution and the properties of 6061 Aluminum composite. *J. Mater. Process. Technol.* **2005**, *170*, 586–592.
8. Zhou, J.; Drużdżel, A.T.; Duszczyk, J. The effect of extrusion parameters on the fretting wear resistance of Al-based composites produced via powder metallurgy. *J. Mater. Sci.* **1999**, *34*, 5089–5097.
9. Rokni, M.R.; Zarei-Hanzaki, A.; Abedi, H.R.; Haghdadi, N. Microstructure evolution and mechanical properties of backward thixoextruded 7075 Aluminum alloy. *Mater. Des.* **2012**, *36*, 557–563.
10. Wen, C.E.; Yasue, K.; Yamada, Y. Fabrication of TiAl by blended elemental powder semisolid forming. *J. Mater. Sci.* **2001**, *36*, 1741–1745.
11. Tzimas, E.; Zavaliangos, A. A comparative characterization of near-equiaxed microstructures as produced by spray casting, magnetohydrodynamic casting and the stress induced, melt activated process. *Mater. Sci. Eng. A* **2000**, *289*, 217–227.

12. Zoqui, E.J.; Shehata, M.T.; Paes, M.; Kao, V.; Es-Sadiqi, E. Morphological evolution of SSM A356 during partial remelting. *Mater. Sci. Eng. A* **2002**, *325*, 38–53.
13. Wu, Y.F.; Kim, G.-Y.; Anderson, I.E.; Lograsso, T.A. Fabrication of Al6061 composite with high SiC particle loading by semi-solid powder processing. *Acta Mater.* **2010**, *58*, 4398–4405.
14. Cheng, Y.S.; Luo, S.J. Strengthening and toughening forming of Al/Al_2O_3 composite in pseudo-semi-solid state. *Acta Metall. Sin.* **2010**, *23*, 21–26.
15. Luo, S.J.; Cheng, Y.S.; Wang, P.X. Pseudo-semi-solid thixoforging of cup shell with Al/Al_2O_3. *Trans. Nonferrous Met. Soc. China* **2006**, *16*, 772–775.
16. Rokni, M.R.; Widener, C.A.; Crawford, G.A. Microstructural evolution of 7075 Al gas atomized powder and high-pressure cold sprayed deposition. *Surf. Coat. Technol.* **2014**, *251*, 254–263.
17. Rokni, M.R.; Widener, C.A.; Crawford, G.A.; West, M.K. An investigation into microstructure and mechanical properties of cold sprayed 7075 Al deposition. *Mater. Sci. Eng. A* **2015**, *625*, 19–27.
18. Schueller, R.D.; Wawner, F.E.; Sachdev, A.K. Nucleation mechanism of the cubic σ phase in squeeze-cast aluminium matrix composites. *J. Mater. Sci.* **1994**, *29*, 424–435.
19. Okamoto, H. *Desk Handbook: Phase Diagrams for Binary Alloys*, 1st ed.; ASM International: Materials Park, OH, USA, 2000; pp. 29–30.
20. Chen, T.J.; Jiang, X.D.; Ma, Y.; Li, Y.D.; Hao, Y. Microstructural evolution and phase transformations during partial remelting of AZ91D magnesium alloy refined by SiC. *J. Alloys Compd.* **2010**, *497*, 147–154.
21. Chen, T.J.; Jiang, X.D.; Ma, Y.; Li, Y.D.; Hao, Y. Effect of initial as-cast microstructure of AZ91D magnesium alloy on its semisolid microstructure. *J. Alloys Compd.* **2010**, *505*, 476–482.
22. Hong, T.W.; Kim, S.K.; Ha, H.S.; Kim, M.G.; Lee, D.B.; Kim, Y.J. Microstructural evolution and semisolid forming of SiC particulate reinforced AZ91HP magnesium composites. *Mater. Sci. Technol.* **2000**, *16*, 887–892.
23. Anderson, I.E.; Foley, J.C. Determining the role of surfaces and interfaces in the powder metallurgy processing of Aluminum alloy powders. *Surf. Interface Anal.* **2001**, *31*, 599–608.
24. Liu, Z.Y.; Sercombe, T.B.; Schaffer, G.B. The effect of particle shape on the sintering of Aluminum. *Metall. Mater. Trans. A* **2007**, *38*, 1351–1357.
25. Ch'ng, H.N.; Pan, J. Sintering of particles of different sizes. *Acta Mater.* **2007**, *55*, 813–824.
26. Snyder, V.A.; Alkemper, J.; Voorhees, P.W. The development of spatial correlations during Ostwald ripening: A test of theory. *Acta Mater.* **2000**, *48*, 2689–2701.
27. Xie, C.L.; Hailat, M.; Wu, X.; Newaz, G.; Taya, M.; Raju, B. Development of short fiber-reinforced NiTi/Al 6061 composite. *J. Eng. Mater. Technol.* **2007**, *129*, 69–76.
28. Jiang, H.T.; Li, M.Q. Effects of isothermal heat treatment on microstructural evolution of semisolid Al-4Cu-Mg alloy. *J. Mater. Eng. Perform.* **2004**, *13*, 488–492.

8

Effects of Annealing on the Martensitic Transformation of Ni-Based Ferromagnetic Shape Memory Heusler Alloys and Nanoparticles

Tina Fichtner [1], Changhai Wang [1], Aleksandr A. Levin [1], Guido Kreiner [1], Catalina Salazar Mejia [1], Simone Fabbrici [2,3], Franca Albertini [2] and Claudia Felser [1,*]

[1] Max-Planck-Institute for Chemical Physics of Solids (MPI CPFS), Noethnitzer Str. 40, DE-01187 Dresden, Germany; E-Mails: tina.fichtner@cpfs.mpg.de (T.F.); changhai.wang@cpfs.mpg.de (C.W.); alexander.levin@cpfs.mpg.de (A.A.L.); guido.kreiner@cpfs.mpg.de (G.K.); catalina.salazar@cpfs.mpg.de (C.S.M.)

[2] Institute of Materials for Electronics and Magnetism, National Research Council (IMEM-CNR), Parco Area delle Scienze 37/A, I-43124 Parma, Italy; E-Mails: simone.fabbrici@imem.cnr.it (S.F.); franca.albertini@imem.cnr.it (F.A.)

[3] Laboratory of Micro and Submicro Enabling Technologies for Emilia-Romagna (MIST E-R), via Gobetti 101, I-40129 Bologna, Italy

* Author to whom correspondence should be addressed; E-Mail: claudia.felser@cpfs.mpg.de

Academic Editor: Kurt R. Ziebeck

Abstract: We report on the effects of annealing on the martensitic phase transformation in the Ni-based Heusler system: $Mn_{50}Ni_{40}Sn_{10}$ and $Mn_{50}Ni_{41}Sn_9$ powder and $Co_{50}Ni_{21}Ga_{32}$ nanoparticles. For the powdered $Mn_{50}Ni_{40}Sn_{10}$ and $Mn_{50}Ni_{41}Sn_9$ alloys, structural and magnetic measurements reveal that post-annealing decreases the martensitic transformation temperatures and increases the transition hysteresis. This might be associated with a release of stress in the $Mn_{50}Ni_{40}Sn_{10}$ and $Mn_{50}Ni_{41}Sn_9$ alloys during the annealing process. However, in the case of $Co_{50}Ni_{21}Ga_{32}$ nanoparticles, a reverse phenomenon is observed. X-ray diffraction analysis results reveal that the as-prepared $Co_{50}Ni_{21}Ga_{32}$ nanoparticles do not show a martensitic phase at room temperature. Post-annealing followed by ice quenching, however, is found to trigger the formation of the martensitic phase. The presence of the martensitic transition is attributed to annealing-induced particle growth and the stress introduced during quenching.

Keywords: phase transitions; magnetism; X-ray and electron scattering

1. Introduction

Heusler alloys that belong to the ferromagnetic shape memory alloys (FSMAs) have been a fascinating class of materials since they were reported by Ullakko *et al.* [1] two decades ago. FSMAs are ferromagnetic and show a magnetic field-induced martensitic transformation. Most shape memory Heusler alloys contain Ni. Those with Mn and Ga, In, Sn and Sb [2–6] are the most investigated compounds. They are characterized by large magnetic field-induced strain [7,8], large magneto-resistance [9,10] and a significant magnetocaloric effect [11–13].

Mn-rich Ni- and Mn-based Heusler alloys have recently become the focus of research owing to their high localized magnetic moment of 4.0 μ_B at the Mn atom and their potential for enhanced magnetocaloric properties [11,14]. Mn-rich $Mn_{50}Ni_{50-x}Sn_x$ Heusler compounds were reported to show a significant exchange bias field [15], large magnetic entropy change [16,17], small thermal hysteresis and large magnetization change at the martensitic transformation [18]. However, little attention has been paid to the effect of annealing on those compounds.

The influence of post-annealing on the martensitic transformation of Ni-rich Ni-Mn-Sn alloys has been studied by Schagel, McCallum and Lograsso [19] on bulk material and Xuan *et al.* [20] on ribbons. Schagel, McCallum and Lograsso [19] reported that the magnetization change at the martensitic transformation is dependent on the annealing time. After an annealing time of approximately four weeks at 1223 K, they observed the highest magnetization change during the martensitic transformation and no overlap of the transition temperatures. Xuan *et al.* [20], on the other hand, reported that annealing of ribbons resulted in an increase in the reverse martensitic transition temperature (T_A) from 230 K (melt-spun) to 265 K (annealed at 1173 K). Concomitantly, the Curie temperature of the austenite T_C^A increased from T_C^A = 270 K (melt-spun) to T_C^A = 275 K (annealed at 1173 K). The reason for the increase of T_A and T_C^A is attributed to stress and structural relaxation during post-annealing.

In addition to these well-investigated Ni- and Mn-based compounds, other FSMAs have been studied, *i.e.*, Ni-Fe-Ga [21–23], Co-Ni-Al [24,25] and Co-Ni-Ga [26–29]. The Co-Ni-Ga Heusler system was intensively studied as a promising alternative to Ni-Mn-Ga alloys, especially for high-temperature shape memory device applications [30–37]. Co-Ni-Ga compounds generally exhibit a dual-phase microstructure: a parent phase (β) and a non-transformable secondary phase (γ). The ductile Ni-rich γ-phase enhances the overall workability of the Co-Ni-Ga alloys. Owing to the absence of a modulated or layered structure in martensite, the magnitude of the magnetic field-induced strain (MFIS) in Co-Ni-Ga alloys is normally much smaller than that of Ni-Mn-Ga alloys [38]. Even so, Co-Ni-Ga compounds are still promising candidates as classical thermally-triggered shape memory materials.

The effects of annealing on the structural and magnetic phase transformation of Co-Ni-Ga compounds have been reported for bulk and ribbon samples [27,39,40]. Liu *et al.* [27] studied the effects of annealing temperature and cooling conditions on the structure and martensitic transformation of bulk $Co_{46}Ni_{27}Ga_{27}$. It was found that the martensitic transformation temperature increased monotonically with increasing annealing temperature and quenching rate. The precipitation of the non-transformable γ-phase can be avoided if the annealing temperature is higher than 1393 K. Sarma *et al.* [39] observed a similar correlation between the annealing temperature and the martensitic transformation in $Co_{47}Ni_{23}Ga_{30}$ ingots. It was also found that a higher annealing temperature significantly increased the Curie temperature and moderately decreased the amount of the γ-phase. Dai *et al.* [40] found that annealing of the $Co_{50}Ni_{20}Ga_{30}$ ribbons at 673 K suppressed the martensitic transformation and significantly influenced their magnetic properties. Post-annealing at higher temperatures (>873 K) and quenching recovered the martensitic phase transition of the $Co_{50}Ni_{20}Ga_{30}$ ribbons. This was attributed to the elimination of the precipitates and weight fractional variations of the transformable β-phase and the non-transformable γ-phase by the post-annealing.

To our knowledge, no study on the effect of annealing on the martensitic phase transformation of nanostructured $Co_{50}Ni_{21}Ga_{32}$ nanoparticles, and $Mn_{50}Ni_{40}Sn_{10}$ and $Mn_{50}Ni_{41}Sn_9$ powder has been conducted. Herein, we report on the synthesis and annealing-dependent martensitic phase transformation of $Co_{50}Ni_{21}Ga_{32}$ nanoparticles, and powdered $Mn_{50}Ni_{40}Sn_{10}$ and $Mn_{50}Ni_{41}Sn_9$ alloys.

This paper is divided into two parts. First, we present the results for the $Mn_{50}Ni_{40}Sn_{10}$ and $Mn_{50}Ni_{41}Sn_9$ powder, followed by the results for the $Co_{50}Ni_{21}Ga_{32}$ nanoparticles. In the following, the as-prepared powder is that which resulted from the melting of the elements, performing a heat treatment with quenching and grinding in a mortar. The post-annealed powder is that subjected to a subsequent annealing procedure without quenching. CNG_AS denotes the as-prepared $Co_{50}Ni_{21}Ga_{32}$ nanoparticles and CNG_AQ the annealed and quenched $Co_{50}Ni_{21}Ga_{32}$ nanoparticles.

2. Results and Discussion

2.1. $Mn_{50}Ni_{40}$ Sn_{10} and $Mn_{50}Ni_{41}$ Sn_9 Powder

2.1.1. Microscopic Analysis

In order to investigate the effect of annealing on the martensitic transformation in Mn-rich Ni-based FSMAs, the alloys $Mn_{50}Ni_{40}Sn_{10}$ and $Mn_{50}Ni_{41}Sn_9$ were synthesized and characterized by optical and scanning electron microscopy (SEM) with energy dispersive X-ray spectroscopy (EDXS). Figure 1 shows the bright field and back-scattered electron micrographs of $Mn_{50}Ni_{40}Sn_{10}$. Both methods exhibit a matrix with an embedded secondary phase, which indicates a two-phase material; the composition of the main phase is $Mn_{47.0}Ni_{41.8}Sn_{11.2}$, and that of the secondary phase is $Mn_{68.9}Ni_{28.8}Sn_{2.3}$. The errors of the SEM/EDX analysis are estimated with ± 0.5 at%. This secondary phase, which contains a high amount of Mn, is described as the γ-phase in the literature [41]. Liu *et al.* [41] reported the presence of this phase in Ni-Fe-Co-Ga alloys and ascribed the absence of martensitic transformation in the altered austenitic matrix to the γ-phase.

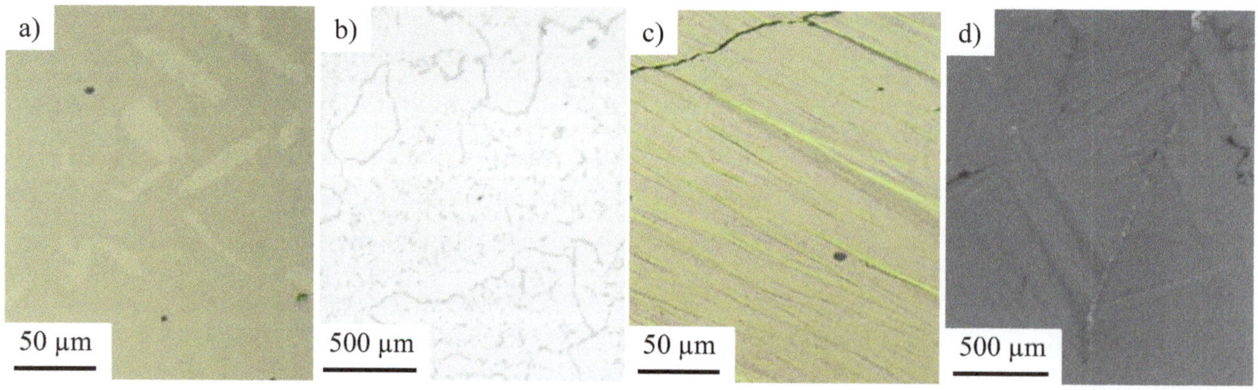

Figure 1. Optical and SEM/EDXS-micrographs of $Mn_{50}Ni_{40}Sn_{10}$: (**a**) bright field and (**b**) back-scattered electron images; (**c**) bright field and (**d**) secondary electron images of $Mn_{50}Ni_{41}Sn_9$.

In the current study, it was found that a tiny variation of the composition of approximately 1 at% changes the microstructure completely, as shown for $Mn_{50}Ni_{41}Sn_9$ in the bright field image in Figure 1c and the secondary electron image in Figure 1d. This microstructure, with its martensitic domains and needle-like shapes, reveals a martensitic structure at room temperature (r.t.), which confirms a martensitic transformation. For this alloy, SEM/EDXS analysis revealed a composition of $Mn_{48.9}Ni_{41.7}Sn_{9.4}$, which is close to the nominal composition.

Additionally, inductively-coupled plasma-optical emission spectroscopy (ICP-OES) analysis was performed to verify the powder compositions. The results showed that the mean analytical composition of the alloy $Mn_{50}Ni_{40}Sn_{10}$ was $Mn_{49.9(1)}Ni_{40.2(1)}Sn_{10.0(1)}$, which is close to the nominal composition. For the compound $Mn_{50}Ni_{41}Sn_9$, an analytical composition of $Mn_{49.7(1)}Ni_{41.3(1)}Sn_{9.0(1)}$ was obtained. Therefore, the analytical compositions of these alloys are in good agreement with their nominal compositions.

2.1.2. Phase Analysis

To support the results from the microstructure analysis of $Mn_{50}Ni_{40}Sn_{10}$, X-ray diffraction (XRD) analysis was performed. Figure 2 shows the r.t. XRD patterns of the as-prepared and the post-annealed powders. Most of the Bragg reflections of the XRD pattern of Figure 2a can be attributed to a cubic Heusler-type crystal structure with the space group (sg) 225 ($Fm\bar{3}m$) and a lattice parameter a = 5.997(2) Å. This result indicates an austenitic crystal structure at r.t.

However, in Figure 2a, a second phase could be referred to the diffraction pattern, which belongs to the fcc γ-phase of Mn-Ni with a lattice parameter of a = 3.724(4) Å. Both determined crystal structures support the EDXS analysis, which resulted in analytical compositions of $Mn_{47.0}Ni_{41.8}Sn_{11.2}$ and $Mn_{68.9}Ni_{28.8}Sn_{2.3}$. This γ-phase is known to be present in Ni-based Heusler alloys, as reported in the literature [41–43], appearing as a secondary phase next to the austenite. Thus, the γ phase is attributed to a reduction of the magnetization change during the martensitic transformation and to a vanishing of the martensitic transformation.

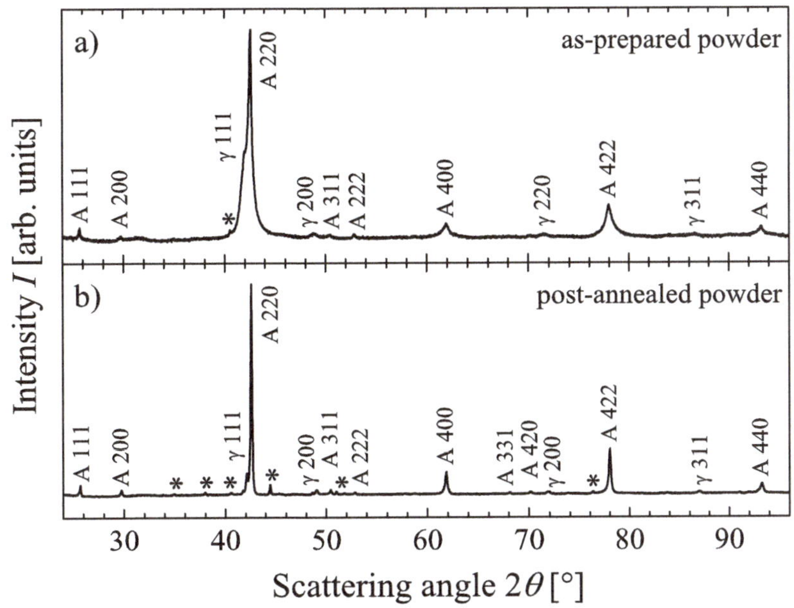

Figure 2. XRD pattern of $Mn_{50}Ni_{40}Sn_{10}$ at r.t. for (**a**) as-prepared and (**b**) post-annealed powders. "A" indicates the austenite and "γ" the fcc Mn-Ni phase. Mn-rich impurities are labeled by asterisks.

After annealing the powder in order to reduce the stress, the Bragg reflections shown in Figure 2b are much sharper, resulting again in a cubic Heusler-type crystal structure $Fm\bar{3}m$ with a lattice parameter of a = 5.993(1) Å and an fcc γ-phase with a lattice parameter of a = 3.711(1) Å.

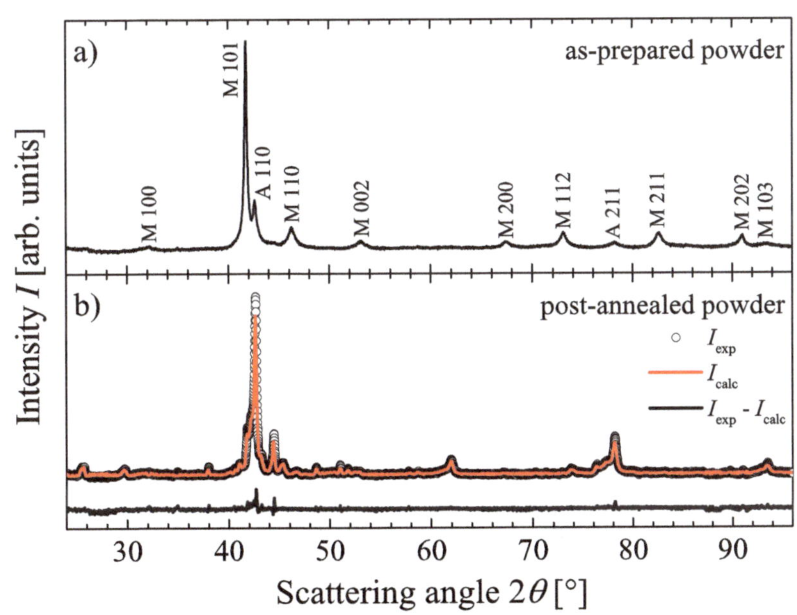

Figure 3. XRD pattern of $Mn_{50}Ni_{41}Sn_9$ at r.t. for (**a**) as-prepared and (**b**) post-annealed powders. "A" indicates the austenite and "M" the martensite. Results of Le Bail fitting are shown in (**b**) (R_{wp} = 10%).

Concerning $Mn_{50}Ni_{41}Sn_9$ (Figure 3a), the Bragg reflections are significantly different from those of the $Mn_{50}Ni_{40}Sn_{10}$ alloy, probably due to a slightly lower Sn-content. At r.t. $Mn_{50}Ni_{41}Sn_9$ reveals a martensitic tetragonal crystal structure ($P4/mmm$, CuAu structure type, sg: 123) with lattice parameters of a = 2.773(1) Å and c = 3.444(2) Å. The corresponding c/a-ratio is approximately 1.24. In addition to this martensite, there is also an austenitic phase with a lattice parameter of a = 2.990(1) Å ($Im\bar{3}m$, tungsten structure type, sg: 229). A crystal structure with a higher order (CsCl structure type, $Pm\bar{3}m$, sg: 221) is indicated by a tiny Bragg reflection at $2\theta = 53.72°$, which is not indexed to the tungsten structure type.

After recrystallization of the post-annealed powder, the diffraction pattern becomes more complex. The diffraction pattern could be indexed to a $7M$ superstructure cell with lattice parameters of a = 4.304(5) Å, b = 5.984(5) Å, and c = 42.223(7) Å and β = 93.97(2)°, using a Le Bail fitting, which is shown in Figure 3b. These lattice parameters are similar to those of the superstructure cell in the Ni-Mn-Ga FMSA [44–46]. It should be noted that owing to strong reflection overlap in the Le Bail fitting, it is not possible to exclude the presence of the austenitic phase at r.t. in this sample.

2.1.3. Magnetic Properties

The magnetic properties were determined by magnetometry. Figure 4a shows the temperature-dependent magnetization analysis of the as-prepared $Mn_{50}Ni_{40}Sn_{10}$ powder with applied fields, $\mu_0 H$, of 0.1, 2 and 7 T in the field cooling and field heating modes, starting from 400 K. Concerning $Mn_{50}Ni_{40}Sn_{10}$ at $\mu_0 H = 0.1$ T, three phase transformations are shown (from low to high temperature): (i) a magnetic transformation in the martensite with a Curie temperature, T_C^M; (ii) a magnetostructural transformation with a martensitic start temperature, M_s, a martensitic finish temperature, M_f, and an inflection point of the martensitic transition, T_M; and (iii) a magnetic transformation in the austenite with a Curie temperature, T_C^A. Thus, $Mn_{50}Ni_{40}Sn_{10}$ possesses martensitic transformation temperatures of M_s = 221 K, M_f = 175 K and T_M = 202 K and two Curie temperatures for the martensite and austenite phases of T_C^M = 144 K and T_C^A = 274 K, respectively. The martensitic transformation temperatures are below r.t. Therefore, the austenite phase found at r.t. from the XRD analysis is verified.

Additionally, the magnetization change during the martensitic transformation is 30 $Am^2\ kg^{-1}$, which indicates a different magnetic order between the martensite and austenite. Thus, the martensite orders ferrimagnetically, whereas the austenite orders ferromagnetically.

The magnetization change at $\mu_0 H = 7$ T is increased by 55% and T_M decreases with further increasing of $\mu_0 H$ from $T_{M,0.1T}$ = 202 K to $T_{M,7T}$ = 189 K, which is ascribed to a field-induced first order magnetostructural transformation. A reduced thermal hysteresis ΔT_{Hyst} of 7 K, as compared to other Heusler alloys [11,14], was obtained, which is comparable with the results published by Han [12] and Ma *et al.* [18]. However, ΔT_{Hyst} remains constant with increasing of $\mu_0 H$, resulting in a possible reduction in energy consumption in magnetocaloric applications, contrary to Ni-rich Ni- and Mn-based Heusler alloys [11,14].

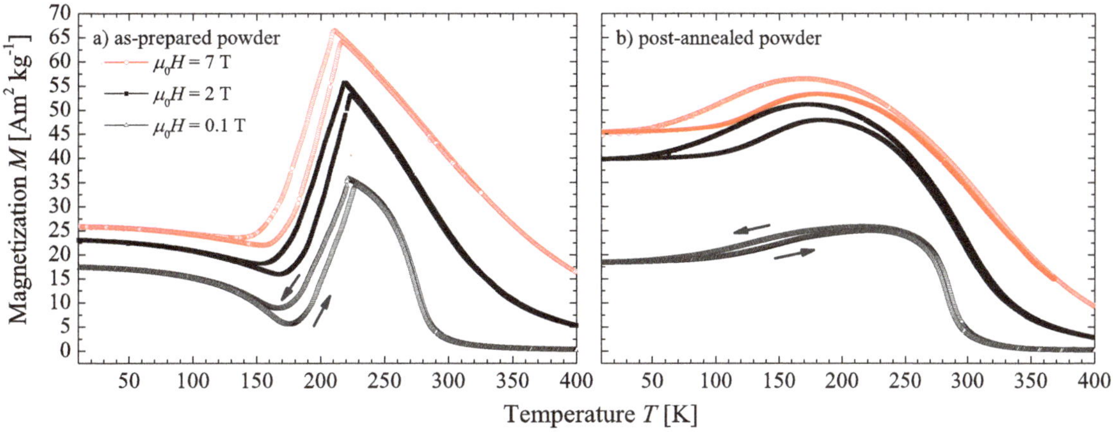

Figure 4. Temperature dependence of magnetization of $Mn_{50}Ni_{40}Sn_{10}$ in applied fields of μ_0H = 0.1, 2 and 7 T during heating and cooling modes for (**a**) as-prepared and (**b**) post-annealed powders.

Figure 4b displays the thermomagnetic curves of this alloy as a post-annealed powder using conditions similar to those of the as-prepared powder. Comparing both at μ_0H = 0.1 T, the martensitic transformation temperatures are very sensitive to the annealing method. Thus, M_s, M_f and T_M are decreasing, whereas T_C^A is increasing (for more details, see the Supplementary Information,Tables S1 and S2). Comparable results are known in Ni-rich Ni-and Mn-based Heusler alloys [19,20]. A similar feature has been found for Ni-Co-Mn-In alloys and is attributed to the modification of the atomic sites (order-disorder), the Mn-Mn distances and the Mn-Mn exchange coupling [47–50]. Additionally, based on this, the Fermi surface and the Brillouin zone boundary can be affected [50–53]. Therefore, for the post-annealed powder, the magnetization change during the martensitic transformation is decreased by one-third and remains constant with increasing μ_0H up to 7 T, indicating similar magnetic order of the martensite and austenite, *i.e.*, ferromagnetism.

At μ_0H = 0.1 T, ΔT_{Hyst} is increased significantly by 17 K, which is comparable to Ni-rich Ni- and Mn-based Heusler alloys [11,14]. However, for Ni-rich Ni-Mn-Sn Heusler alloys reported by Schlagel, McCallum and Lograsso [19] and Xuan *et al.* [20], the effect of annealing increased the magnetization change during the martensitic transformation. This was also observed for Ni-rich Ni-Mn-Co-In Heusler alloys, reported by Ito *et al.* [47]. Therefore, the impact of atomic ordering on annealing in terms of the martensitic transformation are different between Mn-rich Ni- and Mn-based Heusler alloys because of atomic site occupation of the Ni and Mn atoms. A recently reported study of the crystal structure of Mn-rich $Mn_{1.8}Ni_{1.2}Sn$ by our group [54] identified the disorder on the Wyckoff positions $4d$ and $4c$, by Ni and Mn atoms and their effects on the magnetic and transport properties.

Concerning the temperature-dependent magnetization of $Mn_{50}Ni_{41}Sn_9$, shown in Figure 5a, a small variation in composition of 1 at% increases M_s, M_f and T_M to higher temperatures (see the Supplementary Information, Tables S3 and S4). Thus, at r.t., the martensite and austenite are found to be present, verifying the crystal structures of the as-prepared powders shown in Figures 2a and 3a as determined by XRD analysis at r.t.

Comparing the magnetization of the austenite and martensite during the martensitic transformation at μ_0H = 0.1 T, the differences are quite small, which means that the martensite shows paramagnetism and the austenite orders antiferromagnetically. Thus, the main contribution during the martensitic transformation comes from the structural distortion. At $\mu_0 H$ = 2 T, the magnetization change reaches a value of 20 Am^2 kg^{-1}, showing a sharp martensitic transformation.

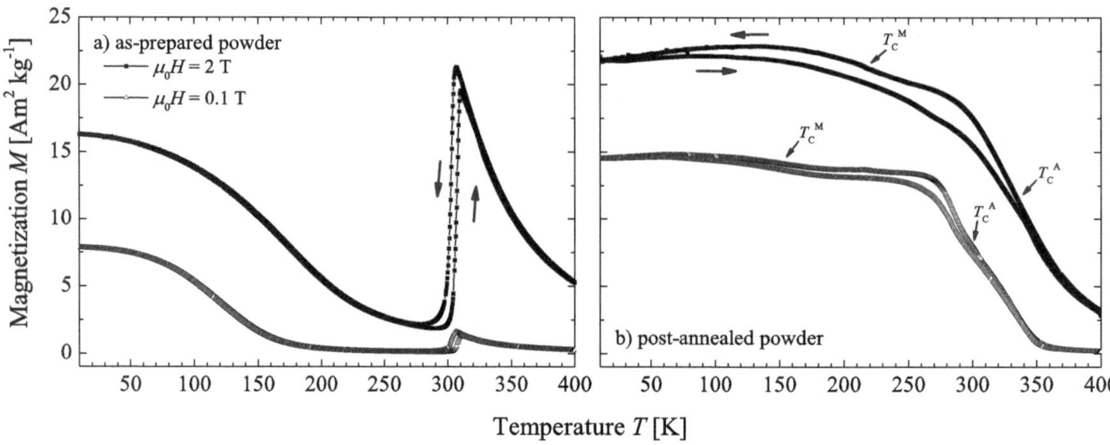

Figure 5. Temperature dependence of magnetization analysis of $Mn_{50}Ni_{41}Sn_9$ in applied fields of 0.1 and 2 T during heating and cooling for (**a**) as-prepared and (**b**) post-annealed powders.

As shown in Figure 5b, the temperature-dependent magnetization of the post-annealed powder is different from that of the as-prepared powder owing to a modified atomic site phenomena, which is verified by the variation of the crystal structures of both powders (see Figure 5).

However, for the post-annealed powder shown in Figure 5b, no clear magnetization change during the martensitic transformation was observed, indicating a similar magnetic coupling of the Mn atoms of the austenite and martensite. Additionally, the ferromagnetic ordering of the austenite increases, which indicates stabilization of the austenitic phase. Therefore, the martensitic transformation merges with T_C of the martensite and austenite. A similar phenomenon on the magnetization curve could be observed for Ni-Mn-Sn thin films [55]. The Curie temperatures of the austenite, T_C^A, and martensite, T_C^M, increase with increasing applied field from $\mu_0 H$ = 0.1 T to 2 T ($T^M_{C,0.1\ T}$ = 153 K *vs.* $T^M_{C,2\ T}$ = 217 K and $T^A_{C,0.1\ T}$ = 303 K *vs.* $T^A_{C,2\ T}$ = 331 K).

2.2. *$Co_{50}Ni_{21}Ga_{32}$ Nanoparticles*

2.2.1. Microscopic Analysis

Figure 6a,b shows the TEM micrograph and particle-size distribution, respectively, of the as-prepared $Co_{50}Ni_{21}Ga_{32}$ nanoparticles. The average particle size is 120 nm, with a broad size distribution in the range of 50–200 nm. As shown in Figure 6c, there are also some $Co_{50}Ni_{21}Ga_{32}$ nanorods and nanowires co-existing with the spherical nanoparticles.

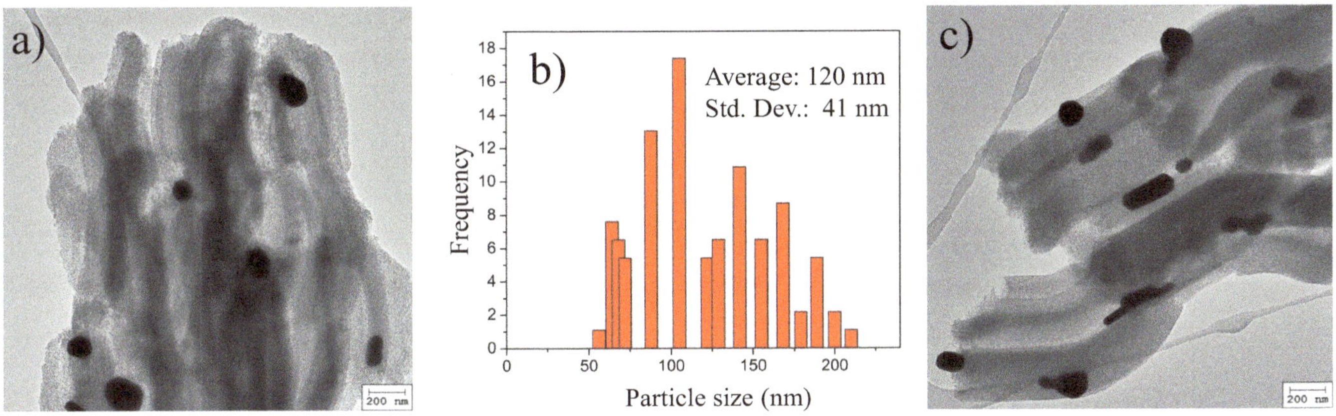

Figure 6. (**a**,**c**) TEM micrographs and (**b**) particle-size distribution of the as-prepared $Co_{50}Ni_{21}Ga_{32.5}$ nanoparticles CNG_AS.

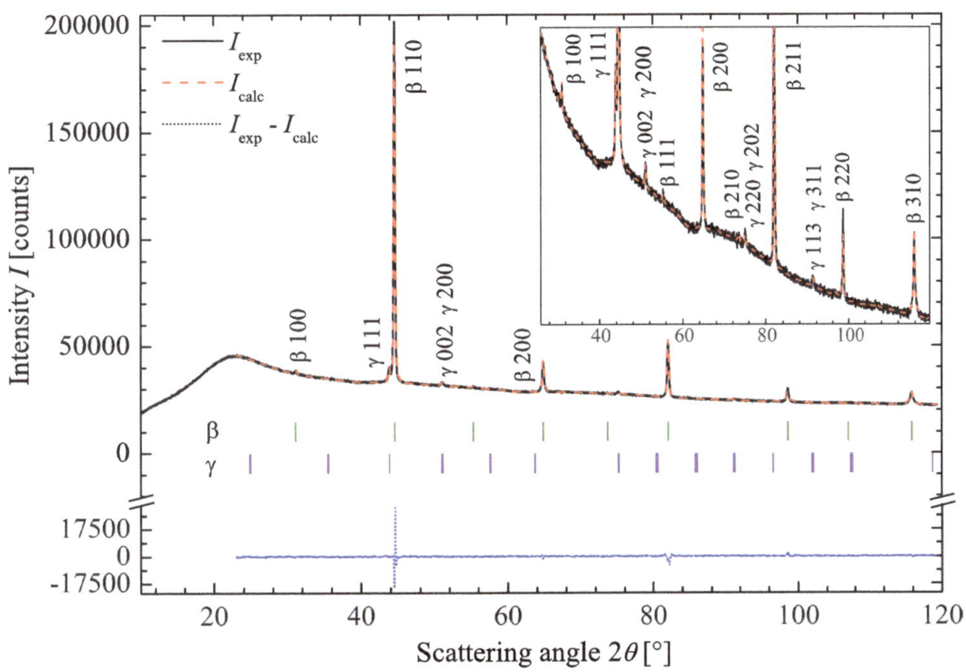

Figure 7. Experimental and the Rietveld-fitted XRD patterns of as-prepared $Co_{50}Ni_{21}Ga_{32}$ nanoparticles. The central-angle 2θ positions of the Bragg reflections according to the refined unit cell parameters of the phases are indicated by bars at the bottom. The difference curve is also shown at the bottom. Miller indices, hkl, of the selected observed reflections of the β ($B2$-type ordering) and γ phases are indicated. The inset shows the scaled XRD pattern for better visualization of the fitting of the weak reflections.

2.2.2. Crystal Structure of $Co_{50}Ni_{21}Ga_{32}$ Nanoparticles

Figures 7 and 8 show the XRD patterns of CNG_AS and CNG_AQ $Co_{50}Ni_{21}Ga_{32}$, respectively. The XRD-derived structural parameters (unit cell parameters, volume and weight content of the phases) and Rietveld fitting quality factors are compiled in Table 1. The XRD patterns exhibit an amorphous halo due to the amorphous mesoporous SBA-15 silica. According to phase analysis and quantitative Rietveld fitting, two modifications with the Co_2NiGa nominal composition are present in the as-prepared nanoparticles (Figures 7). The main phase observed (94.5(1.6) wt% content) is a cubic β phase with a $B2$ partial ordering ($Pm\overline{3}m$, sg: 221, Inorganic Crystal Structure Database (ICSD) Code 169729 [35]) and a unit cell parameter of a = 2.8752(1) Å. As a minor phase with a content of 5.5(1.1) wt%, a fully-ordered γ phase with a slightly tetragonally distorted structure ($P/4mmm$, sg: 123, ICSD Code 157788 [34]) and unit cell parameters of a = 3.580(3) Å and c = 3.567(5) Å is detected (Table 1). In the β Co_2NiGa structure with $B2$-type ordering, the Ni and Ga atoms occupy the same position, whereas in the tetragonal γ-phase, different atoms are placed in different positions [34,35] and the unit cell volume is approximately two-times larger than that of the β phase. The $B2$-type ordering in the β phase is clearly indicated by the presence of the weak reflection, with Miller indices hkl = 100 at $2\theta = 31.12°$ and the distinct absence of the superlattice reflection at $2\theta = 26.87°$, which is characteristic of the β-phase with $L2_1$ atomic ordering (hkl = 111 for the $L2_1$-type β-structure).

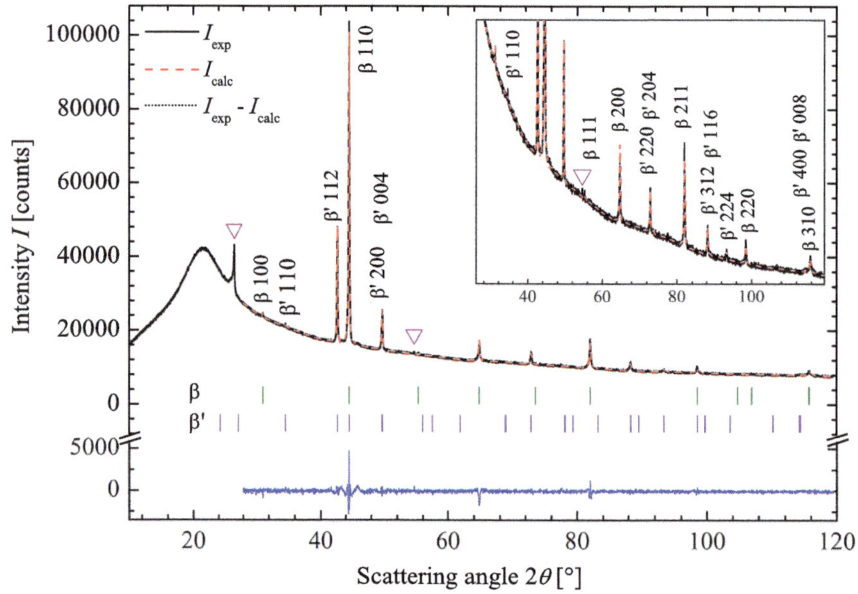

Figure 8. Experimental and Rietveld-fitted XRD patterns of annealed/quenched $Co_{50}Ni_{21}Ga_{32}$ nanoparticles. The central-angle 2θ positions of the Bragg reflections according to refined unit cell parameters of the phases are indicated by bars at the bottom. The difference curve is also shown at the bottom. Miller indices hkl of the selected observed reflections of the β ($B2$-type ordering) and β' phases are indicated. The inset shows the scaled XRD pattern for better visualization of the weak reflections. Non-indexed reflections are marked by triangles (probably 112 and 224 reflections of the Ga_7Ni_3 phase).

As shown in Figure 8, considerable changes occur in the crystalline phase composition and content after heat treatment. The minor tetragonal γ-phase disappears, and the content of the $B2$-type ordered Co_2NiGa β-phase decreases to 72.2(1.4) wt%, whereas its unit cell parameter increases to a = 2.8763(1) Å. The formation of a rather large amount of the highly ordered Co_2NiGa martensite phase (β') is observed (27.8(3.5) wt%) according to the results of the Rietveld quantitative analysis. In the β'-phase (sg: 139, $I/4mmm$, ICSD Code 169733 [35]), the Co, Ni and Ga atoms are placed in different atomic positions, similar to what occurs in the γ-phase, but with approximately twice the unit cell volume, owing to the two-times larger tetragonal unit cell parameter c. Recently, the structure of the Co_2NiGa (β') phase was theoretically predicted [35] by means of density functional theory (DFT) calculations. According to Arróyave *et al.* [35], the tetragonal Co_2NiGa (β') phase exhibits the greatest structural stability among Co_2NiGa modifications. Thus, annealing results in a partial $\beta \longrightarrow \beta$' phase transformation from the partially ordered β phases ($B2$-type of ordering) to the highly ordered tetragonal β' that has a unit cell volume approximately four-times larger than that of the β phase. The presence of the different Co_2NiGa phases (β, β' and γ) is confirmed experimentally by the good quality of the Rietveld fitting of the XRD patterns (Figures 7 and 8) reflected in small values of the fitting agreement factors and structural Bragg factors (see Table 1).

Table 1. Structural parameters (unit cell parameters and weight content of the phases) and the Rietveld fitting quality factors of $Co_{50}Ni_{21}Ga_{32}$ nanoparticles. The estimated standard deviations (e.s.d.s) are indicated in round brackets.

ID	Phase	Space Group	Unit Cell Parameters		Wt Fraction [a] (wt%)	Rietveld Fitting Quality Factors			
			a (Å)	c (Å)		R_{wp} [b] (%)	cR_{wp} [c] (%)	R_B [d] (%)	$M_{e.s.d}$ [e]
CNG_AS	β	$Pm\bar{3}m$	2.8752(1)	a	94.5(1.6)	1.73	15.25	2.10	3.14
	γ	$P4/mmm$	3.580(3)	3.567(5)	5.5(1.1)			6.85	
CNG_AQ	β	$Pm\bar{3}m$	2.8763(1)	a	72.2(1.4)	1.26	10.56	2.12	2.46
	β'	$P4/mmm$	3.6691(4)	7.331(2)	27.8(3.5)			6.26	

[a] the amorphous SBA-15 phase and non-identified crystalline phase (for CNG_AQ) are not taken into account; [b] weighted profile factor; [c] background-corrected weighted profile factor; [d] Bragg factor; [e] Berrar's factor.

The crystallite size was analyzed using the Scherrer equation. It was revealed that the crystallite size of the β ($B2$-ordered) phase (D = 45(8) nm) is larger than that of its γ counterpart (D = 33(6) nm) for CNG_AS, whereas annealing resulted in crystallites of nearly the same size for the β ($B2$-type ordered) and β' crystallites, *i.e.*, 54 nm. The crystallite size of the unknown phase in CNG_AQ was D = 69(16) nm. A comparison with TEM results shows that the $Co_{50}Ni_{21}Ga_{32}$ nanoparticles are composed of either two or three nanocrystallites. This fact shows that the crystallite size derived from powder XRD analysis corresponds to the size of the coherent domains and not that of the particles observed by TEM.

2.2.3. Magnetism of the $Co_{50}Ni_{21}Ga_{32}$ Alloys

The magnetic properties of $Co_{50}Ni_{21}Ga_{32}$ nanoparticles changed significantly after annealing and quenching. As shown in Figure 9, large differences are observed in the thermomagnetic behavior of the two samples. The field-cooling (FC) and field-heating (FH) curves of the CNG_AS sample are dissimilar, and the divergence is visible across the entire measured temperature range. The origin of the observed anomaly is unclear, but it is expected to be unrelated to the martensite transformation, since a transformation hysteresis was not observed. It is also probably associated with the presence of the γ phase, because a similar thermomagnetic feature was observed in $Co_{50}Ni_{21}Ga_{32}$ nanoparticles of a single γ phase. Further investigations are required to clarify this point. In comparison, the FC and FH curves of the CNG_AQ sample almost match, indicating the absence of martensite transformation across the measured temperature range. As pointed out by Brown [32], the martensite transformation in Co-Ni-Ga alloys does not occur abruptly, and it normally covers a broad temperature interval (even larger than 200 K). Therefore, in addition to the low weight fraction of the $Co_{50}Ni_{21}Ga_{32}$ phase in the nanoparticles, it is expected that thermomagnetic measurements might not be appropriate for probing the martensitic transformation in shape memory Heusler nanoparticles, such as $Co_{50}Ni_{21}Ga_{32}$. Instead, temperature-dependent XRD or neutron diffraction analysis might be a more suitable method. Furthermore, the magnetization of the CNG_AQ sample was much lower than that of the CNG_AS sample. This might be due to the absence of a γ phase after annealing, which contributes significantly to the magnetization of the two-phase $Co_{50}Ni_{21}Ga_{32}$ nanoparticles.

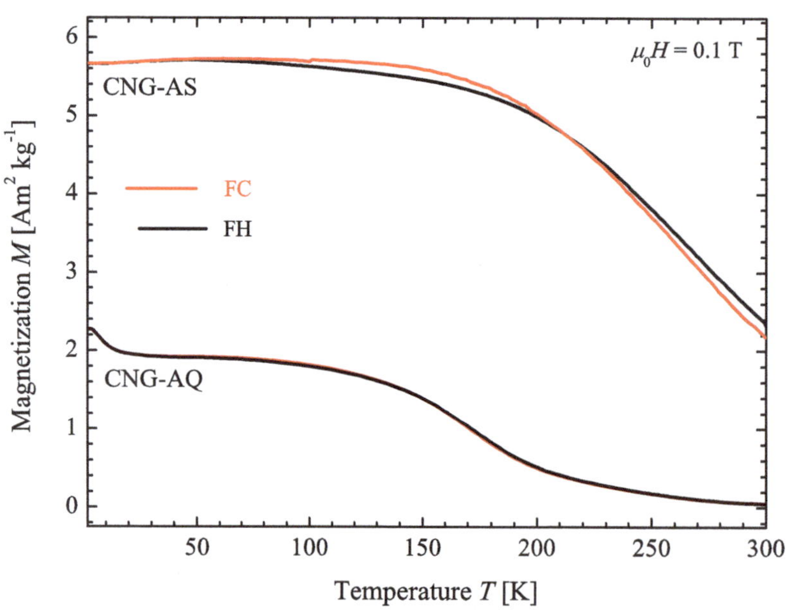

Figure 9. Temperature-dependent magnetization (M - T) curves of $Co_{50}Ni_{21}Ga_{32}$ nanoparticles before and after annealing. FC, field cooling; FH, field heating. Applied magnetic field: 0.1 T.

3. Materials and Methods

3.1. Experimental Details of the $Mn_{50}Ni_{40}Sn_{10}$ and $Mn_{50}Ni_{41}Sn_9$ Alloys

Bulk $Mn_{50}Ni_{40}Sn_{10}$ and $Mn_{50}Ni_{41}Sn_9$ alloys with a total mass of 5 g were prepared by inductive melting of the elements (Ni foil, 99.9%, ChemPur; Mn pieces, 99.99%, ChemPur; Sn granules, 99.999%, ChemPur) in Al_2O_3 crucibles (Ø = 10 mm, Friatec, Al23). The total mass loss was $\leq$ 1 wt%. A subsequent heat treatment for 3 weeks was performed by encapsulating the alloys in an arc-welded tantalum ampoule at 0.3 bar Ar and jacketing them in an evacuated fused quartz glass ampoule.

The $Mn_{50}Ni_{40}Sn_{10}$ sample was annealed at 973 K (onset of melting at T_o = 1188 K from DSC analysis) and the $Mn_{50}Ni_{41}Sn_9$ sample at 1073 K (T_o = 1,205 K). Lastly, both alloys were quenched in ice-water by shattering the silica ampoule.

For post-annealing, the powders were ground in a WC/Co mortar and an agate mortar and sieved to grain sizes, d, of $\leq$ 50 μm for $Mn_{50}Ni_{40}Sn_{10}$ and 100 μm for $Mn_{50}Ni_{41}Sn_9$. Subsequently, the powders were enclosed in an arc-welded tantalum ampoule at 0.3 bar Ar, which, in turn, was jacketed by an evacuated fused quartz glass ampoule. Finally, both ampoules were annealed for 24 h at 873 K and cooled to r.t. over 48 h.

To characterize the alloys, powder XRD measurements were performed at r.t. using an image-plate Huber G670 Guinier camera, Rimsting, Germany equipped with a Ge(111) monochromator. The compounds were analyzed by using Cu-$K\alpha_1$ radiation ($\lambda = 1.540598$ Å) in the range $10° \leq \theta \leq 100°$. A small amount of powder was homogeneously dispersed on Mylar foil. The WinXPOW package [56] was used to determine the lattice parameters. The Le Bail fitting was carried out by Jana, 2006 [57].

Thermal analysis was performed with a differential scanning calorimeter (DSC/TG Netzsch STA 449C, Selb, Germany) at a heating rate of 10 K/min from r.t. to a maximum temperature of 1673 K.

ICP-OES by means of a spectrometer VISTA (Varian Inc., Palo Alto, CA, USA) was used to analyze the final composition after annealing. Chemical analysis by EDXS was performed, and SEM images of secondary electrons and back-scattered electrons were taken using a Philips XL30 system, (SEMTech Solutions, Amsterdam, The Netherlands) with a LaB_6 cathode and an XFLASH® detector, (Bruker, Billerica, MA, USA) at 15 kV. Wavelength dispersive X-ray spectroscopy analysis was done using a Cameca SX-100 electron probe analyzer, (CAMECA, Gennevilliers, France) and elemental standards.

Magnetic properties were investigated by means of a commercial system (VSM, quantum design, San Diego, CA, USA). The temperatures for the magnetic properties measurements were in the range from 1.8 to 400 K.

3.2. Experimental Details of the $Co_{50}Ni_{21}Ga_{32}$ Nanoparticles

All chemicals were purchased from Sigma-Aldrich, (St. Louis, MO, USA) or Alfa Aesar, (Ward Hill, MA, USA) and used as-received. The nonionic triblock copolymer, Pluronic P123 (MW 5800), and tetraethoxysilane (TEOS) were used as a structure-directing surfactant and silicon source, respectively. Furthermore, the precursors for the $Co_{50}Ni_{21}Ga_{32}$ nanocrystals were $CoCl_2 \times 6H_2O$ (99.9%), $Ni(NO_3)_2 \times 6H_2O$ (99.999%) and $Ga(NO_3) \times H_2O$ (99.9%).

First, mesoporous SBA-15 silica was prepared as the template for the $Co_{50}Ni_{21}Ga_{32}$ nanoparticles. Four grams of Pluronic P123 were dissolved into a solution consisting of 30 g of water and 120 mL of 2 M HCl at 310 K and stirred overnight. Subsequently, 8.5 g of TEOS was vigorously stirred for 24 h into the clear solution at the aforementioned temperature. The milky suspension was then transferred to a 250-mL polypropylene storage bottle and hydrothermally treated at 353 K for 24 h. The solid product was recovered, washed with deionized water and air-dried at r.t. in the chemical hood. The dried SBA-15 powders were then slowly heated at 1 K/min to 773 K and annealed for 6 h in order to remove the structure-directing agents.

The $Co_{50}Ni_{21}Ga_{32}$ nanoparticles were prepared using 1 g of SBA-15 as the template. The amounts of precursor salts were 0.234 g (0.98 mmol) of $CoCl_2 \times 6H_2O$, 0.144 g (0.49 mmol) of $Ni(NO_3)_2 \times 6H_2O$ and 0.22 g (0.65 mmol) of $Ga(NO_3) \times H_2O$. First, the precursor salts were dissolved in 50 mL of methanol and then mildly ultrasonically vibrated for 10 min. Subsequently, 1 g of SBA-15 silica was added to the precursor solution, and the suspension was ultrasonically vibrated for another 20 min. The methanol was removed by a rotor evaporator, and the dried powder (200 mg) was heated at 10 K/min, under H_2 atmosphere to 1123 K and held at this temperature for 6 h.

The chemical composition of $Co_{50}Ni_{21}Ga_{32}$ and the metal-to-silica weight ratios were obtained by a chemical analysis method using ICP-OES by means of the VISTA spectrometer (Varian Inc., Palo Alto, CA, USA). The as-prepared Co-Ni-Ga nanoparticles were found to have a $Co_{50}Ni_{21}Ga_{32}$ composition, and the weight fraction of $Co_{50}Ni_{21}Ga_{32}$ in the $Co_{50}Ni_{21}Ga_{32}$/silica nanocomposites was 13.45 wt%.

Suspensions of SBA-15-supported $Co_{50}Ni_{21}Ga_{32}$ nanoparticles were used for TEM sample preparation. Several drops of the suspensions were loaded onto a carbon-coated copper grid and transferred to the microscope after being completely dried. The average particle size was evaluated by counting more than 100 individual particles. A Tecnai 10 TEM system (FEI, Eindhoven, The Netherlands) equipped with a LaB_6-source at an acceleration voltage of 100 kV was used to investigate particle morphology, size distribution and electron diffraction. Images were recorded with a F224HD 2k slow-scan charge-coupled device (CCD) camera (Tietz Video and Image Processing Systems, Gauting, Germany).

The crystal structure of the $Co_{50}Ni_{21}Ga_{32}$/SBA-15 nanocomposite system was investigated by powder XRD technique. An X'Pert PRO diffractometer designed in Bragg–Brentano geometry and supplied with a solid-state X'Celerator linear detector was used (PANalytical B.V., Almelo, The Netherlands). The Cu-$K\alpha_1$-radiation of the Cu anode of the X-ray tube (λ = 1.540598 Å) monochromatized by a primary Johansson-type Ge(111) monochromator was utilized. Additionally, XRD measurements of the powder samples mixed with the Si 640c powder internal standard (National Institute of Standards, Gaithersburg, MD, USA) were performed. The recorded XRD patterns of the samples were corrected to the $\Delta 2\theta_{zero}$ shift using these data as an external standard. The contribution of the substrate was corrected by subtracting the XRD pattern of the substrate measured with high counting statistics.

The microstructure calculations for $Co_{50}Ni_{21}Ga_{32}$ samples were carried out by means of the program *SIZECR* [58]. First, an appropriate correction of the instrumental broadening according to the pseudo-Voigt profile type of the reflections observed [59] was performed. The inconsistency of the observed XRD reflection broadening with the presence of the microstrains in the $Co_{50}Ni_{21}Ga_{32}$ crystallites was established. Finally, the average sizes of XRD crystallites of the $Co_{50}Ni_{21}Ga_{32}$ phases were calculated from Scherrer's equation assuming zero microstrain contribution.

Using the recorded XRD patterns and ICSD [60], the phase analysis was carried out. To confirm the results of the phase analysis and to obtain the quantitative characteristics of the observed crystalline phases, Rietveld fitting of the XRD patterns was carried out by means of the Rietveld program *WinCSD* [61]. The weight scheme $w_i = 1/y_i$ was used in the Rietveld fitting, where y_i is the intensity recorded at the 2θ-step i of the XRD pattern. The isotropic overall temperature factors $B_{iso,overall}$ of the atoms were refined for each phase. To neglect the contribution of the amorphous SiO_2 (SBA-15), the XRD patterns were cut at diffraction angles 2θ of less than 23°, where no XRD reflections of the crystalline phases were observed. At higher angles, the residual amorphous contribution was modeled by a background function. The estimated standard deviations (e.s.d.s) of the crystalline phase parameters were obtained in the Rietveld refinement and were underestimated owing to serial correlations. These were corrected by multiplication on the correcting coefficients, $M_{e.s.d.}$ (see Table 1), calculated by means of Berrar's procedure [62] implemented in $WinCSD$.

Additionally, for the CNG_AQ sample, two non-indexed reflections of an unknown phase were detected at 2θ = 25.88° and 53.22° (marked by triangles in Figure 8). The first one is more intensive, whereas the second one is very weak. According to the 2θ values, these reflections can be attributed to an intermetallic phase with a nominal composition of Ga_7Ni_3 ($Immm$, space group No. 229, ICSD Code 408313 [63]) with a calculated unit cell parameter of a = 8.225(4) Å. This phase is probably characterized by a highly preferred orientation along [112], resulting in the observation of only the two reflections mentioned above (with hkl = 112 and 224, respectively). Owing to a small amount of this phase and the limited number of the reflections observed, attempts to take this phase into account during the Rietveld fitting failed. Finally, for the CNG_AQ sample, the cutting was done at $2\theta = 28°$ in order to exclude the contribution of the most intensive reflection of this phase.

In order to investigate the magnetic properties of $Co_{50}Ni_{21}Ga_{32}$ nanoparticles before and after annealing, temperature-dependent magnetization ($M-T$) measurements were conducted. These measurements were performed under a small magnetic field of 0.1 T from 300 to 1.8 K (field cooling) and then from 1.8 to 300 K (field heating) using the commercial magnetometers Quantum Design MPMS-XL-7 and MPMS-3. The actual magnetization of the $Co_{50}Ni_{21}Ga_{32}$ nanoparticles was calculated based on the weight fraction of the nanoparticles in the nanocomposites. The nanoparticle samples were fixed in a quartz capillary with molten paraffin wax in order to avoid unwanted sample movement during the measurements.

4. Conclusions

In conclusion, we studied the effect of annealing on powdered $Mn_{50}Ni_{40}Sn_{10}$ and $Mn_{50}Ni_{41}Sn_9$ alloys and on $Co_{50}Ni_{21}Ga_{32}$ nanoparticles. Concerning the as-prepared powders of $Mn_{50}Ni_{40}Sn_{10}$ and $Mn_{50}Ni_{41}Sn_9$, a slight change in the Sn-content increased the martensitic transformation temperatures, resulting in an austenitic phase for $Mn_{50}Ni_{40}Sn_{10}$ and a martensitic phase for $Mn_{50}Ni_{41}Sn_9$ at r.t. The annealing of both alloys resulted in changes to the crystal structure of the martensite of $Mn_{50}Ni_{41}Sn_9$ owing to a stabilization of the austenitic phase. It also resulted in a decrease of the martensitic transformation temperatures and an increase in the thermal hysteresis during the martensitic transformation. This might be associated with a release of stress during the annealing process.

We also investigated the annealing effect on $Co_{50}Ni_{21}Ga_{32}$ nanoparticles. XRD analysis showed the absence of a martensitic phase in the as-prepared $Co_{50}Ni_{21}Ga_{32}$ nanoparticles at r.t. After the heat treatment, a martensitic phase with a tetragonal crystal structure formed together with a parent austenitic phase. Thermomagnetic measurements revealed a pronounced decrease in magnetization of the annealed nanoparticles, which is consistent with the variations of the crystal structure. The intrinsic features in the martensite transformation of $Co_{50}Ni_{21}Ga_{32}$ compounds make it difficult to probe the phase transition by temperature-dependent magnetic measurements.

Acknowledgments

This work was financially supported by the ERC Advanced Grant No. 291472 "Idea Heusler" and DFG Project No. 1.2-A in the research unit FOR 1464 ASPIMATT. Additionally, the contributions of S. Kostmann (microstructure analysis), S. Scharsach (DSC/TG), U. Schmidt, A. Völzke and S. Schwinger (ICP-OES) are gratefully acknowledged.

Author Contributions

Tina Fichtner and Changhai Wang performed the synthesis of the alloys $Mn_{50}Ni_{40}Sn_{10}$ and $Mn_{50}Ni_{41}Sn_9$ or $Co_{50}Ni_{21}Ga_{32}$, respectively. Aleksandr A. Levin carried out the XRD measurements of the $Co_{50}Ni_{21}Ga_{32}$ samples for Rietveld fitting and performed quantitative analysis and crystallite size calculations. Guido Kreiner characterized the $Mn_{50}Ni_{40}Sn_{10}$ and $Mn_{50}Ni_{41}Sn_9$ alloys. Catalina Salazar Mejia analyzed the magnetic properties of the $Mn_{50}Ni_{40}Sn_{10}$ and $Mn_{50}Ni_{41}Sn_9$ alloys. Simone Fabbrici and Franca Albertini discussed the analysis and the results. Claudia Felser supervised the work and discussed the results and the analysis with the other authors.

Conflicts of Interest

The authors declare no conflict of interest.

References

1. Ullakko, K.; Huang, J.K.; Kantner, C.; O'Handley, R.C.; Kokorin, V.V. Large magnetic-field-induced strains in Ni_2MnGa single crystals. *Appl. Phys. Lett.* **1996**, *69*, 1966–1968.
2. Brück, E. Developments in magnetocaloric refrigeration. *J. Phys. D Appl. Phys.* **2005**, *38*, R381–R391.
3. Planes, A.; Manosa, L.; Acet, M. Magnetocaloric effect and its relation to shape-memory properties in ferromagnetic Heusler alloys. *J. Phys. Condes. Matter* **2009**, *21*, 233201:1–233201:29.
4. Buchelnikov, V.D.; Sokolovskiy, V.V. Magnetocaloric effect in Ni-Mn-X (X = Ga, In, Sn, Sb) Heusler alloys. *Phys. Met. Metallogr.* **2011**, *112*, 633–665.
5. Franco, V.; Blázquez, J.; Ingale, B.; Conde, A. The Magnetocaloric Effect and Magnetic Refrigeration Near Room Temperature: Materials and Models. *Annu. Rev. Mater. Res.* **2012**, *42*, 305–342.
6. Roy, S.B. Chapter two—Magnetocaloric effect in intermetallic compounds and alloys. In *Handbook of Magnetic Materials*; Buschow, K., Ed.; Elsevier: Amsterdam, The Netherlands, 2014; Volume 22, pp. 203–316.
7. Sozinov, A.; Likhachev, A.A.; Lanska, N.; Ullakko, K. Giant magnetic-field-induced strain in NiMnGa seven-layered martensitic phase. *Appl. Phys. Lett.* **2002**, *80*, 1746–1748.
8. Müllner, P.; Chernenko, V.; Kostorz, G. Large magnetic-field-induced deformation and magneto-mechanical fatigue of ferromagnetic Ni-Mn-Ga martensites. *Mater. Sci. Eng. A* **2004**, *387–389*, 965–968.

9. Koyama, K.; Okada, H.; Watanabe, K.; Kanomata, T.; Kainuma, R.; Ito, W.; Oikawa, K.; Ishida, K. Observation of large magnetoresistance of magnetic Heusler alloy $Ni_{50}Mn_{36}Sn_{14}$ in high magnetic fields. *Appl. Phys. Lett.* **2006**, *89*, 182510:1–182510:3.
10. Yu, S.Y.; Liu, Z.H.; Liu, G.D.; Chen, J.L.; Cao, Z.X.; Wu, G.H.; Zhang, B.; Zhang, X.X. Large magnetoresistance in single-crystalline $N_{50}Mn_{50-x}In_x$ alloys (x = 14 – 16) upon martensitic transformation. *Appl. Phys. Lett.* **2006**, *89*, 162503:1–162503:3.
11. Krenke, T.; Duman, E.; Acet, M.; Wassermann, E.F.; Moya, X.; Manosa, L.; Planes, A. Inverse magnetocaloric effect in ferromagnetic Ni-Mn-Sn alloys. *Nat. Mater.* **2005**, *4*, 450–454.
12. Han, Z.D.; Wang, D.H.; Zhang, C.L.; Tang, S.L.; Gu, B.X.; Du, Y.W. Large magnetic entropy changes in the $Ni_{45.4}Mn_{41.5}In_{13.1}$ ferromagnetic shape memory alloy. *Appl. Phys. Lett.* **2006**, *89*, 182507:1–182507:3.
13. Liu, J.; Gottschall, T.; Skokov, K.; Moore, J.; Gutfleisch, O. Giant magnetocaloric effect driven by structural transitions. *Nat. Mater.* **2012**, *11*, 620–626.
14. Han, Z.D.; Wang, D.H.; Zhang, C.L.; Xuan, H.C.; Gu, B.X.; Du, Y.W. Low-field inverse magnetocaloric effect in $Ni_{50-x}Mn_{39+x}Sn_{11}$ Heusler alloys. *Appl. Phys. Lett.* **2007**, *90*, 042507:1–042507:3.
15. Ma, L.; Wang, W.H.; Lu, J.B.; Li, J.Q.; Zhen, C.M.; Hou, D.L.; Wu, G.H. Coexistence of reentrant-spin-glass and ferromagnetic martensitic phases in the $Mn_2Ni_{1.6}Sn_{0.4}$ Heusler alloy. *Appl. Phys. Lett.* **2011**, *99*, 182507:1–182507:3.
16. Wu, Z.; Liu, Z.; Yang, H.; Liu, Y.; Wu, G. Martensitic and magnetic transformation behaviours in $Mn_{50}Ni_{42-x}Sn_8Co_x$ polycrystalline alloys. *J. Phys. D Appl. Phys.* **2011**, *44*, 385403:1–385403:7.
17. Ghosh, A.; Mandal, K. Large magnetic entropy change and magnetoresistance associated with a martensitic transition of Mn-rich $Mn_{50.5-x}Ni_{41}Sn_{8.5+x}$ alloys. *J. Phys. D Appl. Phys.* **2013**, *46*, 435001:1–435001:7.
18. Ma, L.; Wang, S.Q.; Li, Y.Z.; Zhen, C.M.; Hou, D.L.; Wang, W.H.; Chen, J.L.; Wu, G.H. Martensitic and magnetic transformation in $Mn_{50}Ni_{50-x}Sn_x$ ferromagnetic shape memory alloys. *J. Appl. Phys.* **2012**, *112*, 083902:1–083902:4.
19. Schlagel, D.L.; McCallum, R.W.; Lograsso, T.A. Influence of solidification microstructure on the magnetic properties of Ni-Mn-Sn Heusler alloys. *J. Alloys Compd.* **2008**, *463*, 38–46.
20. Xuan, H.C.; Xie, K.X.; Wang, D.H.; Han, Z.D.; Zhang, C.L.; Gu, B.X.; Du, Y.W. Effect of annealing on the martensitic transformation and magnetocaloric effect in $Ni_{44.1}Mn_{44.2}Sn_{11.7}$ ribbons. *Appl. Phys. Lett.* **2008**, *92*, 242506:1–242506:3.
21. Oikawa, K.; Ota, T.; Ohmori, T.; Tanaka, Y.; Morito, H.; Fujita, A.; Kainuma, R.; Fukamichi, K.; Ishida, K. Magnetic and martensitic phase transitions in ferromagnetic Ni-Ga-Fe shape memory alloys. *Appl. Phys. Lett.* **2002**, *81*, 5201–5203.
22. Liu, Z.H.; Zhang, M.; Cui, Y.T.; Zhou, Y.Q.; Wang, W.H.; Wu, G.H.; Zhang, X.X.; Xiao, G. Martensitic transformation and shape memory effect in ferromagnetic Heusler alloy Ni_2FeGa. *Appl. Phys. Lett.* **2003**, *82*, 424–426.
23. Hamilton, R.; Efstathiou, C.; Sehitoglu, H.; Chumlyakov, Y. Thermal and stress-induced martensitic transformations in NiFeGa single crystals under tension and compression. *Scr. Mater.* **2006**, *54*, 465–469.

24. Oikawa, K.; Wulff, L.; Iijima, T.; Gejima, F.; Ohmori, T.; Fujita, A.; Fukamichi, K.; Kainuma, R.; Ishida, K. Promising ferromagnetic Ni-Co-Al shape memory alloy system. *Appl. Phys. Lett.* **2001**, *79*, 3290–3292.
25. Karaca, H.; Karaman, I.; Lagoudas, D.; Maier, H.; Chumlyakov, Y. Recoverable stress-induced martensitic transformation in a ferromagnetic CoNiAl alloy. *Scr. Mater.* **2003**, *49*, 831–836.
26. Wuttig, M.; Li, J.; Craciunescu, C. A new ferromagnetic shape memory alloy system. *Scr. Mater.* **2001**, *44*, 2393–2397.
27. Liu, J.; Xia, M.; Huang, Y.; Zheng, H.; Li, J. Effect of annealing on the microstructure and martensitic transformation of magnetic shape memory alloys CoNiGa. *J. Alloys Compd.* **2006**, *417*, 96–99.
28. Dai, X.F.; Liu, G.D.; Liu, Z.H.; Wu, G.H.; Chen, J.L.; Meng, F.B.; Liu, H.Y.; Yan, L.Q.; Qu, J.P.; Li, Y.X.; *et al.* Superelasticity of CoNiGa:Fe single crystals. *Appl. Phys. Lett.* **2005**, *87*, 112504:1–112504:3.
29. Dogan, E. Co-Ni-Ga High Temperature Shape Memory Alloys. Master's Thesis, Texas A&M University, College Station, TX, USA, 2010.
30. Sato, M.; Okazaki, T.; Furuya, Y.; Wuttig, M. Magnetostrictive and Shape Memory Properties of Heusler Type Co_2NiGa Alloys . *Mater. Trans.* **2003**, *44*, 372–376.
31. Sato, M.; Okazaki, T.; Furuya, Y.; Kishi, Y.; Wuttig, M. Phase transformation and magnetic property of heusler type Co_2NiGa alloys. *Mater. Trans.* **2004**, *45*, 204–207.
32. Brown, P.J.; Ishida, K.; Kainuma, R.; Kanomata, T.; Neumann, K.U.; Oikawa, K.; Ouladdiaf, B.; Ziebeck, K.R. Crystal structures and phase transitions in ferromagnetic shape memory alloys based on Co-Ni-Al and Co-Ni-Ga. *J. Phys. Condens. Matter* **2005**, *17*, 1301–1310.
33. Oikawa, K.; Ota, T.; Imano, Y.; Omori, T.; Kainuma, R.; Ishida, K. Phase equilibria and phase transformation of Co-Ni-Ga ferromagnetic shape memory alloy system. *J. Phase Equilib. Diffus.* **2006**, *27*, 75–82.
34. Dai, X.; Liu, G.; Li, Y.; Qu, J.; Li, J.; Chen, J.; Wu, G. Structure and magnetic properties of highly ordered Co_2NiGa alloys. *J. Appl. Phys.* **2007**, *101*, 09N503:1–09N503:3.
35. Arróyave, R.; Junkaew, A.; Chivukula, A.; Bajaj, S.; Yao, C.Y.; Garay, A. Investigation of the structural stability of Co_2NiGa shape memory alloys via ab initio methods. *Acta Mater.* **2010**, *58*, 5220–5231.
36. Dogan, E.; Karaman, I.; Chumlyakov, Y.; Luo, Z. Microstructure and martensitic transformation characteristics of CoNiGa high temperature shape memory alloys. *Acta Mater.* **2011**, *59*, 1168–1183.
37. Dadda, J.; Maier, H.; Karaman, I.; Karaca, H.; Chumlyakov, Y. Pseudoelasticity at elevated temperatures in [001] oriented $Co_{49}Ni_{21}Ga_{30}$ single crystals under compression. *Scr. Mater.* **2006**, *55*, 663–666.
38. Li, Y.X.; Liu, H.Y.; Meng, F.B.; Yan, L.Q.; Liu, G.D.; Dai, X.F.; Zhang, M.; Liu, Z.H.; Chen, J.L.; Wu, G.H.; *et al.* Magnetic field-controlled two-way shape memory in CoNiGa single crystals. *Appl. Phys. Lett.* **2004**, *84*, 3594–3596.
39. Sarma, S.; Srinivasan, A. Influence of Annealing Temperature on the Properties of Co-Ni-Ga Ferromagnetic Shape Memory Alloy. *Adv. Mater. Res.* **2008**, *52*, 63–68.

40. Dai, X.F.; Wang, H.Y.; Liu, G.D.; Wang, Y.G.; Duan, X.F.; Chen, J.L.; Wu, G.H. Effect of heat treatment on the properties of $Co_{50}Ni_{20}Ga_{30}$ ferromagnetic shape memory alloy ribbons. *J. Phys. D Appl. Phys.* **2006**, *39*, 2886–2889.
41. Liu, J.; Scheerbaum, N.; Hinz, D.; Gutfleisch, O. A high-temperature coupling of martensitic and magnetic transformations and magnetic entropy change in Ni-Fe-Ga-Co alloys. *Scr. Mater.* **2008**, *59*, 1063–1066.
42. Yu, S.; Yan, S.; Kang, S.; Tang, X.; Qian, J.; Chen, J.; Wu, G. Magnetic field-induced martensite-austenite transformation in Fe-substituted NiMnGa ribbons. *Scr. Mater.* **2011**, *65*, 9–12.
43. Tao, Q.; Han, Z.D.; Wang, J.J.; Qian, B.; Zhang, P.; Jiang, X.F.; Wang, D.H.; Du, Y.W. Phase stability and magnetic-field-induced martensitic transformation in Mn-rich NiMnSn alloys. *AIP Adv.* **2012**, *2*, 042181, doi:10.1063/1.4772626.
44. Khachaturyan, A.; Shapiro, S.; Semenovskaya, S. Adaptive phase formation in martensitic transformation. *Phys. Rev. B* **1991**, *43*, 10832–10843.
45. Kaufmann, S.; Rößler, U.K.; Heczko, O.; Wuttig, M.; Buschbeck, J.; Schultz, L.; Fähler, S. Adaptive Modulations of Martensites. *Phys. Rev. Lett.* **2010**, *104*, 145702:1–145702:4.
46. Li, Z. Study on Crystallographic Features of Ni-Mn-Ga Ferromagnetic Shape Memory Alloys. Ph.D. Thesis, University de Lorraine and Northeastern University, Metz, France, 10 September 2011.
47. Ito, W.; Nagasako, M.; Umetsu, R.Y.; Kainuma, R.; Kanomata, T.; Ishida, K. Atomic ordering and magnetic properties in the $Ni_{45}Co_5Mn_{36.7}In_{13.3}$ metamagnetic shape memory alloy. *Appl. Phys. Lett.* **2008**, *93*, 232503:1–232503:3.
48. Ito, W.; Ito, K.; Umetsu, R.Y.; Kainuma, R.; Koyama, K.; Watanabe, K.; Fujita, A.; Oikawa, K.; Ishida, K.; Kanomata, T.; *et al.* Kinetic arrest of martensitic transformation in the NiCoMnIn metamagnetic shape memory alloy. *Appl. Phys. Lett.* **2008**, *92*, 021908:1–021908:3.
49. Kustov, S.; Corró, M.L.; Pons, J.; Cesari, E. Entropy change and effect of magnetic field on martensitic transformation in a metamagnetic Ni-Co-Mn-In shape memory alloy. *Appl. Phys. Lett.* **2009**, *94*, 191901:1–191901:3.
50. Chen, L.; Hu, F.X.; Wang, J.; Shen, J.; Sun, J.R.; Shen, B.G.; Yin, J.H.; Pan, L.Q.; Huang, Q.Z. Effect of post-annealing on martensitic transformation and magnetocaloric effect in $Ni_{45}Co_5Mn_{36.7}In_{13.3}$ alloys. *J. Appl. Phys.* **2011**, *109*, 07A939:1–07A939:3.
51. Webster, P.J.; Ziebeck, K.R.A.; Town, S.L.; Peak, M.S. Magnetic order and phase transformation in Ni_2MnGa. *Philos. Mag. B* **1984**, *49*, 295–310.
52. Sato, H.; Toth, R. Effect of additional elements on the period of CuAu II and the origin of the long-period superlattice. *Phys. Rev.* **1961**, *124*, 1833–1847.
53. Entel, P.; Buchelnikov, V.D.; Khovailo, V.V.; Zayak, A.T.; Adeagbo, W.A.; Gruner, M.E.; Herper, H.C.; Wassermann, E.F. Modelling the phase diagram of magnetic shape memory Heusler alloys. *J. Phys. D Appl. Phys.* **2006**, *39*, 865–889.
54. Fichtner, T.; Kreiner, G.; Chadov, S.; Fecher, G.H.; Schnelle, W.; Hoser, A.; Felser, C. Magnetic and transport properties in the Heusler series $Ni_{2-x}Mn_{1+x}Sn$ affected by chemical disorder. *Intermetallics* **2015**, *57*, 101–112.

55. Auge, A.; Teichert, N.; Meinert, M.; Reiss, G.; Hütten, A.; Yüzüak, E.; Dincer, I.; Elerman, Y.; Ennen, I.; Schattschneider, P.; *et al.* Thickness dependence of the martensitic transformation, magnetism, and magnetoresistance in epitaxial Ni-Mn-Sn ultrathin films. *Phys. Rev. B* **2012**, *85*, 214118:1–214118:8.
56. *STOE WinXPOW*, Version 2.25; Stoe & Cie GmbH: Darmstadt, Germany, 2009.
57. *Jana2006*; Institute of Physics: Praha, Czech Republic, 2006.
58. Levin, A.A.; Levichkova, M.; Hildebrandt, D.; Klisch, M.; Weiss, A.; Wynands, D.; Elschner, C.; Pfeiffer, M.; Leo, K.; Riede, M.; *et al.* Effect of film thickness, type of buffer layer, and substrate temperature on the morphology of dicyanovinyl-substituted sexithiophene films. *Thin Solid Films* **2012**, *520*, 2479–2487.
59. Rehani, B.R.; Joshi, P.B.; Lad, K.N.; Pratap, A. Crystallite size estimation of elemental and composite silver nano-powders using XRD principles. *Indian J. Pure Appl. Phys.* **2006**, *44*, 157–161.
60. Belsky, A.; Lynn, V. Research papers New developments in the Inorganic Crystal Structure Database (ICSD): Accessibility in support of materials research and design research papers. *Acta Cryst. B* **2002**, *58*, 364–369.
61. Akselrud, L.; Grin, Y. WinCSD : Software package for crystallographic calculations (Version 4). *J. Appl. Cryst.* **2014**, *47*, 803–805.
62. Bérar, J.F.; Lelann, P. E.s.d.'s and estimated probable error obtained in Rietveld refinements with local correlations. *J. Appl. Crystallogr.* **1991**, *24*, 1–5.
63. Häussermann, U.; Elding-ponte, M.; Svensson, C.; Lidin, S. Compounds with the Ir_3Ge_7 structure type : Interpenetrating frameworks with flexible Bonding Properties. *Chem. Eur. J.* **1998**, *4*, 1007–1015.

Influence of Particulate Reinforcement and Equal-Channel Angular Pressing on Fatigue Crack Growth of an Aluminum Alloy

Lisa Köhler *, Kristin Hockauf and Thomas Lampke

Institute of Materials Science and Engineering, Technische Universität Chemnitz, Erfenschlager Str. 73, 09125 Chemnitz, Germany; E-Mails: kristin.hockauf@mb.tu-chemnitz.de (K.H.); thomas.lampke@mb.tu-chemnitz.de (T.L.)

* Author to whom correspondence should be addressed; E-Mail: lisa.koehler@mb.tu-chemnitz.de

Academic Editor: Heinz Werner Höppel

Abstract: The fatigue crack growth behavior of unreinforced and particulate reinforced Al 2017 alloy, manufactured by powder metallurgy and additional equal-channel angular pressing (ECAP), is investigated. The reinforcement was done with 5 vol % Al_2O_3 particles with a size fraction of 0.2–2 μm. Our study presents the characterization of these materials by electron microscopy, tensile testing, and fatigue crack growth measurements. Whereas particulate reinforcement leads to a drastic decrease of the grain size, the influence of ECAP processing on the grain size is minor. Both reinforced conditions, with and without additional ECAP processing, exhibit reduced fatigue crack growth thresholds as compared to the matrix material. These results can be ascribed to the well-known effect of the grain size on the crack growth, since crack deflection and closure are directly affected. Despite their small grain size, the thresholds of both reinforced conditions depend strongly on the load ratio: tests at high load ratios reduce the fatigue threshold significantly. It is suggested that the strength of the particle-matrix-interface becomes the critical factor here and that the particle fracture at the interfaces dominates the failure behavior.

Keywords: Al 2017 alloy; Al_2O_3 particulate reinforcement; fatigue crack growth; equal-channel angular pressing (ECAP)

---

1. Introduction

Aluminum matrix composites (AMCs) are designed to meet the requirements of advanced engineering applications. Reinforcement with ceramic particles provides high strength and stiffness, and enhances wear resistance, thermal stability, and creep resistance [1,2]. It has been shown that methods of severe plastic deformation (SPD) can help to homogenize the particle distribution [3], which improves the mechanical properties. Equal-channel angular pressing (ECAP) has emerged as one of the most frequently used SPD methods for processing particulate reinforced AMCs [4–7]. For unreinforced materials, ECAP processing leads to strain-hardening of the material and decreases the grain size by the rearrangement of dislocations to subgrain boundaries and cell walls [8–10].

Focusing on potential future applications for AMCs produced by ECAP, fatigue crack growth is an important field of interest. To the best of our knowledge, previous studies only focused on crack propagation either in ultrafine-grained materials produced by ECAP or in AMCs. These investigations show the well-known effect of the grain or particle size on the fatigue crack growth. The grain refinement through ECAP leads to a minimized amount of crack deflection and roughness-induced crack closure, which results in higher crack propagation rates and, therefore, in minor thresholds [11–18]. In reinforced materials, particle sizes larger than 2–5 μm have a contrary effect on crack propagation: by increasing crack deflection and roughness-induced crack closure, crack propagation rates are lowered [19–24].

The purpose of the present study is to investigate the influence of particulate reinforcement with a particle size less than 2 μm and additional ECAP on the near-threshold fatigue crack growth of an aluminum alloy. This small particle size was chosen in order to minimize local stress concentrations in the matrix material [25]. It is well-known that small particles cause significantly higher strain hardening and that they are less susceptible to cracking as compared to coarser particles [26,27]. As matrix material, the Al 2017 alloy has been chosen here and in preliminary studies [28], because it combines attractive properties, such as good fracture toughness in the underaged condition, high resistance against crack propagation, and excellent high-temperature strength, which enables future applications for lightweight structures, e.g., for the aerospace and aeronautical industries.

2. Experimental Section

2.1. Material

A commercially available, gas-atomized, spherical Al 2017 powder alloy was used as matrix material. The particle size fraction of the powder, with a chemical composition such as that presented in Table 1, was below 100 μm. Al_2O_3 powder with a particle size fraction of 0.2–2 μm was used as reinforcing component. The composite powder, consisting of 95 vol % matrix material and 5 vol % Al_2O_3 particles, was processed in a high-energy ball mill and hot-degassed at 450 °C and 0.06 bar for 4 h. Afterwards, compaction was performed by hot isostatic pressing at 450 °C and 1100 bar for 3 h. The manufacturing process is described in more detail in [28].

Table 1. Chemical composition of the Al 2017 powder alloy.

Element	Al	Cu	Mn	Mg	Fe	Si	Zn
wt %	>94.1	4.1	0.8	0.6	0.2	0.1	traces

The non-ECAP-processed conditions, unreinforced, and reinforced, were solid-solution treated at 505 °C for 60 min, water-quenched, and subsequently naturally aged at room temperature for one month in order to adjust an underaged condition. For the ECAP-processed conditions, as described in [28], a pre-ECAP aging at 140 °C was carried out for 10 min to increase the workability. Immediately after the pre-ECAP aging, the material was ECAP-processed at 140 °C for one pass. The ECAP processing was performed in a device with an internal angle of 120 and a cross-section of 15 × 15 mm^2, equipped with movable walls and a bottom slider. This processing resulted in an equivalent strain of 0.67 in each passage [29]. After ECAP, the material was peak-aged at 140 °C for 360 min to optimize the ductility. With this combined treatment of ECAP and aging, both elevated strength and moderate ductility can be achieved [30,31].

2.2. Methods of Mechanical Testing and Electron Microscopy

Quasi-static tensile tests were performed in a Zwick-Roell servohydraulic testing machine (Zwick, Ulm, Germany) at a strain rate of $10^{-3}\cdot s^{-1}$ at room temperature. Cylindrical specimens were used with a cross section of 3.5 mm and a gauge length of 10.5 mm.

Fatigue crack propagation measurements were performed ΔK-controlled in a Rumul testronic resonant testing machine (Russenberger Prüfmaschinen AG, Neuhausen am Rheinfall, Switzerland) at room temperature and at three constant load ratios: R = 0.1, 0.4 and 0.7. Single-edge-notched bend (SEB) specimens with a geometry according to ASTM E399-12e3 [32] and a thickness of 5 mm were used. Figure 1 shows the position in the ECAP billet out of which the SEB specimens were extracted. In the choice of the extraction direction, it was considered that the crack is most likely sidetracked in the direction of the last ECAP shear plane [12]. The crack length was measured continuously according to the principle of the indirect electric potential method with crack measurement foils.

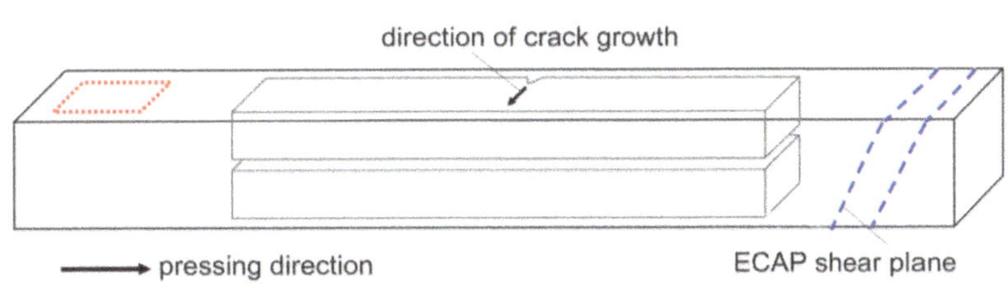

Figure 1. Position of two single-edge-notched bend specimens within the ECAP billet. The ideal direction of crack growth is parallel to the ECAP shear plane. The plane from which samples for microstructural analyses were taken is marked in red.

After precracking at ΔK = 2–7 MPam$^{1/2}$ (depending on material condition and load ratio), ΔK was lowered in small steps until the fatigue threshold ΔK_{th} was reached. Then, 0.5 mm crack growth were realized at this cyclic stress intensity to avoid any load history effect from the threshold region. In the following, ΔK was raised again in small steps up to a cyclic stress intensity of about 11 MPam$^{1/2}$ and a crack length of 5 mm.

Samples for microstructural analyses were extracted parallel to the extrusion direction, as shown in Figure 1, and electro-polished followed by a final ion milling polishing step. These samples were analyzed by scanning transmission electron microscopy (STEM) and quadrant back scatter diffraction (QBSD) at an accelerating voltage of 30 kV, as well as by electron back scatter diffraction (EBSD, EDAX TSL

OIM 5.2, AMETEK GmbH, Wiesbaden, Germany) at 15 kV using a Zeiss Neon 40 field emission microscope (Carl Zeiss MicroImaging GmbH, Jena, Germany).

3. Results and Discussion

3.1. Tensile Tests

The tensile properties of the investigated material conditions are shown in Table 2. The unreinforced aluminum matrix (in an underaged heat treatment condition) exhibits the lowest yield strength with 252 MPa just as the highest ductility, which shows in a uniform elongation of 20%. ECAP processing (with subsequent aging as described in the experimental section) leads to a significant increase in yield strength (487 MPa, which is an increase of 93%) in combination with a moderate ductility (11% uniform elongation). The strategy of combining ECAP and aging, which is used here and which is referred to as "optimizing heat treatment" in [33], enables this favorable combination of strength and ductility: ECAP processing results in a severely strain hardened material with a high dislocation density. In the following process of thermal aging, numerous finely-dispersed precipitates form and strengthen the matrix. Simultaneously, thermal recovery takes place and increases the material's ductility.

Table 2. Tensile properties of the unreinforced and reinforced conditions.

Material	Yield Strength in MPa	Ultimate Tensile Strength in MPa	Uniform Elongation in %	Elongation to Fracture in %
Al 2017 unreinforced	252	435	20	29
Al 2017 unreinforced after 1 ECAP pass	487	566	11	17
Al 2017 with 5 vol % Al_2O_3	328	511	11	13
Al 2017 with 5 vol % Al_2O_3 after 1 ECAP pass	509	580	5	8

The dispersion strengthening, which is caused by the reinforcement with 5 vol % particles, leads to an improvement in strength, which is less compared to the strengthening, which is achieved by ECAP of the unreinforced material. ECAP processing of the particulate reinforced material increases the yield strength from 328 MPa to 509 MPa (which is an increase of 55%). This shows, that the effect of strain hardening during ECAP is less pronounced than in the unreinforced material. Both reinforced conditions exhibit minor ductility, which is ascribed to the particles which act as interior notches [28].

3.2. Microstructure

One ECAP pass of the unreinforced matrix material leads to a bimodal grain structure with inhomogeneously distributed refined grains. The initial grain size is significantly reduced in the shear bands, which is shown in Figure 2. In the outside parts, almost no grain refinement has taken place.

The reinforced material exhibits elongated areas with a width of 100–200 μm and a height of about 7 μm without reinforcement components. These areas are preferentially orientated parallel to the direction of extrusion. In Figure 3, micrographs of the reinforced areas are shown. The Al_2O_3 particles are finely dispersed and exhibit an intact interface to the aluminum matrix. In these micrographs, the presence of the coarse intermetallic phase Al_2Cu can also be seen. Additional ECAP does not lead to significant grain refinement (see Figure 4).

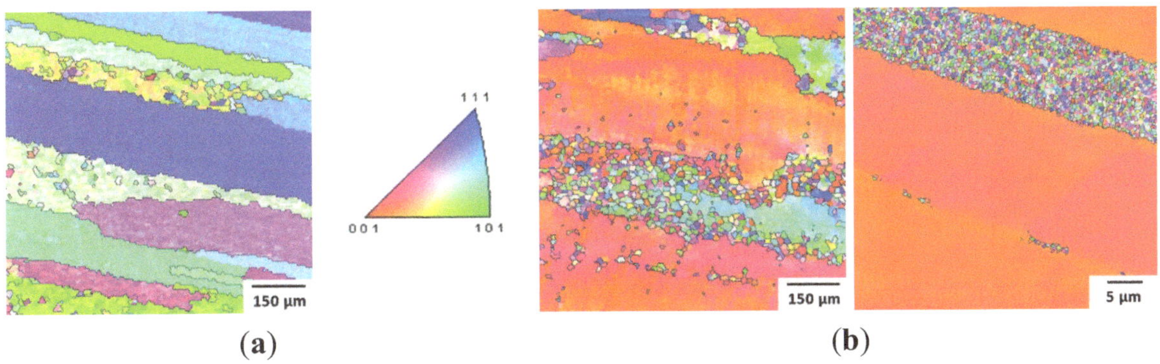

Figure 2. Electron back scatter diffraction (EBSD) color-coded orientation maps of the unreinforced matrix material conditions: (**a**) non-ECAP-processed and (**b**) after one ECAP pass. In the shear bands, the grain size is considerably reduced after one ECAP pass. By contrast, the initial grain size outside the shear bands is almost maintained.

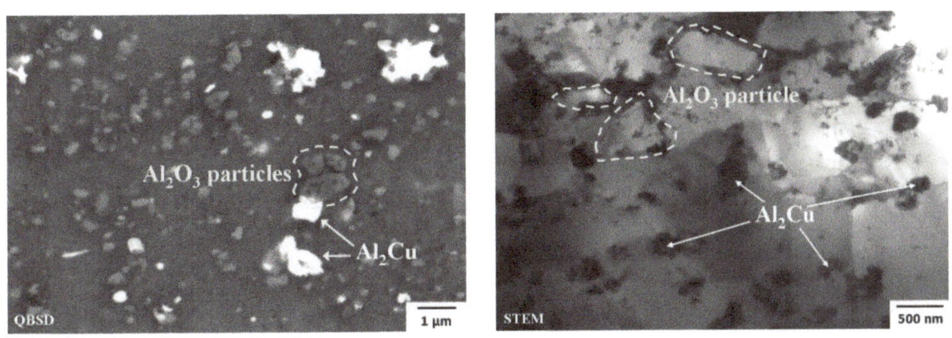

Figure 3. Quadrant backscatter diffraction (QBSD, **left**) and scanning transmission electron microscopy (STEM, **right**) micrographs of Al 2017 with 5 vol % Al_2O_3 particles with a size of 0.2–2 µm. Al_2O_3 particles are marked with broken lines and the Al_2Cu precipitates with arrows. The particles are finely dispersed and matrix and particles exhibit an intact interface.

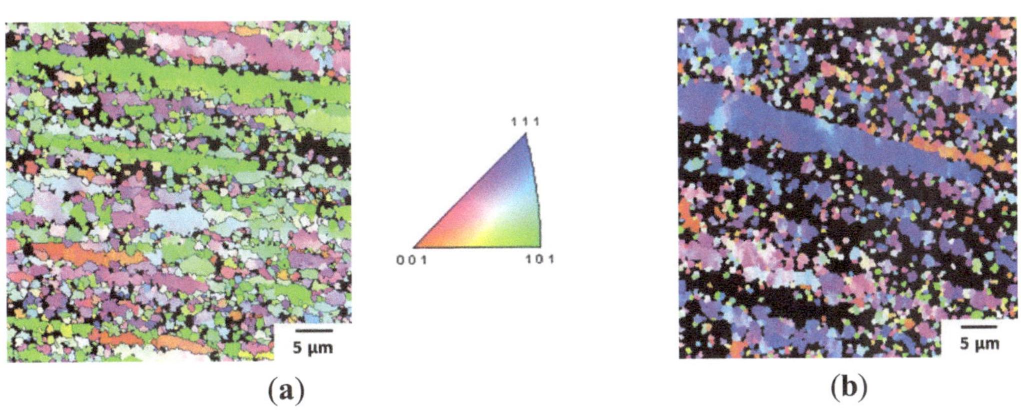

Figure 4. EBSD color-coded orientation maps of the reinforced conditions: (**a**) non-ECAP-processed and (**b**) after additional ECAP. For the reinforced material, ECAP does not lead to a significant grain refinement.

3.3. Fatigue Crack Surface Analysis

For the unreinforced conditions the fatigue crack surfaces are shown in Figure 5. The non-ECAP-processed condition exhibits a strongly faceted surface with pronounced band-typed steps. This fissured crack surface can be explained by the large grain size of this condition and thus by a high amount of crack deflection. For the material after one ECAP pass, plateaus are present on the crack surface. Additionally, striations are clearly visible. Such signs of "micro-damage" have also been observed for other ECAP-processed aluminum alloys [16]. They were interpreted as the severely strain-hardened material is not being able to compensate the stresses at the crack tip by deforming plastically. Compared to the unprocessed material, the crack path after ECAP is less tortuous and more straight-forward. Presumably, the crack propagates preferably along or within the shear bands.

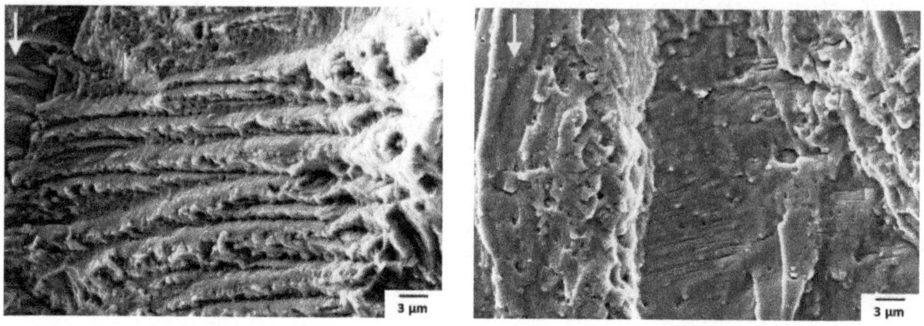

(a) (b)

Figure 5. Fatigue crack surfaces of the unreinforced conditions: (**a**) non-ECAP-processed and (**b**) after one ECAP pass near the fatigue threshold. The crack propagation direction is marked with an arrow. The surface of the non-ECAP-processed condition exhibits a band-type faceted appearance, whereas ECAP leads to plateaus and clearly visible striations.

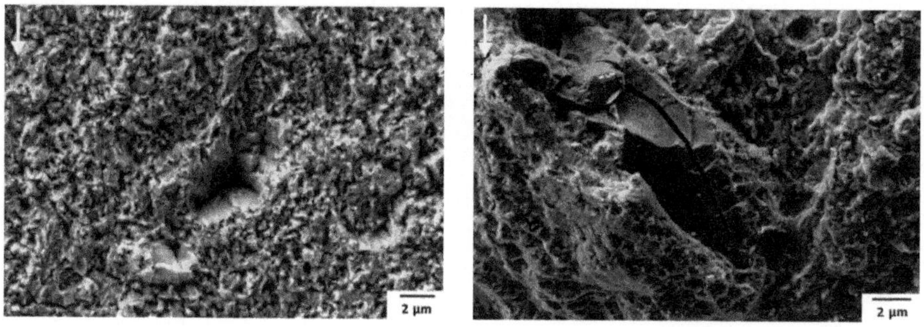

(a) (b)

Figure 6. Fatigue crack surfaces of the reinforced conditions: (**a**) non-ECAP-processed and (**b**) after one ECAP pass near the fatigue threshold. The crack propagation direction is marked with an arrow. The surface is less rough and barely fissured as compared to the unreinforced conditions. Through additional ECAP, the particle failure is more likely to fracture than to break out.

In Figure 6, the fatigue crack surfaces of the reinforced conditions are shown. In contrast to the unreinforced conditions, the crack surfaces are characterized by a minor roughness. Hardly any faceting is noticeable. Due to the small grain and particle sizes of these conditions, crack deflection is reduced to a minimum. This outcome is in good agreement with the findings in [19,34]. It is noticeable that, in the ECAP-processed condition, particle fracture is more pronounced, whereas the initial reinforced condition exhibits a higher rate of particle break outs. It cannot be concluded with certainty if the particle fracture proceeded due to fatigue loading or whether it was caused by ECAP.

3.4. Fatigue Crack Propagation

Fatigue crack growth curves near the threshold for the tested load ratios are shown in Figure 7. The matrix material exhibits the highest thresholds and the lowest fatigue crack growth rates for all tested conditions. Due to the major grain size of the matrix material, the amount of crack deflection and roughness-induced crack closure and, therefore, the strongly branched crack path lead to decreased fatigue crack growth rates, which results in major thresholds. One ECAP pass results in higher crack growth rates and lower thresholds. Presumably, due to the preferred crack propagation along the shear bands and the finer grains in this area, less crack deflection and closure occurs. Supposedly, also the high dislocation density induced by ECAP inhibits stress relief and, therefore, early damaging at the crack tip takes place [13].

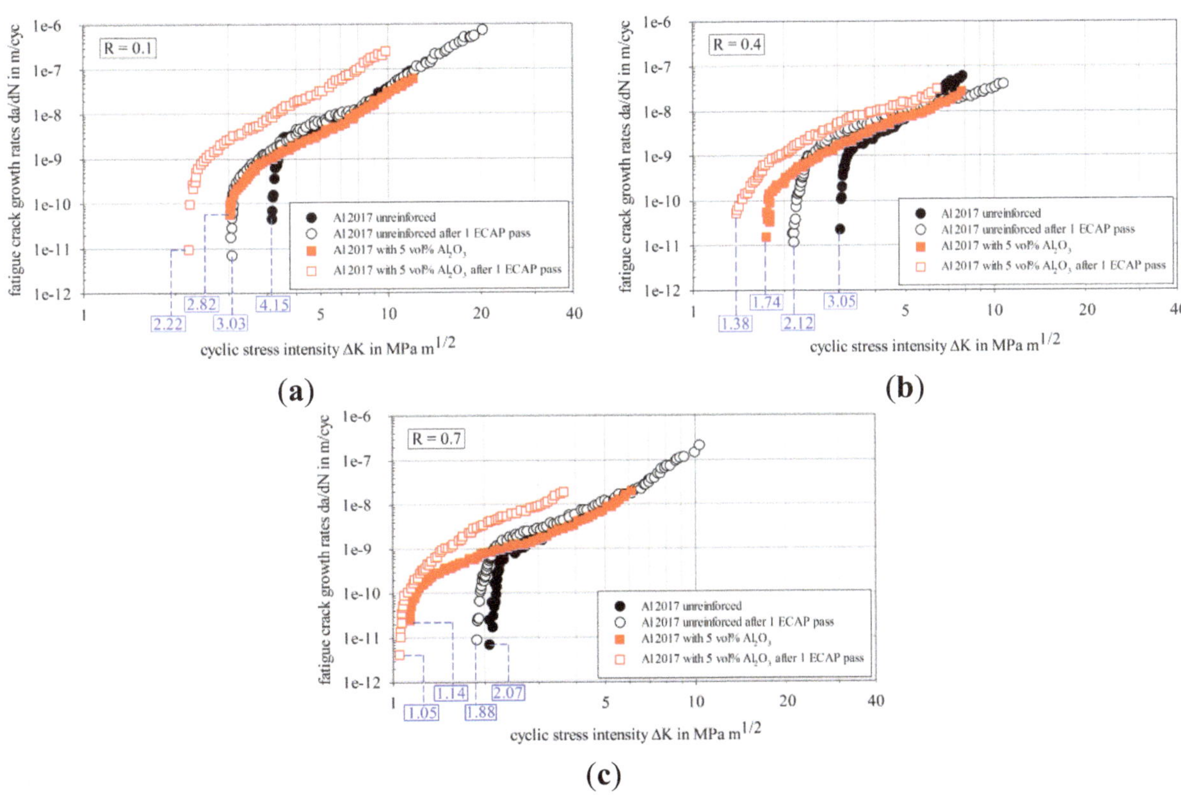

Figure 7. Fatigue crack growth curves of Al 2017 in unreinforced and reinforced conditions with additional ECAP at load ratios of: (**a**) $R = 0.1$, (**b**) $R = 0.4$, and (**c**) $R = 0.7$. The thresholds for the reinforced conditions are lower than for the matrix material conditions. With the increase of the load ratio, this effect becomes more pronounced. The crack growth rates for the reinforced ECAP-processed conditions are generally the highest.

For the reinforced conditions, at low cyclic stress intensities, the crack propagation rates are higher as compared to the matrix material conditions. As the near threshold regime is highly sensitive to the microstructure, the main factor for this behavior is presumably the significantly smaller grain size and the resulting minimized effects of crack deflection and roughness-induced crack closure. For the ECAP-processed conditions, fatigue crack growth thresholds are further reduced as compared to their corresponding not-processed conditions. This is probably due to the high dislocation density in these materials, which leads to a diminished deformation ability at the crack tip.

Figure 8 presents the fatigue threshold depending on the load ratio for all tested conditions. The matrix material and both reinforced conditions exhibit a strong dependency on the load ratio. Although the curve progression is nearly parallel for these three conditions, the underlying mechanisms are different. The matrix material shows the highest thresholds and also a strong load ratio dependency. This can be explained by the well-known effect of the grain size on the crack growth at different load ratios [11–13,15,35]. The amount of crack deflection is constant for all load ratios, whereas the effect of roughness induced crack closure is minimized by increasing the load ratio. The fatigue thresholds of the ECAP-processed matrix material show a minor dependency on the load ratio. Supposedly, the small grain size in the shear bands and therefore the generally low amount of crack deflection and closure lead here to minimized load ratio effects.

However, despite their small grain size, both reinforced conditions exhibit a strong dependency on the load ratio. In this case, assumedly, the thresholds depend critically on the interaction between the softer aluminum matrix where crack propagation takes place and the interface where decohesion or brittle particle facture can occur [36]. At higher load ratios, the susceptibility to cracking of the particle-matrix-interface is strongly increased. On that account, fatigue thresholds further decrease for higher load ratios. This, in turn, results in a higher load-ratio-dependency. For the reinforced condition, additional ECAP does not influence the load ratio dependency.

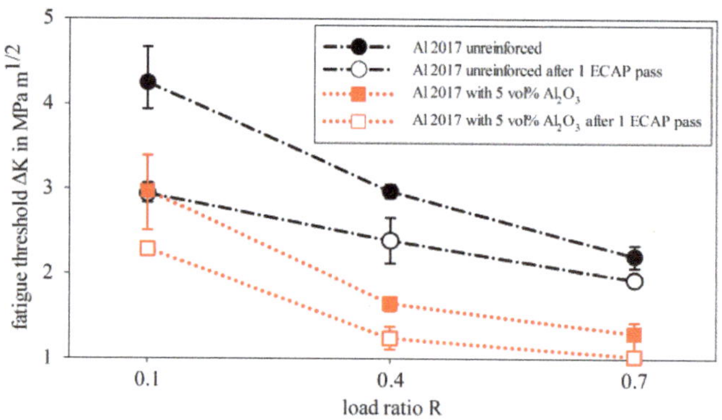

Figure 8. Fatigue thresholds of Al 2017 in unreinforced and reinforced conditions with additional ECAP at different load ratios. The ECAP-processed matrix material exhibits the least dependency on the load ratio. In contrast, the matrix material and the reinforced conditions depend strongly on the load ratio and display almost the same amount for the three conditions.

4. Conclusions

In this paper, the effect of particulate reinforcement and equal-channel angular pressing on the fatigue crack growth in an aluminum alloy produced by powder metallurgy is investigated. For the Al 2017 powder alloy, reinforced with 5 vol % Al_2O_3 with a size fraction of 0.2–2 μm, the fatigue properties are compared to the unreinforced material. Two different conditions for both materials are chosen: non-processed and after one pass of ECAP.

(1) One ECAP pass of the unreinforced material results in an inhomogeneous microstructure with shear bands. These shear bands are characterized by a considerably reduced grain size. In the surrounding areas, the grains are strongly elongated, but not refined. The grain size of the reinforced material is significantly lower, as compared to the unreinforced. Additional ECAP does not lead to a significant grain refinement, here.

(2) The yield strength of both materials, unreinforced and reinforced, is increased by one pass of ECAP. However, the strain hardening effect of ECAP is more pronounced in the unreinforced material.

(3) Due to the major grain size of the unreinforced matrix material, the high amount of crack deflection and roughness-induced crack closure lead to a tortuous crack path and therefore to minor crack propagation rates. The grain refinement through one ECAP pass reduces these effects. Supposedly because crack propagation occurred in the finer-grained shear bands, the fatigue thresholds and their dependency on the load ratio are diminished for the unreinforced material after ECAP.

(4) For the reinforced material, the fatigue thresholds are drastically reduced due to the significantly smaller grain size of these conditions. Despite their small grain size, the reinforced materials exhibit a strong dependency on the load ratio. Here, the interaction between the aluminum matrix and the interface is perceived as crucial: the susceptibility to cracking of the particle-matrix-interface is strongly increased at higher load ratios. This leads to a "particle-failure-induced" load ratio dependency.

Acknowledgments

The authors gratefully acknowledge funding of the Collaborative Research Center SFB 692 received from the German Research Foundation (Deutsche Forschungsgemeinschaft DFG).

Author Contributions

L.K. performed and analyzed the experiments and is the primary author of the paper. K.H. conceived and designed the experiments and discussed the results and analysis with the author. T.L. supervised the work.

Conflicts of Interest

The authors declare no conflict of interest.

References

1. Chawla, N.; Shen, Y.L. Mechanical behavior of particle reinforced metal matrix composites. *Adv. Eng. Mater.* **2001**, *3*, 357–370.
2. Mani, B.; Paydar, M.H. Application of forward extrusion-equal channel angular pressing (FE-ECAP) in fabrication of aluminum metal matrix composites. *J. Alloys Compd.* **2010**, *492*, 116–121.

3. Tan, M.J.; Zhang, X. Powder metal matrix composites: Selection and Processing. *Mater. Sci. Eng. A* **1998**, *244*, 80–85.
4. Ramu, G.; Bauri, R. Effect of equal channel angular pressing (ECAP) on microstructure and properties of Al-SiC_p composites. *Mater. Des.* **2009**, *30*, 3554–3559.
5. Sabirov, I.; Kolednik, O.; Valiev, R.Z.; Pippan, R. Equal channel angular pressing of metal matrix composites: Effect on particle distribution and fracture toughness. *Acta Mater.* **2005**, *53*, 4919–4930.
6. Valiev, R.Z.; Islamgaliev, R.K.; Kuzmina, N.F.; Li, Y.; Langdon, T.G. Strengthening and grain refinement in an Al-6061 metal matrix composite through intense plastic straining. *Scr. Mater.* **1999**, *40*, 117–122.
7. Muñoz-Morris, M.A.; Calderón, N.; Guitierrez-Urrutia, I.; Morris, D.G. Matrix grain refinement in Al-TiAl composites by severe plastic deformation: Influence of particle size and processing route. *Mater. Sci. Eng. A* **2006**, *425*, 131–137.
8. Segal, V.M.; Reznikov, V.I.; Drobyshevskiy, A.E.; Kopylov, V.I. Plastic working of metals by simple shear (Russian translation). *Russ. Metall.* **1981**, *1*, 99–105.
9. Valiev, R.Z.; Langdon, T.G. Principles of equal-channel pressing as a processing tool for grain refinement. *Prog. Mater. Sci.* **2006**, *51*, 881–981.
10. Furukawa, M.; Horita, Z.; Langdon, T.G. Developing ultrafine grain sizes using severe plastic deformation. *Adv. Eng. Mater.* **2001**, *3*, 121–125.
11. Pao, P.S.; Jones, H.N.; Cheng, S.F.; Feng, C.R. Fatigue crack propagation in ultrafine grained Al-Mg alloy. *Int. J. Fatigue* **2005**, *27*, 1164–1169.
12. Hübner, P.; Kiessling, R.; Biermann, H.; Hinkel, T.; Jungnickel, W.; Kawalla, R.; Höppel, H.-W.; May, J. Static and cyclic crack growth behavior of ultrafine-grained Al produced by different severe plastic deformation methods. *Metall. Mater. Trans. A* **2007**, *38*, 1926–1933.
13. Vinogradov, A. Fatigue limit and crack growth in ultra-fine grain metals produced by severe plastic deformation. *J. Mater. Sci.* **2007**, *42*, 1797–1808.
14. Cavaliere, P. Fatigue properties and crack behavior of ultra-fine and nanocrystalline pure metals. *Int. J. Fatigue* **2009**, *31*, 1476–1489.
15. Estrin, Y.; Vinogradov, A. Fatigue behaviour of light alloys with ultrafine grain structure produced by severe plastic deformation: An overview. *Int. J. Fatigue* **2010**, *32*, 898–907.
16. Meyer, L.W.; Sommer, K.; Halle, T.; Hockauf, M. Crack growth in ultrafine-grained AA6063 produced by equal-channel angular pressing. *J. Mater. Sci.* **2008**, *43*, 7426–7431.
17. Avtokratova, E.V.; Sitdikov, O.S.; Kaibyshev, R.O.; Watanabe, Y. Behavior of a submicrocrystalline aluminum alloy 1570 under conditions of cyclic loading. *Phys. Met. Metallogr.* **2009**, *107*, 291–297.
18. Hockauf, K.; Wagner, M.F.-X.; Halle, T.; Niendorf, T.; Hockauf, M.; Lampke, T. Influence of precipitates on low-cycle fatigue and crack growth behavior in an ultrafine-grained aluminum alloy. *Acta Mater.* **2014**, *80*, 250–263.
19. Chen, Z.Z.; Tokaji, K. Effects of particle size on fatigue crack initiation and small crack growth in SiC particulate-reinforced aluminum alloy composites. *Mater. Lett.* **2004**, *58*, 2314–2321.
20. Levin, M.; Karlsson, B. Influence of SiC particle distribution and prestraining on fatigue crack growth rates in aluminum AA 6061-SiC composite material. *Mater. Sci. Technol.* **1991**, *7*, 596–607.

21. Mason, J.J.; Ritchie, R.O. Fatigue crack growth resistance in SiC particulate and whisker reinforced P/M 2124 aluminum matrix composites. *Mater. Sci. Eng. A* **1997**, *231*, 170–182.
22. Llorca, J. Fatigue of particle- and whisker-reinforced metal-matrix composites. *Prog. Mater. Sci.* **2002**, *47*, 283–353.
23. Oh, K.H.; Han, K.S. Short-fiber/ particle hybrid reinforcement: Effects on fracture toughness and fatigue crack growth of metal matrix composites. *Comput.. Sci. Technol.* **2007**, 67, 1719–1726.
24. Chawla, N.; Ganesh, V.V. Fatigue crack growth of SiC particle reinforced metal matrix composites. *Int. J. Fatigue* **2010**, *32*, 856–863.
25. Huang, M.; Li, Z. Size effects on stress concentration induced by a prolate ellipsoidal particle and void nucleation mechanism. *Int. J. Plast.* **2005**, *21*, 1568–1590.
26. Xue, Z.; Huang, Y.; Li, M. Particle size effect in metallic materials: A study by the theory of mechanism-based strain gradient plasticity. *Acta Mater.* **2002**, *50*, 149–160.
27. Slipenyuk, A.; Kuprin, V.; Milman, Y.; Goncharuk, V.; Eckert, J. Properties of P/M processed particle reinforced metal matrix composites specified by the reinforcement concentration and matrix-to-reinforcement particle size ratio. *Acta Mater.* **2006**, *54*, 157–166.
28. Wagner, S.; Siebeck, S.; Hockauf, M.; Nestler, D.; Podlesak, H.; Wielage, B.; Wagner, M.F.X. Effect of SiC-reinforcement and equal-channel angular pressing on microstructure and mechanical properties of AA2017. *Adv. Eng. Mater.* **2012**, *14*, 388–393.
29. Iwahashi, Y.; Wang, J.T.; Horita, Z.; Nemoto, M.; Langdon, T.G. Principle of equal-channel angular pressing for the processing of ultra-fine grained materials. *Scr. Mater.* **1996**, *35*, 143–146.
30. Kim, J.K.; Jeong, H.G.; Hong, S.I.; Kim, Y.S.; Kim, W.J. Effect of aging treatment on heavily deformed microstructure of a 6061 aluminum alloy after equal channel angular pressing. *Scr. Mater.* **2014**, *45*, 901–907.
31. Hockauf, M.; Meyer, L.W.; Zillmann, B.; Hietschold, M.; Schulze, S.; Krüger, L. Simultaneous improvement of strength and ductility of Al-Mg-Si alloys by combining equal-channel angular extrusion with subsequent high-temperature short-time aging. *Mater. Sci. Eng. A* **2009**, *503*, 167–171.
32. ASTM E399-12e3. *Standard Test Method for Linear-elastic Plane-Strain Fracture Toughness KIc of Metallic Materials.* American Society for Testing and Materials: West Conshohocken, PA, USA, 1970.
33. Hockauf, M.; Schönherr, R.; Wagner, S.; Meyer, L.W.; Podlesak, H.; Mücklich, S.; Wielage, B.; Krüger, L.; Hahn, F.; Weber, D. Equal-channel angular pressing of medium- to high-strength precipitation hardening aluminium wrought alloys. *Mat.-Wiss. Werkstofftech.* **2009**, *40*, 540–550.
34. Shang, J.K.; Ritchie, R.O. On the particle-size dependence of fatigue-crack propagation thresholds in SiC-particulate-reinforced aluminium alloy composites: Role of crack closure and crack trapping. *Acta Metall.* **1989**, *37*, 2267–2278.
35. Hockauf, K.; Halle, T.; Hockauf, M.; Wagner, M.F.-X.; Lampke, T. Near-threshold fatigue crack propagation in an ECAP-processed ultrafine-grained aluminium alloy. *Mater. Sci. Forum* **2011**, *667–669*, 873–878.

36. Shang, J.K.; Yu, W.; Ritchie, R.O. Role of silicon carbide particles in fatigue crack growth in SiC-particulate-reinforced aluminum alloy composites. *Mater. Sci. Eng. A* **1988**, *102*, 181–192.

10

Fatigue Behavior of an Ultrafine-Grained Al-Mg-Si Alloy Processed by High-Pressure Torsion

Maxim Murashkin [1,2,*], Ilchat Sabirov [3], Dmitriy Prosvirnin [4,†], Ilya Ovid'ko [5], Vladimir Terentiev [4,†], Ruslan Valiev [1,2] and Sergey Dobatkin [4,†]

[1] Institute for Physics of Advanced Materials, Ufa State Aviation Technical University, Ufa 450000, Russia; E-Mail: rzvaliev@mail.rb.ru

[2] Laboratory for Mechanics of Bulk Nanostructured Materials, Saint Petersburg State University, Saint-Petersburg 199034 Russia

[3] IMDEA Materials Institute, Getafe, Madrid 28906, Spain; E-Mail: ilchat.sabirov@imdea.org

[4] Baikov Institute for Metallurgy and Materials Science of the Russian Academy of Sciences, Moscow 119991, Russia; E-Mails: imetran@yandex.ru (D.P.); fatig@mail.ru (V.T.); dobatkin@ultra.imet.ac.ru (S.D.)

[5] Research Laboratory for Mechanics of New Nanomaterials, Saint-Petersburg State Polytechnical University, Saint-Petersburg 195251, Russia; E-Mail: ovidko@gmail.com

† These authors contributed equally to this work.

* Author to whom correspondence should be addressed; E-Mail: m.murashkin.70@gmail.com

Academic Editor: Heinz Werner Höppel

Abstract: The paper presents the evaluation of the mechanical and fatigue properties of an ultrafine-grained (UFG) Al 6061 alloy processed by high-pressure torsion (HPT) at room temperature (RT). A comparison is made between the UFG state and the coarse-grained (CG) one subjected to the conventional aging treatment T6. It is shown that HPT processing leads to the formation of the UFG microstructure with an average grain size of 170 nm. It is found that yield strength ($\sigma_{0.2}$), ultimate tensile strength (σ_{UTS}) and the endurance limit (σ_f) in the UFG Al 6061 alloy are higher by a factor of 2.2, 1.8 and 2.0 compared to the CG counterpart subjected to the conventional aging treatment T6. Fatigue fracture surfaces are analyzed, and the fatigue behavior of the material in the high cycle and low cycle regimes is discussed.

Keywords: Al alloy; severe plastic deformation; high pressure torsion; ultrafine-grained microstructure; mechanical strength; fatigue properties

1. Introduction

The enhancement in the level of physical and mechanical properties in Al alloys as a result of ultrafine-grained (UFG) structure formation induced by severe plastic deformation (SPD) is currently of great interest [1–5]. As a rule, SPD processing leads to the formation of the microstructure with a grain/subgrain size in the range of 50 nm–1 μm in bulk Al billets [1–5]. Earlier, it was demonstrated that UFG Al alloys have a mechanical strength 1.2–2.0-times as much as their coarse-grained (CG) counterparts processed by a standard aging treatment [1–4,6]. Numerous studies have also shown that other properties, including ductility [1,7,8], fracture resistance [9] and electrical conductivity [1,10,11], can be enhanced. The fatigue behavior of the UFG Al alloys has been widely studied over the last decade, and a detailed overview of those studies can be found in [12–17]. The outcomes of these studies can be roughly generalized as follows. The high-cycle fatigue (HCF) properties are usually not enhanced as significantly as mechanical strength. An insignificant improvement in fatigue life observed in the UFG Al alloys in the HCF regime was related to a low increase in resistance against crack nucleation, as the fatigue life in the HCF is determined by the resistance of material to the crack nucleation [13]. The low cycle fatigue (LCF) behavior of the UFG Al alloys is very complex, since ultrafine grains have a low ability to sustain cyclic loads in the LCF regime. This was related to the limited ductility on monotonic and cycling deformation promoting early crack initiation, as well as a higher fraction of grain boundaries favorable for crack propagation [13]. Unlike CG Al alloys, their UFG counterparts often show an insignificant increase of the stress amplitude (low cyclic hardening) or even its decrease (cyclic softening) when they are cyclically deformed under a constant strain amplitude [18–20]. Nevertheless, intelligent microstructural design in the UFG Al alloys can significantly improve their HCF and LCF behavior [21,22].

It should be noted that the fatigue behavior of the UFG Al alloys has been studied mainly on specimens processed by equal-channel angular pressing (ECAP) [12–14] or by cryorolling [18,23]. It is well known that the ECAP technique or other ECAP-based methods result in the formation of a UFG microstructure with a grain/subgrain size of 300–500 nm [12–14], whereas a microstructure consisting mainly of cells/subgrains having a similar size is formed during cryorolling [18,23]. However, in the literature, there are no reports on the fatigue properties of the UFG Al alloys with a grain size of <200 nm, which can demonstrate record mechanical strength up to 1,000 MPa and are usually processed by HPT [24,25]. The main limitation for fatigue studies on the HPT-processed metallic materials was the small size of the processed disks. Recent progress in up-scaling of the HPT technique along with miniaturization of the specimens for fatigue testing allows one to investigate the fatigue properties of the HPT processed disks [26]. Therefore, the main objective of the present work is to study the fatigue properties and fatigue behavior of an HPT-processed Al-Mg-Si alloy with a grain size of 170 nm. The Al-Mg-Si alloy was selected due to the wide application of the material in construction, auto- and aero-space engineering [21] and as a conductor in electrical engineering [27].

2. Experimental Section

Commercial-grade Al 6061 alloy having a chemical composition (in wt%) 1.00 Mg, 0.61 Si, 0.12 Cu, 0.38 Cr, 0.60 Fe, 0.02 Ti and balance Al was selected for this investigation. Disks having a diameter of 20 mm and a thickness of 1.4 mm were machined from the as-received hot-pressed rod. The disks were solution-treated at 530 °C for 2 h and water quenched. Some disks after water quenching were subjected to the conventional treatment T6, namely artificial ageing at 160 °C for 12 h.

The disks were HPT processed for 10 rotations at RT under a load of 188 tons (e.g., a pressure of 6 GPa). The as-processed disks had a thickness of 0.9 mm. The microstructure and properties of the HPT-processed alloy were studied in the areas located at a distance of ~5 mm (half of radius) from the center of the disks.

Microstructure studies were performed using the transmission electron microscope (TEM, JEOL, Tokyo, Japan) JEM 2100 at an accelerating voltage of 200 kV. Observations were made in both the bright and the dark field imaging modes, and selected area electron diffraction (SAED) patterns were recorded from areas of interest using an aperture of a 1-μm nominal diameter. The lineal intercept method was used to estimate the grain size. At least 200 grains were analyzed to estimate the average grain size.

Specimens for tensile testing and fatigue testing were machined from the solution-treated material and HPT-processed disks. The central axis of the tensile and fatigue specimens was located at a distance of 5.2 mm from the disk center, as shown in Figure 1, where a drawing of the samples is also presented. Two specimens were machined from each disk. All machined tensile and fatigue specimens were mechanically polished to a mirror-like surface using silica colloidal solution at the final stage. Tensile tests were carried out using an Instron 8862 universal testing machine (JEOL, Tokyo, Japan). Tensile specimens were deformed to failure at room temperature with a constant cross-head speed corresponding to an initial strain rate of 10^{-3} s^{-1}. Three tensile specimens were tested for each material condition, and the results thus obtained were found to be reproducible.

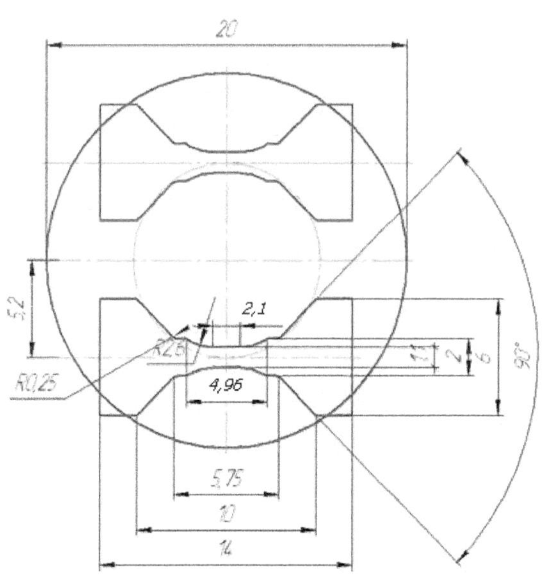

Figure 1. Schematic drawing of specimens machined from disks.

Fatigue testing was carried out using an Instron Electropuls E 3000 testing module with a frequency of 30 Hz under repeated tension conditions with a constant minimum cycle stress of σ_{min} = 100 MPa.

The fatigue fracture surfaces were studied using the scanning electron microscope (SEM, JEOL, Tokyo, Japan) JSM-6490LV operating at an accelerating voltage of 20 kV. The SEM is equipped with the INCA X-sight attachment for energy-dispersive X-ray (EDX) analysis.

3. Results and Discussion

3.1. Effect of HPT Processing on Microstructure of the Al 6061 Alloy

Figure 2 illustrates the microstructure of the Al 6061 alloy before and after HPT processing. Conventional treatment T6 of the as-received alloy resulted in the formation of a CG microstructure with an average grain size of 70 μm (Figure 2a). Particles of primary excess phase forming bands oriented along the pressing direction are observed in the microstructure (Figure 2a). They have a complex chemical composition of $Al_{68}Fe_{21}Si_{6.5}Cr_{2.5}$ (wt%) and are typically formed during the crystallization of slabs. These particles have a spherical shape; their average size is about 1.8 μm, and the volume fraction is about 2.4%. Such a distribution of these particles is usually observed in the hot-pressed Al-Mg-Si alloys. Second phase precipitates having a needle shape are also present in the microstructure (Figure 2b). As is well known, these are the metastable β''-Mg_2Si precipitates, which are typically formed during T6 treatment of the Al-Mg-Si alloys [28,29].

HPT processing of the solution-treated Al 6061 alloy resulted in the formation of a very homogeneous UFG microstructure consisting mainly of equiaxed grains having an average size of 170 nm (Figure 2c,d). This is also confirmed by SAED patterns (Figure 2d). No precipitates, like Guinier–Preston zones and/or second phases, were observed in the processed disks. Such a microstructure is typically formed during HPT processing of the Al-Mg-Si alloys [11,30,31]. It should be also noted that earlier studies using the atom probe tomography (APT) technique revealed segregations of Mg, Cu and Si atoms along grain boundaries in the Al 6061 alloy [30] and in the Al 6060 alloy [32] processed by HPT under similar conditions. HPT processing did not have any effect on the size and volume fraction of the primary excess phases. However, particles changed their orientation and formed a banded structure along the shear plane during HPT processing (Figure 3). It should be noted that in order to avoid the effect of the orientation of the particle bands on the properties of the alloy, specimens for tensile and fatigue testing were cut out, so that particle bands in the gauge sections were oriented along the specimen axis.

3.2. Effect of HPT Processing on Tensile and Fatigue Behavior of the Al 6061 Alloy

The results of mechanical tensile testing of the Al 6061 alloy after conventional T6 treatment and HPT processing are presented in Table 1. It is seen that HPT processing resulted in dramatic improvement of the mechanical strength of the material. The yield strength increased from 276 MPa to 605 MPa, whereas the ultimate tensile strength increased from 365 MPa to 675 MPa. No other SPD processing technique, such as ECAP [33,34], or ECAP combined with cold rolling (CR), or artificial aging (AA) [33], led to such a high mechanical strength in this alloy. Such a significant increase in mechanical strength of the HPT-processed Al 6061 alloy can be related primarily to the formation of a homogeneous UFG structure with a very small grain size, promoting the grain size hardening according to the Hall–Petch relationship [2]. At the same time, segregations of solute atoms in the Al solid solution, which were found earlier in the structure of the HPT-processed Al 6061 alloy in [30], can also provide some additional strengthening.

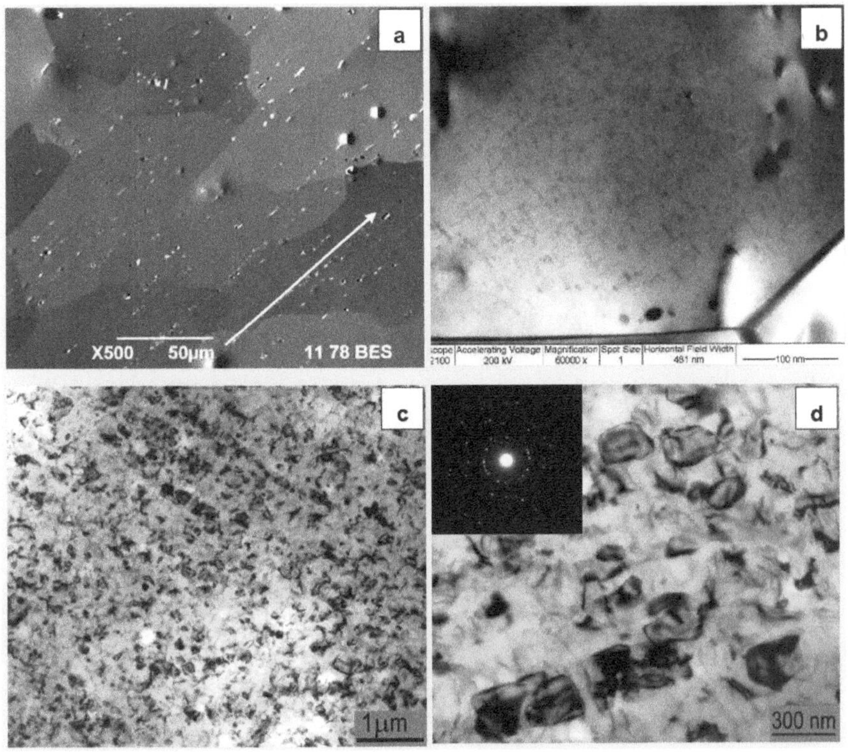

Figure 2. Microstructure of the Al 6061 alloy after conventional T6 treatment (**a**,**b**) and high-pressure torsion (HPT) processing at RT (**c**,**d**). The arrow indicates the pressing direction and the orientation of the bands formed by primary phase particles.

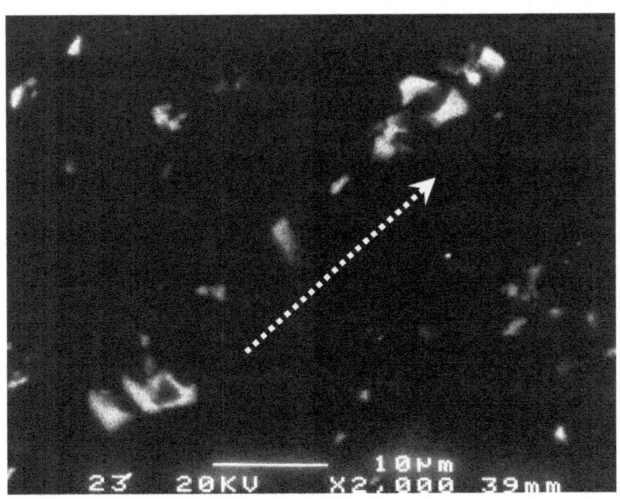

Figure 3. Distribution of the primary excess phases in the Al 6061 alloy processed by HPT at RT (the arrow indicates the shear plane and the orientation of the bands formed by primary phase particles).

Table 1. Mechanical properties of the Al 6061 alloy in the present work in comparison with the mechanical properties reported in earlier studies. ECAP, equal-channel angular pressing; AA, artificial aging; CR, cold rolling; UFG, ultrafine grained; CG, coarse grained.

Processing	State	$\sigma_{0.2}$ (MPa)	σ_{UTS} (MPa)	δ (%)	σ_f (MPa)	σ_f/σ_{UTS}	Reference
T6	CG	276	365	14	100	0.27	present work
HPT for 10 turns at RT	UFG	605	675	5.5	200	0.30	
T6	CG	275	310	12	97	0.31	[34]
ECAP at 125 °C for 1 pass	UFG	310	375	20	80	0.21	[21]
ECAP at 100 °C, 4 passes	UFG	386	434	11	-	-	[33]
ECAP + AA (130 °C for 24 h)	UFG	434	470	10	-	-	
ECAP + CR (15%)	UFG	470	500	8	-	-	

Figure 4 shows the results of fatigue testing of CG T6 treated and HPT-processed material. It is seen that HPT processing has doubled the endurance limit (σ_f) of the material at 10^7 cycles: it increased from 100 MPa in the T6-treated CG alloy to 200 MPa after HPT processing (Figure 4). It should be noted that the σ_f-values obtained after cyclic testing of sub-size specimens with the CG microstructure subjected to the conventional T6 treatment are in very good agreement with the results of the testing of standard reference specimens machined from a similar material [12–14,35]. For example, according to the results of fatigue testing of standard samples, the σ_f-value of the Al 6061 alloy in the T6 state is 97 MPa [13].

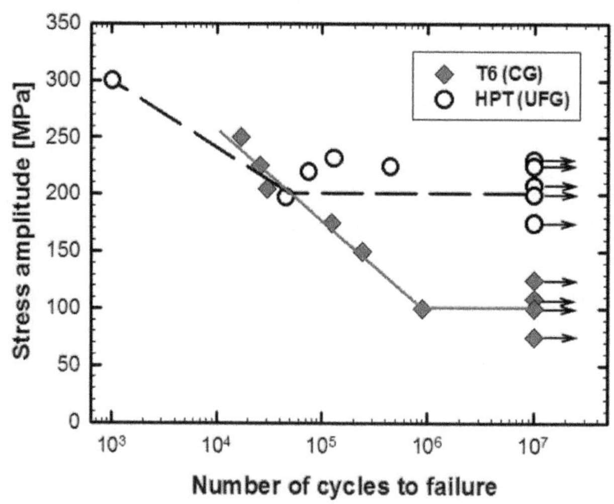

Figure 4. Fatigue curves for the studied Al 6061 alloy.

It should be also noted that the level of σ_f-value achieved in the HPT-processed Al 6061 alloy with an average grain size of 170 nm is also twice the cyclic strength of the UFG material with an average grain size of ~500 nm processed by ECAP in [34].

In Figure 5, the endurance limit is plotted *vs.* the ultimate tensile strength of the CG Al 6061 alloy processed by conventional thermal treatments (O, T4, T6) and by thermo-mechanical treatment (T91) [13,35]. The analysis of the data shows that the ratio of the fatigue limit to the ultimate tensile strength (σ_f/σ_{UTS}) for the given material decreases from ~0.50 to 0.25 with increasing σ_{UTS} (Figure 5).

It should be noted that a lower σ_f/σ_{UTS} ratio of 0.20 was reported for the UFG Al 6061 alloy processed by ECAP in [21]. Such low σ_f/σ_{UTS} ratios were found also in other UFG Al-Mg-Si alloys [12–14]. In our work, grain refinement down to 170 nm in the Al 6061 alloy via HPT processing resulted in an increase of the σ_f/σ_{UTS} ratio from 0.27 to 0.30 and in a record value of σ_f = 200 MPa. As is well known, fatigue life in the HCF regime is determined by the resistance of the material to crack initiation [13]. Unlike UFG Al 6061 alloy produced by ECAP in [34], our UFG alloy shows a very homogeneous grain structure and, therefore, should have also a homogeneous resistance to crack nucleation all over the gauge section without any 'weakest link'. Thus, it can be concluded that further grain refinement down to the nanoscale and the generation of a very homogeneous grain structure can be proposed as a recipe for the improvement of HCF properties in the Al alloys.

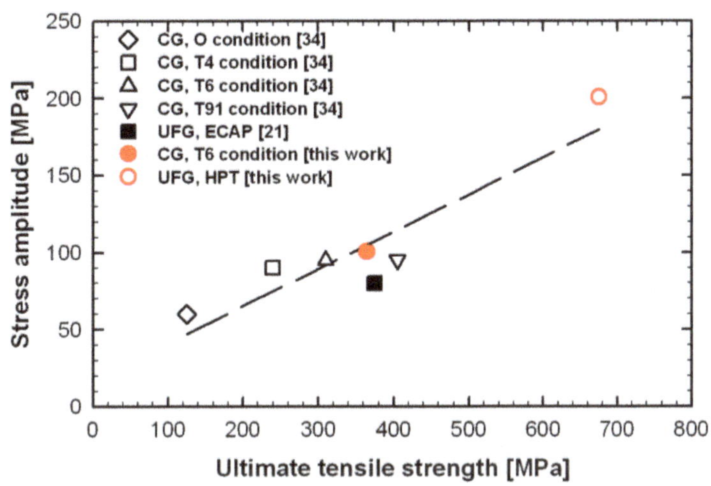

Figure 5. Ultimate tensile strength *vs.* endurance limit for the Al 6061 alloy subjected to various processing treatments.

In the LCF regime, the endurance limit of the HPT-processed Al alloy was somewhat lower than in its CG counterpart (Figure 4). This observation can be rationalized based on the low strain hardening ability of the UFG material, compared to that of the CG T6 one (Table 1), promoting early crack initiation, as well as a much higher volume fraction of grain boundaries favorable for crack propagation [13].

3.3. Analysis of Fatigue Fracture Surfaces

Figures 6 and 7 present SEM images of the fatigue fracture surface of CG T6-treated and HPT-processed Al 6061 alloy, respectively. The area of fatigue crack initiation in the CG T6 material is located on the surface of specimens and is extended into the specimen by ~150 μm (Figure 6a). This is known as Stage I or the short crack growth propagation stage. The increase of the stress intensity factor K as a consequence of crack growth results in the development of slips in different planes close to the crack tip, initiating Stage II, e.g., the stage of stable fatigue crack growth (Figure 6b). An important characteristic of Stage II is the presence of surface ripples, known as "striations". The ductile striations are clearly observed on the fatigue fracture surface (Figure 6c). Their width is in the range of 1–2 μm. The average width of striations is about 3 μm. The most accepted mechanism for the formation of striations is the successive blunting and re-sharpening of the crack tip during cyclic loading [35,36]. Striations are grouped in areas of homogenous propagation separated by tear ridges in which the material fails by shearing, resulting in the apparent continuity of the fatigue advancing front. Finally, Stage III is related to unstable crack growth as K approaches K_{IC} (Figure 6e). At this stage, crack growth is controlled by static modes of failure and is very sensitive to the microstructure, load ratio and stress state. The ductile fracture surface consisting of spherical dimples is observed (Figure 6g). The size of the dimples is in the range of 2–20 μm. Figure 6g shows a fracture pattern within the transition area from a fatigue fracture to the final breakdown.

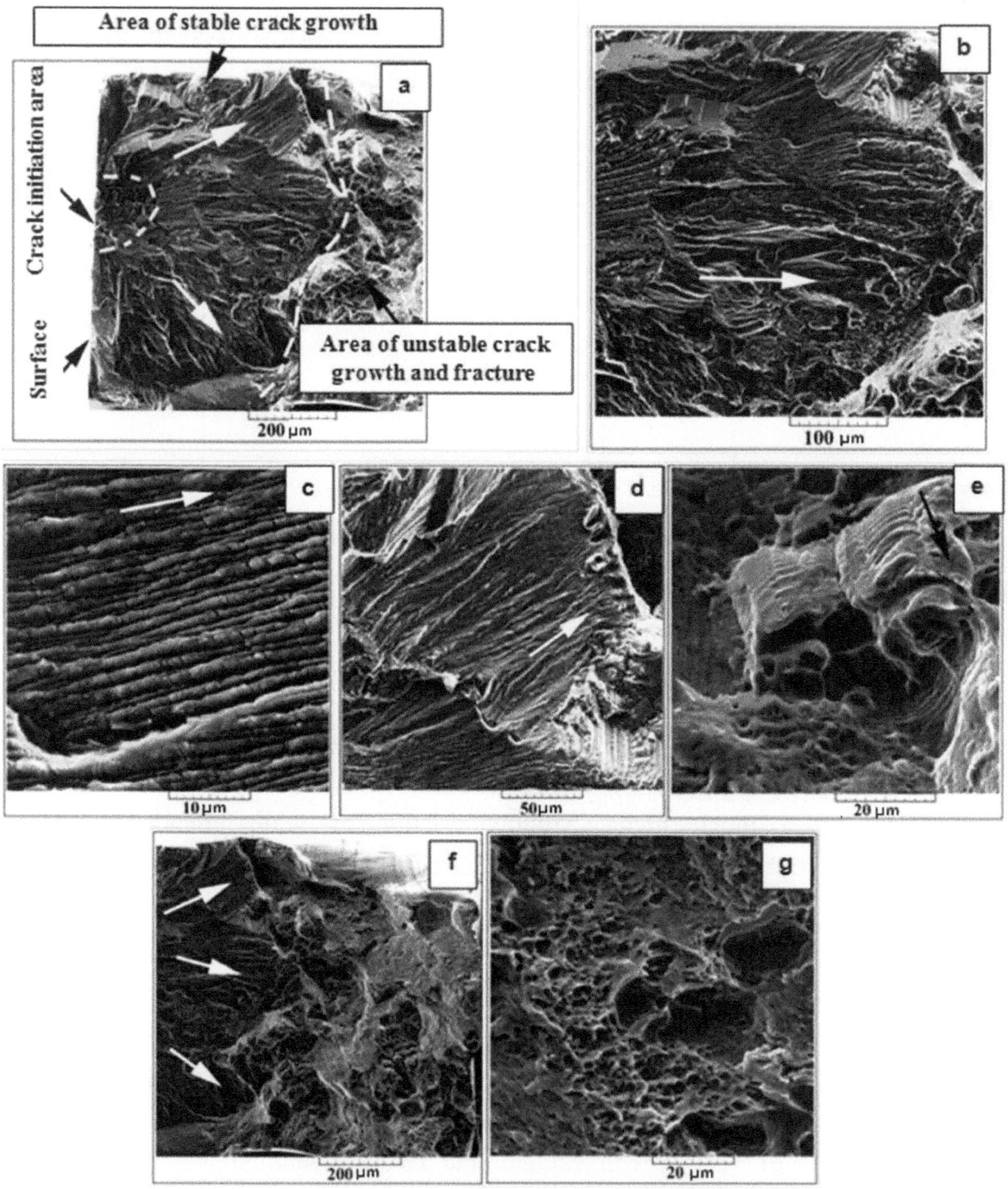

Figure 6. SEM images of the fatigue fracture surface of the CG T6 Al 6061 alloy tested under a stress amplitude of 150 MPa to 2.2×10^5 cycles) (the arrows indicate the direction of fatigue crack propagation).

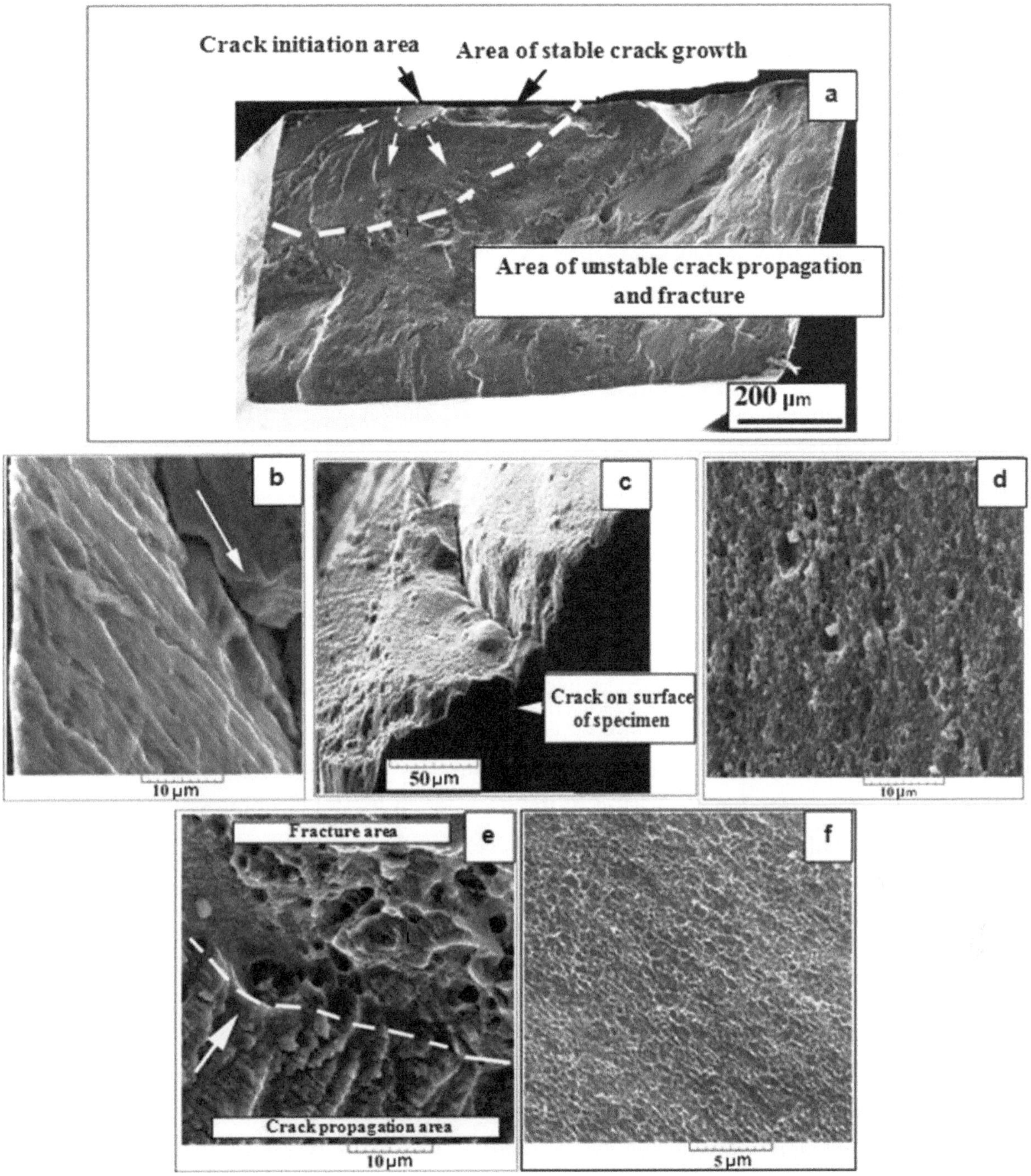

Figure 7. SEM images of the fatigue fracture surface of the UFG Al 6061 alloy: (**a**,**b**,**e**,**f**) tested under a stress amplitude of 225 MPa to 4.4×10^5 cycles; (**c**,**d**) tested under a stress amplitude of 200 MPa to 4.35×10^4 cycles (the arrows indicate the direction of fatigue crack propagation).

Similar stages of fatigue crack initiation and propagation can be also easily defined on the fatigue fracture surface of the HPT-processed alloy (Figure 7). The area of fatigue crack initiation is also located on the surface of specimens, though its size is smaller: ~50 µm (Figure 7a). The fracture surface relief in the areas of stable and unstable crack growth is less developed as compared to that on in the fatigue-tested CG T6 specimens. Brittle striations are observed in the area of stable crack propagation in the UFG material (Figure 7b) [36]. Unlike in the fatigue-tested CG T6 specimens, stable and unstable crack propagation is accompanied by the formation and growth of micro-cracks in the direction parallel to the applied load (Figure 7c). The length of these micro-cracks is 50–100 μm. These micro-cracks should result in energy dissipation, thus delaying the growth of the main fatigue crack [35]. The dimpled ductile fracture surface is observed in the area of unstable crack growth and the final fracture of the specimen (Figure 7d–g). However, the size of these dimples is in the range of 1–2 μm, which is much lower compared to that in the CG T6 material. A significant difference in the morphology of fatigue fracture surfaces of the CG T6-treated and UFG Al 6061 alloy can be related to the grain structure of these material conditions. As is well known, during the loading phase, the crack is opened by normal stress, which activates plastic slips at the tip. At the crack tip of the CG T6 material with a grain size of 70 μm, dislocations are generated in the grain interior and do not necessarily reach the grain boundaries, when the material is cyclically deformed in Stage II. Conventional transgranular fatigue crack growth followed by specimen failure is observed. In the UFG material with a grain size of 170 nm, dislocation pile-ups are definitely formed at grain boundaries near the crack tip, and intragranular slip coupled with grain boundary sliding can occur [37]. Unaccommodated grain boundary sliding might result in local microcracking or nanovoid formation at grain boundaries. Triple junction nanovoids can also appear due to unaccommodated grain boundary sliding. Both transgranular and intragranular fatigue crack growth can take place in Stage II in this case, though this cannot be clearly resolved in SEM due to the very small grain size. In Stage III, nanovoids can grow and partially relieve the constraints on a grain. Individual single crystal ligaments deform extensively and finally experience chisel-point failure. These grain boundary and triple junction voids also act as sites for nucleation of the dimples in the UFG alloy, which are significantly larger than the individual grains, and the rim of these dimples on the fracture surface (1–2 μm) is typically a magnitude larger than the grain size (170 nm) [37].

4. Conclusions

The following conclusions are outlined.

1. High-pressure torsion (HPT) of the Al 6061 alloy at room temperature leads to the formation of a very homogeneous ultrafine-grained (UFG) microstructure with an average grain size of 170 nm. The yield strength and ultimate tensile strength of the HPT processed alloy are increased as compared to the CG counterpart subjected to the conventional T6 heat treatment from 276 MPa to 605 MPa and from 365 MPa to 675 MPa, correspondingly.
2. HPT processing of the Al 6061 alloy improves its endurance limit by a factor of two (from 100 MPa after T6 treatment to 200 MPa after HPT). This is related to the formation of a very homogeneous UFG microstructure with homogeneous resistance to fatigue crack initiation in the HCF regime. In the LCF regime, the UFG alloy shows somewhat lower fatigue resistance due to its lower strain hardening ability.
3. Classical stages of fatigue crack initiation and propagation are clearly observed on the fatigue fracture surfaces of the CG T6-treated alloy and UFG alloy. Ductile striations are observed in the stage of stable crack propagation in the CG T6-treated alloy, whereas brittle striations seem to dominate on the fatigue fracture surface of the UFG alloy. The dimpled fracture surface is

observed at the final stage of crack propagation in both material conditions, with dimples having a smaller size in the UFG alloy.

Acknowledgments

The work has been done under the financial support of the Russian Federal Ministry for Education and Science. R.Z. Valiev and M. Yu. Murashkin gratefully acknowledge the support through Grant No. 14.B25.31.0017, and I. Ovidko acknowledges the support through Programme "5-100-2020". I. Sabirov acknowledges gratefully the Spanish Ministry of Economy and Competitiveness for financial support through the Ramon y Cajal Fellowship.

Author Contributions

Maxim Murashkin formulated the idea of this work, processed the samples, conceived of the workflow and created the initial draft. Ilchat Sabirov carried out mechanical tensile tests and performed microstructural characterization of the processed material. Dmitriy Prosvirnin performed fatigue testing. Vladimir Terentiev performed the SEM studies of the fatigue fracture surfaces. Ilya Ovidko, Ruslan Valiev and Sergey Dobatkin contributed with the overall development of the main concepts presented in this paper. All authors discussed the experimental results, participated in manuscript preparation and approved the final manuscript.

Conflicts of Interest

The authors declare no conflict of interest.

References

1. Valiev, R.Z.; Zhilyaev, A.P.; Langdon, T.G. *Bulk Nanostructured Materials: Fundamentals and Applications.* John Wiley & Sons: Hoboken, NJ, USA, 2014; p. 456.
2. Sabirov, I.; Murashkin, M.Y.; Valiev, R.Z. Nanostructured aluminium alloys produced by severe plastic deformation: New horizons in development. *Mater. Sci. Eng. A* **2013**, *560*, 1–24.
3. Markushev, M.V.; Murashkin, M.Y. Mechanical properties of submicrocrystalline Al alloys processed by equal-channel angular pressing. *Phys. Met. Metall.* **2000**, *90*, 506–515.
4. Roven, H.J.; Nesboe, H.; Werenskiold, J.C.; Seibert, T. Mechanical properties of aluminum alloys processed by SPD: Comparison of different alloy systems and possible product areas. *Mater. Sci. Eng. A* **2005**, *410–411*, 426–429.
5. Lyakishev, N.P.; Alymov, M.I.; Dobatkin, S.V. Structural bulk nanomaterials. *Russ. Metall.* **2003**, *3*, 191–202.
6. Dobatkin, S.V.; Zakharov, V.V.; Vinogradov, A.Y.; Kitagava, N.; Krasilnikov, N.A.; Rostova, T.D.; Bastrash, E.N. Nanocrystalline structure in Al-Mg-Sc alloys by severe plastic deformation. *Russ. Metall.* **2006**, *6*, 533–540.
7. Zhao, Y.; Liao, X.; Cheng, S.; Ma, E.; Zhu, Y. Simultaneously increasing the ductility and strength of nanostructured alloys. *Adv. Mater.* **2006**, *18*, 2280–2283.

8. Horita, Z.; Ohashi, K.; Fujita, T.; Kaneko, K.; Langdon, T.G. Achieving high strength and high ductility in precipitation-hardened alloys. *Adv. Mater.* **2005**, *17*, 1599–1603.
9. Markushev, M.V.; Bampton, C.C.; Murashkin, M.Y.; Hardwick, D.A. Structure and properties of ultra-fine grained Al alloys produced by severe plastic deformation. *Mater. Sci. Eng. A.* **1997**, *234–236*, 927–931.
10. Murashkin, M.; Sabirov, I.; Kazykhanov, V.; Bobruk, E.; Dubravina, A.; Valiev, R.Z. Enhanced mechanical properties and electrical conductivity in ultra-fine grained Al alloy processed via ECAP-PC. *J. Mater. Sci.* **2013**, *48*, 4501–4509.
11. Valiev, R.Z.; Murashkin, M.Y.; Sabirov, I. A nanostructural design to produce high-strength Al alloys with enhanced electrical conductivity. *Scr. Mater.* **2014**, *76*, 13–16.
12. Vinogradov, A.Y.; Khasimoto, S. Fatigue in ultrafine-grained materials processed by equal-channel angular pressing. *Russ. Metall.* **2004**, *1*, 42–51.
13. Estrin, Y.; Vinogradov, A. Fatigue behaviour of light alloys with ultrafine grain structure produced by severe plastic deformation: An overview. *Int. J. Fatigue* **2010**, *32*, 898–907.
14. Estrin, Y.; Vinogradov, A. Extreme grain refinement by severe plastic deformation: A wealth of challenging science. *Acta Mater.* **2013**, *61*, 782–817.
15. Höppel, W.; Kautz, M.; Murashkin, M.Y.; Xu, C.; Langdon, T.G.; Valiev, R.Z.; Mughrabi, H. An overview: Fatigue Behavior of Ultrafine-Grained Metals and Alloys. *Int. J. Fatigue* **2006**, *28*, 1001–1010.
16. Höpel, H.W.; Göken, M. Fatigue Behavior in Nanostructured Metals. In *Nanostructured Metals and Alloys: Processing, Microstructure, Mechanical Properties and Applications*; Whang, S.H., Ed.; Woodhead Publishing Limited: Suite, PA, USA, 2011; pp. 507–541.
17. Mughrabi, H.; Höppel, H.W.; Kautz, M. Fatigue and microstructure of ultra-fine grained metals produced by severe plastic deformation. *Scr. Mater.* **2004**, *51*, 807–812.
18. Malekjani, S.; Hodgson, P.D.; Cizek, P.; Sabirov, I.; Hilditch, T.B. Cyclic deformation response of UFG 2024 Al alloy. *Int. J. Fatigue* **2011**, *33*, 700–709.
19. Canadinc, D.; Maier, H.J.; Gabor, P.; May, J. On the cyclic deformation response of ultrafine-grained Al–Mg alloys at elevated temperatures. *Mater. Sci. Eng. A* **2008**, *496*, 114–120.
20. Vinogradov, A.; Washikita, A.; Kitagawa, K.; Kopylov, V.I. Fatigue life of fine-grain Al-Mg-Sc alloys produced by equal-channel angular pressing. *Mater. Sci. Eng. A* **2003**, *349*, 318–326.
21. Chung, C.S.; Kim, J.K.; Kim, H.K.; Kim, W.J. Improvement of high-cycle fatigue life in a 6061 Al alloy produced by equal channel angular pressing. *Mater. Sci. Eng. A* **2002**, *337*, 39–44.
22. Lapovok, R.; Loader, C.; Dalla Torre, F.H.; Semiatin, S.L. Microstructure evolution and fatigue behavior of 2124 aluminum processed by ECAE with back pressure. *Mater. Sci. Eng. A* **2006**, *425*, 36–46.
23. Malekjani, S.; Hodgson, P.D.; Cizek, P.; Hilditch, T.B. Cyclic deformation response of ultra-fine pure Al. *Acta Mater.* **2011**, *59*, 5358–5367.
24. Valiev, R.Z.; Enikeev, N.A.; Murashkin, M.Y.; Kazykhanov, V.U.; Sauvage, X. On the origin of the extremely high strength of ultra-fine grained Al alloys. *Scr. Mater.* **2010**, *63*, 949–952.
25. Liddicoat, P.V.; Liao, X.Z.; Zhao, Y.; Zhu, Y.; Murashkin, M.Y.; Lavernia, E.J.; Valiev, R.Z.; Ringer, S.P. Nanostructural hierarchy increases the strength of aluminium alloys. *Nat. Comm.* **2010**, *1*, 63–70.

26. Ruffing, C.; Ivanisenko, Y.; Kerscher, E. Fatigue behavior of ultrafine grained medium Carbon steel processed by severe plastic deformation. *IOP Conf. Series: Mater. Sci. Eng.* **2014**, *63*, 012163.
27. Polmear, I.J. Light Alloys. In *From Traditional Alloys to Nanocrystals*; Butterworth-Heinemann: Oxford, UK, 2006; p. 417.
28. Mondolfo, L.F. *Aluminum Alloys: Structure and Properties*; Butterworth: Oxford, UK, 1976; p. 971.
29. Murayama, M.; Hono, K. Pre-precipitate clusters and precipitation processes in Al-Mg-Si alloys. *Acta Mater.* **1999**, *47*, 1537–1548.
30. Nurislamova, G.; Sauvage, X.; Murashkin, M.; Islamgaliev, R.; Valiev, R. Nanostructure and related mechanical properties of an Al-Mg-Si alloy processed by severe plastic deformation. *Phil. Mag. Lett.* **2008**, *88*, 459–466.
31. Moreno-Valle, E.C.; Sabirov, I.; Perez-Prado, M.T.; Murashkin, M.Y.; Bobruk, E.V.; Valiev, R.Z. Effect of the grain refinement via severe plastic deformation on strength properties and deformation behavior of an Al 6061 alloy at room and cryogenic temperatures. *Mater. Lett.* **2011**, *65*, 2917–2919.
32. Sha, G.; Tugcu, K.; Liao, X.Z.; Trimby, P.W.; Murashkin, M.Y.; Valiev, R.Z.; Ringer, S.P. Strength, grain refinement and solute nanostructures of an Al-Mg-Si alloy (AA6060) processed by high-pressure torsion. *Acta Mater.* **2014**, *63*, 169–179.
33. Murashkin, M.Y.; Markushev, M.V.; Ivanisenko, Y.V.; Valiev, R.Z. Strength of commercial aluminum alloys after equal channel angular pressing and post-ECAP processing. *Solid State Phen.* **2006**, *114*, 91–96.
34. Hatch, J.E. *Aluminium: Properties and Physical Metallurgy*; ASM International: Materials Park, OH, USA, 1984; p. 636.
35. Suresh, S. *Fatigue of Materials*; Cambridge University Press: Cambridge, UK, 1998; p. 679.
36. Milella, P.P. *Fatigue and Corrosion in Metals*; Springer: Berlin, Germany, 2013; p. 763.
37. Dalla Torre, F.; van Swygenhoven, H.; Victoria, M. Nanocrystalline electrodeposited Ni: Microstructure and tensile properties. *Acta Mater.* **2002**, *50*, 3957–3970.

Investigation of Material Performance Degradation for High-Strength Aluminum Alloy Using Acoustic Emission Method

Yibo Ai [1], Chang Sun [1,†], Hongbo Que [2] and Weidong Zhang [1,†,*]

1 National Center for Materials Service Safety, University of Science and Technology Beijing, Beijing 100083, China; E-Mails: ybai@ustb.edu.cn (Y.A.); sunchang2009@163.com (C.S.)
2 Qishuyan Institute Co., Ltd., China South Locomotive & Rolling Stock Corporation Limited, Changzhou 213011, China; E-Mail: quehbff@126.com

† These authors contributed equally to this work.

* Author to whom correspondence should be addressed; E-Mail: zwd@ustb.edu.cn

Academic Editor: Hugo F. Lopez

Abstract: Structural materials damages are always in the form of micro-defects or cracks. Traditional or conventional methods such as micro and macro examination, tensile, bend, impact and hardness tests can be used to detect the micro damage or defects. However, these tests are destructive in nature and not in real-time, thus a non-destructive and real-time monitoring and characterization of the material damage is needed. This study is focused on the application of a non-destructive and real-time acoustic emission (AE) method to study material performance degradation of a high-strength aluminum alloy of high-speed train gearbox shell. By applying data relative analysis and interpretation of AE signals, the characteristic parameters of materials performance were achieved and the failure criteria of the characteristic parameters for the material tensile damage process were established. The results show that the AE method and signal analysis can be used to accomplish the non-destructive and real-time detection of the material performance degradation process of the high-strength aluminum alloy. This technique can be extended to other engineering materials.

Keywords: AE; material performance degradation; relative analysis; aluminum alloy; non-destructive; real-time

1. Introduction

With the continuous increase in speed of high-speed train, the material widely used is high-aluminum alloy because of its lightweight and higher strength and good fatigue properties [1,2]. The gearbox of high-speed train is one of the important parts of its drive system, and is of great importance for operation safety of high-speed train. High-strength aluminum alloy A356, a key material of the gearbox, which has excellent properties of rigidity, strength, ductility, fatigue, casting, and shrinkage tendency [3]. At present, the development of this material has almost been completed, but performance analysis of service process has not been done [4,5]. As the operational safety is important for transportation purposes, a method of non-destructive and real-time detection for the material is urgently required.

Traditional method for materials research is microscopic observation, but this method is always accompanied by destructive testing and not in real time. Among the common non-destructive method such as acoustic emission (AE), X-ray and ultrasound, AE is a dynamic non-destructive testing method and not limited by materials [6]. Lugo *et al.* [7] studied the damage evolution in a 7075 aluminum alloy by using acoustic emissions to quantify the evolution of damage, but their research is based on the destructive experiment. Behrens *et al.* [8] presented an approach for online crack detection during cold forging of steel components using the acoustic emission technique, but the data was obtained by off-line laboratory experiments. Han *et al.* [9] investigated the effects of the change in macro-structures due to heat treatment and weld on fatigue and corresponding acoustic emission behaviors of Q345 steels, but they only considered AE counts possibility in their work. Maillet *et al.* [10] used an energy-based procedure to select AE events generated by damage in composite materials, but it was only appropriate for the composite materials and the AE signal of metal materials is different. Therefore, the technical feasibility and relative analysis method of the AE method used for non-destructive testing should be further studied especially in real-time detection and characterization for materials.

Material performance degradation is an important direction for life prediction and service safety assessment. Since the structural dimension and the whole life degradation data are difficult to be achieved in limited time and cost, the method of performance degradation can be used. In a service process, although the direct degradation data cannot be achieved, the indirect degradation data such as strength, amplitude and event count can be detected and used to characterize the degradation process. Recorded data at different times can be analyzed and used to establish failure threshold and degradation model. By extrapolation, it can also be used to estimate the remaining life of structural material before failure. Roca *et al.* [11,12] used neural networks and acoustic emission signals to estimate the stability of gas metal arc welding (GMAW) process, and it is a good method to analyze the AE signals, although the characteristics of AE signal of high-speed train gearbox shell is different from that of GMAW. Cholette *et al.* [13] presented a novel data-driven degradation modeling and monitoring methodology based on hidden Markov models; but the method needs a large data set. Hong *et al.* [14] put forward a novel performance degradation assessment method based on ensemble empirical mode composition and Gaussian mixture model, however, their data was not in real-time. Savage *et al.* [15] investigated a methodology to provide the cumulative failure distribution for degrading, uncertain, and dynamic systems, but this uncertainty method was not appropriate for engineering applications. Dobrica *et al.* [16] developed a hydrodynamic model to simulate the operation of scratched journal bearings, but the parametric study presented in their paper corresponds to a particular journal bearing configuration. Presently, there is still no life prediction data and the traditional testing method, which is expensive and time-consuming for high-speed train gearbox materials.

The aim of this work is to investigate the tensile damage of gearbox shell material of the high-speed train using an AE detection system. The AE system was implemented to obtain the AE signals during the tensile test process, and relative analysis method was then used to determine the performance degradation characteristic parameters and failure criteria.

2. Materials and Methods

2.1. Materials Tensile Damage

The results of the tensile test process of ductile metallic materials can be divided into four stages as elastic, yield, plasticity and fracture [17]. These four stages depend on the different characteristics of the tensile test process as identified on the axial load-axial elongation curve of the materials studied which is shown in Figure 1. Based on the axial load-axial elongation curve, according to different characteristics of the four stages of tensile process, the four stages can be clearly divided.

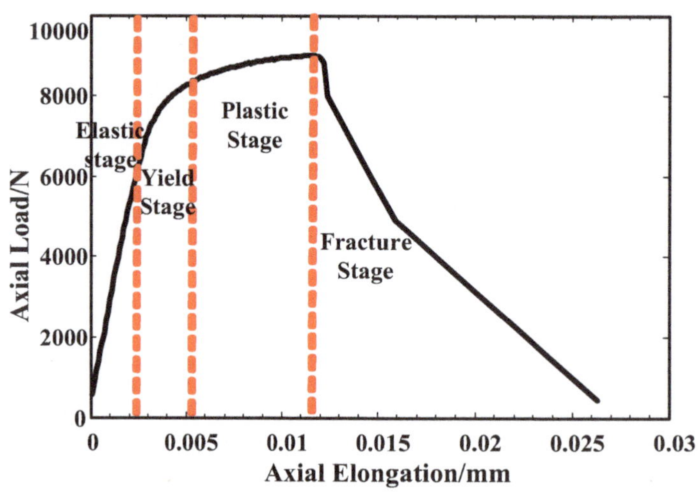

Figure 1. The axial load-axial elongation curve and four stages of tensile process.

Elastic stage: the material deformation is elastic, and the force and elongation is proportional. If withdraw the force, the elongation disappears, and no residual elongation.

Yield stage: In this stage, a lot of dislocation and metal bond cleavage occurs inside the metal material, the plastic deformation begins, which accompanied by a large number of acoustic emission phenomena [18].

Plastic deformation stage: the material is deformed under the action of external forces, and when the external force is removed, the original material cannot be recovered.

Fracture stages: stress increased while strain decreased, and the material eventually fractured.

For the material service, when the material is in the elastic stage, it is considered safe. When the material enters the yield stage, final fracture will be attained quickly. In this paper, the transition point from elastic stage to yield stage is the failure point of the tensile process.

2.2. Acoustic Emission Technology

Material deformation and crack propagation under stress are an important mechanism of structural failure. Acoustic Emission (AE) is a phenomenon of rapid release of energy and generation of transient elastic wave from a localized source of the material. This elastic wave is also known as the stress wave emission [18]. The elastic wave emitted from the sound emission source can reach the surface of the material, and can cause the displacement which can be detected by acoustic emission sensors. The mechanical vibration of the material which is caught by the detectors is converted to an electrical signal, is then amplified, processed and recorded.

AE instrument with high sensitivity can detect micro crack propagation with the change of micrometer. AE diagnostic methods are used to diagnose the crack generation and propagation due to plastic deformation, fatigue and wear. Acoustic emission parameter analysis method is used to analyze and interpret AE signal based on waveform characteristics of the acoustic emission signal. Commonly used parameters of AE are energy, ring-down count, duration, amplitude, effective voltage and so on. For the same acoustic emission source, these parameters are correlated.

2.3. System of Tensile Test with Acoustic Emission Instrument

The system of tensile test with acoustic emission testing includes tensile testing machine, acoustic emission instrument and workstation. The acoustic emission probes are fixed on the test samples. When tensile testing machine is working, acoustic emission instrument is synchronously triggered. The acoustic emission instrument parameters can be adjusted through the workstation, and the acoustic emission signals are stored. Figure 2 shows the tensile test system with acoustic emission instrument.

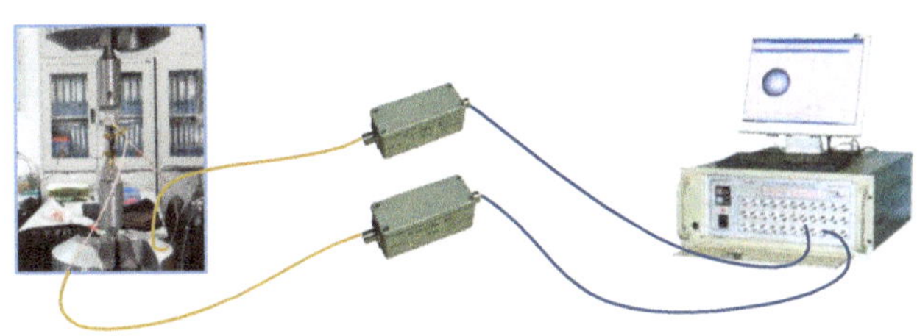

Figure 2. The system of tensile test with acoustic emission instrument.

Material tensile is performed on MTS810 material testing machine, MTS632.12C-20. The strain rate was 10^{-4}/s and test environment was 27 °C and 40% RH. Three test samples were cut from the gearbox shell, and the parameters of sample size are shown in Figure 3.

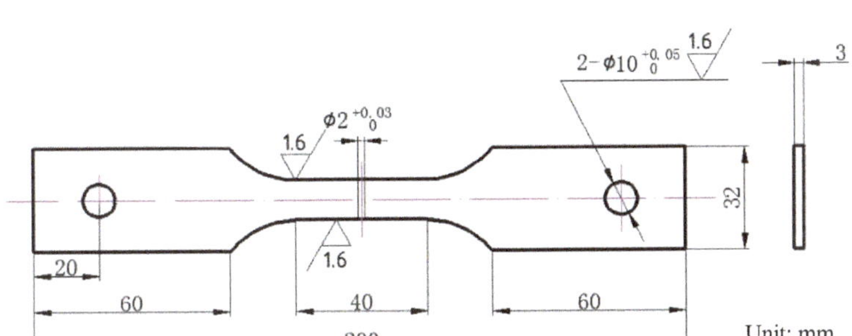

Figure 3. The sample size of aluminum alloy.

Acoustic emission instrument used in the test was SAMOS acoustic emission system of PAC Company. The sensors type was R15A; resonant frequency was 150 kHz; pre-gain amplifier was 40 dB; threshold was 45 dB; peak identification time (PDT) was 300 μs; impact identification time (HDT) was 600 μs; hit lockout time (HLT) was 1000 μs; sampling frequency was 1 MHz; and the crash file length was 2 kB.

3. Results and Discussion

3.1. Characteristic of AE Signal for the Tensile Test Process

Form Figure 4, the axial load-axial elongation curves of multiple samples are drawn in a graph. It can be seen that although there are some differences in the axial load-axial elongation curves, the transition point from elasticity to yield is more concentrated. It indicated that this process is of high reproducibility and consistent rules, so it is available to use statistical analysis method to develop the research.

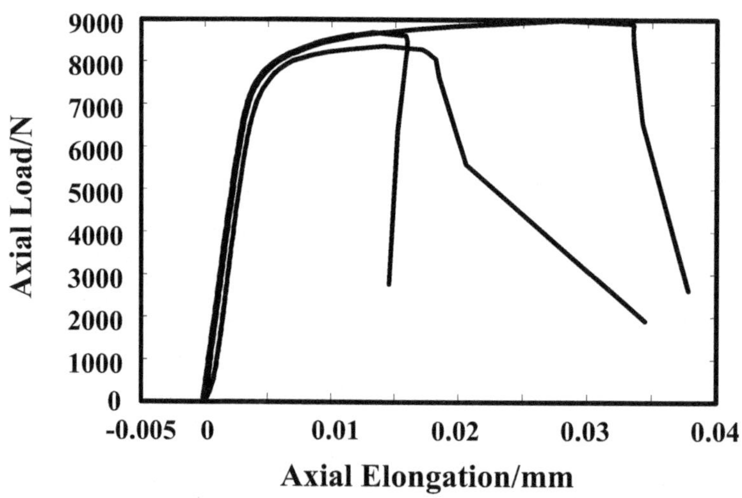

Figure 4. The axial load-axial elongation curve of multiple samples.

In the same tensile process, the AE signal and axial elongation were shown in Figure 5. It can be seen that the AE signals increased significantly at the point from the elastic stage to the yield stage, so the AE signals can characterize the process from the elastic stage to the yield stage of the material tensile damage. But the AE signal boundary between elastic stage and yield stage is not clear, and the AE signals of different samples have the same tendency but different values. The point from the elastic stage to the yield stage cannot be determined by using actual measured AE signals. But the characteristic of the AE signals of different samples at the point are having statistical regularity, and it is considered appropriate to use statistical analysis method to find the characteristic parameters of the material performance degradation.

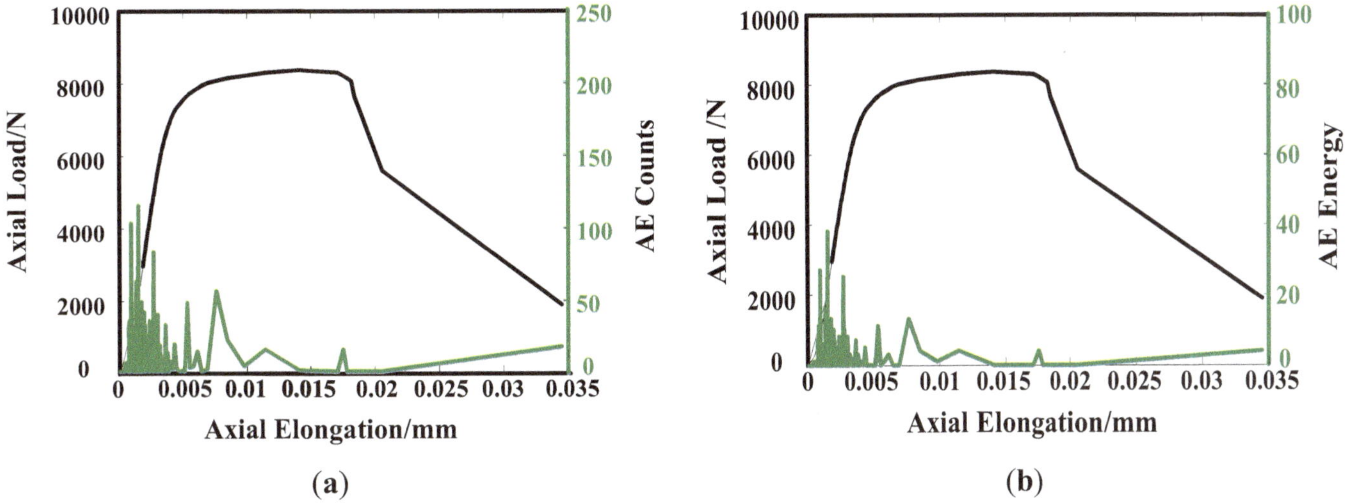

Figure 5. Acoustic emission (AE) signal and axial elongation curve. (**a**) Ring-down count and axial elongation curve; (**b**) Energy and axial elongation curve.

3.2. Characteristic Parameters of the Material Performance Degradation

The AE signals were collected from the tensile tests, and the data was built into a series, which is named C as characteristic parameter set.

$$C(1), C(2), C(3), \cdots, C(t) \tag{1}$$

$$C(t) = [c_1(t)\ c_2(t)\ c_3(t)\ c_4(t)\ c_5(t)\ c_6(t)] \tag{2}$$

$c_1(t) \sim c_6(t)$ were the AE signals at *t* time as ring-down count, energy, amplitude, rise-time, duration, and peak-frequency. For a test process, the characteristic parameter set is as follows:

$$\begin{aligned} \mathrm{C} = [\mathrm{C}(1)\ \mathrm{C}(2)\ \mathrm{C}(3)\ \cdots\ \mathrm{C}(t)\ \cdots]^T &= \begin{bmatrix} c_1(1) & c_2(1) & c_3(1) & c_4(1) & c_5(1) & c_6(1) \\ c_1(2) & c_2(2) & c_3(2) & c_4(2) & c_5(2) & c_6(2) \\ \vdots & \vdots & \vdots & \vdots & \vdots & \vdots \\ c_1(t) & c_2(t) & c_3(t) & c_4(t) & c_5(t) & c_6(t) \\ \vdots & \vdots & \vdots & \vdots & \vdots & \vdots \end{bmatrix} \\ &= [c_1\ c_2\ c_3\ c_4\ c_5\ c_6] \end{aligned} \tag{3}$$

$$c_i = [c_i(1)\ c_i(2)\ \cdots\ c_i(t)\ \cdots]^T,\ i = 1,2,3,4,5,6 \tag{4}$$

In theory, the correlation coefficient series for the same sample should be larger at beginning, and with the arrival of a bundle of data the correlation coefficient has a tendency to adjust, and then with the advent of large amounts of data the correlation coefficient would be gradually stabilized.

For the six parameters of AE signal, the correlation coefficients between two AE signals has the tendency from vibration to decline to smooth, but these result cannot be correlated with the tensile process. So the correlation coefficient series of acoustic emission signals need to be integrated to reduce the volatility of the correlation coefficient series trends. The mean and median correlation coefficient series C_{mean} and C_{median} were calculated as follows:

$$\mathrm{C}_{\mathrm{mean}} = [\mathrm{C}_{mean}(1)\ \mathrm{C}_{mean}(2)\ \mathrm{C}_{mean}(3)\ \cdots\ \mathrm{C}_{mean}(t)\ \cdots]^T \quad (5)$$

$$\mathrm{C}_{mean}(t) = \frac{\sum_{i=1}^{6}(c_{corr}^{i}(t))}{6} \quad (6)$$

$$c_{corr}^{i}(t) = \sum_{j\in C_R} \rho(c_i(t), c_j(t)),\ \mathrm{i} = 1,2,3,4,5,6,\ j \in C_R = \{1,2,3,4,5,6, \text{and } j \neq i\} \quad (7)$$

$$\rho\left(c_i(t), c_j(t)\right) = \frac{E[(c_i^t - E(c_i^t))(c_j^t - E(c_j^t))]}{D(c_i)D(c_j)},\ c_i^t = [c_i(1)\ c_i(2)\ \cdots\ c_i(t)\ ]^T,\quad c_j^t = \left[c_j(1)\ c_j(2)\ \cdots\ c_j(t)\ \right]^T \quad (8)$$

$$\mathrm{C}_{\mathrm{median}} = [\mathrm{C}_{\mathrm{median}}(1)\ \mathrm{C}_{\mathrm{median}}(2)\ \mathrm{C}_{\mathrm{median}}(3)\ \cdots\ \mathrm{C}_{\mathrm{median}}(\mathrm{t})\ \cdots]^T \quad (9)$$

$$\mathrm{C}_{\mathrm{median}}(t) = \mathrm{median}(c_{corr}^{i}(t)),\ i = 1,2,3,4,5,6 \quad (10)$$

Then the C_{mean} and C_{median} curve are in the Figure 6:

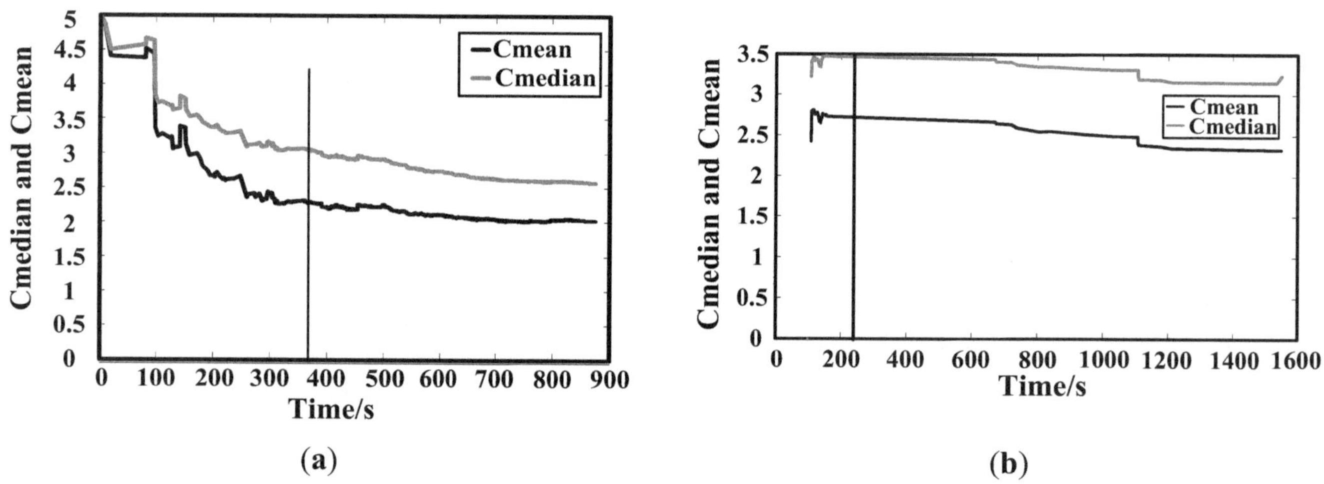

Figure 6. C_{mean} and C_{median} curve for AE signal of the tensile samples. (**a**) Sample a; (**b**) Sample b.

It can be seen that the C_{mean} and C_{median} curves have the same tendency, and the value at the transition point from elastic stage to yield stage are closer. For the 15 samples, these tendencies are same and the transition point values are closer. Therefore, the mean and median correlation coefficient series C_{mean} and C_{median} can characterize the process of the material tensile damage just as the characteristic parameter of material performance degradation, while the closely values can be used to conceive the failure criteria of the material performance degradation.

3.3. Failure Criteria of the Material Performance Degradation

To determine the failure criteria, 15 tensile tests with AE instrument have been done. Some of the statistical results of AE signal were shown in Table 1.

Table 1. The statistical results of AE signal.

Sample	C_{mean} at Transition Point	C_{median} at Transition Point
Sample 1	2.34	2.82
Sample 2	2.28	3.03
Sample 3	2.47	3.06
Sample 4	2.70	3.45
Sample 5	2.14	2.75
Sample 6	2.82	3.12

It can be seen that C_{mean} and C_{median} are at different value levels, but the each value of C_{mean} and C_{median} at the point from elastic stage to yield stage is relatively concentrated seperately. The difference value of each row of C_{mean} and C_{median} is related to the number of samples and the material dispersion of different position of casting structure. For normal distribution is a common method in the research for stress strength, the normal distribution analysis has been done for C_{mean} and C_{median} at transition point. The normal probability density functions were shown in Figure 7.

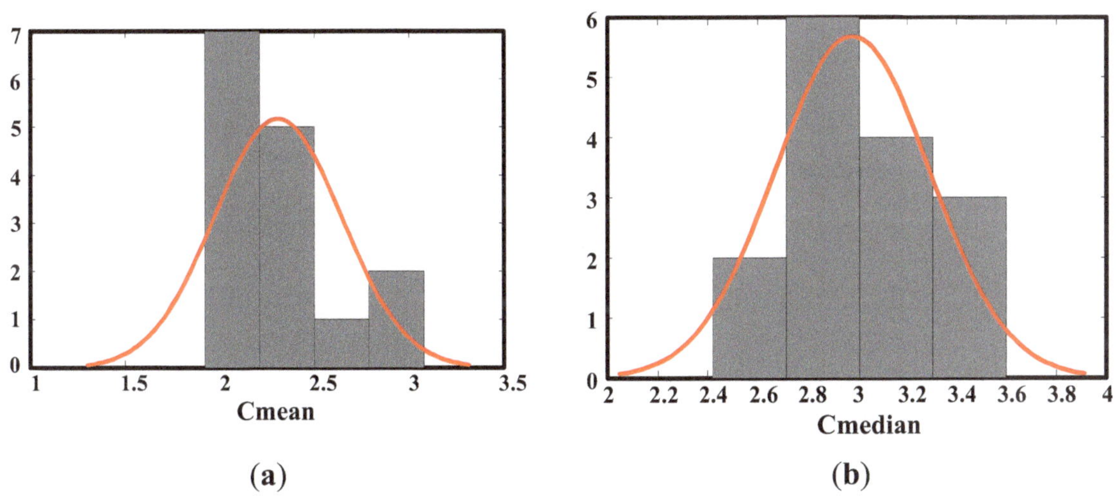

Figure 7. Histogram and normal distribution analysis of C_{mean} and C_{median} at the transition point. (**a**) Histogram and normal distribution analysis of C_{mean}; (**b**) Histogram and normal distribution analysis of C_{median}.

$$\mu_{mean} = 2.3087, \sigma^2_{mean} = 0.113, \mu_{median} = 2.9813, \sigma^2_{median} = 0.0968 \tag{11}$$

It can be seen that the value of C_{median} is more concentrated than the value of C_{mean}, so the value of C_{median} at the point from elastic stage to yield stage was chosen to be the failure criteria, which named as $C_{T\text{point}}$. $f_{T\text{point}}(C_{T\text{point}})$ is the normal probability density function, and $F_{T\text{point}}(C_{T\text{point}})$ is the normal distribution function. $f_{T\text{point}}(C_{T\text{point}})$ and $F_{T\text{point}}(C_{T\text{point}})$ were shown in Figure 8.

$$f_{T\text{point}}\left(C_{T\text{point}}\right) = \frac{1}{\sigma_{\text{median}}\sqrt{2\pi}} \exp\left[-\frac{1}{2}\left(\frac{C_{T\text{point}} - \mu_{\text{median}}}{\sigma_{\text{median}}}\right)^2\right] \tag{12}$$

$$F_{T\text{point}}\left(C_{T\text{point}}\right)=\frac{1}{\sigma_{\text{median}}\sqrt{2\pi}}\int_{-\infty}^{C_{T\text{point}}}\exp\left[-\frac{1}{2}\left(\frac{x-\mu_{\text{median}}}{\sigma_{\text{median}}}\right)^{2}\right]\text{d}x \tag{13}$$

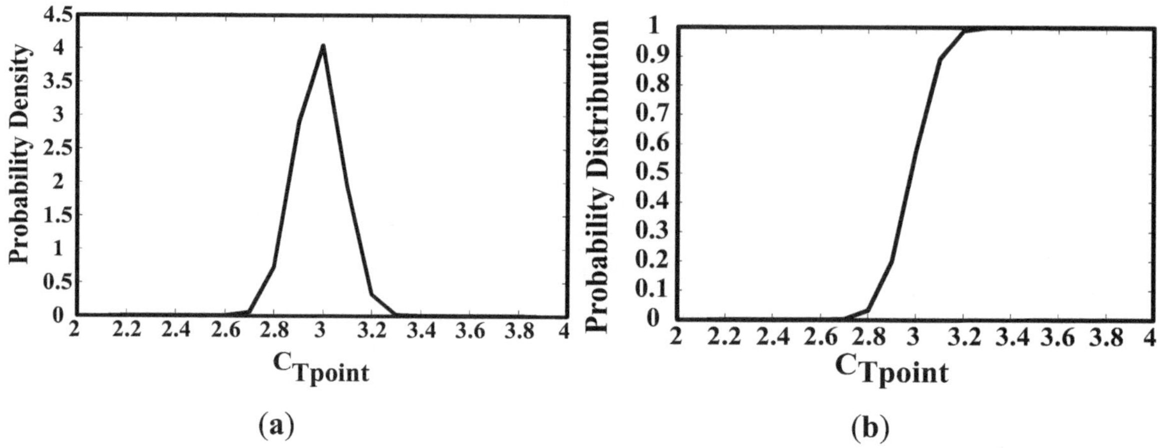

Figure 8. The normal probability density function and the normal distribution function of $C_{T\text{point}}$. (**a**) The normal probability density function of $C_{T\text{point}}$; (**b**) The normal distribution function of $C_{T\text{point}}$.

For the C_{median} series are declining with vibration at the point from elastic stage to yield stage, the probability distribution of tensile damage, named $F_T(C_{T\text{point}})$, is as follows:

$$F_T\left(C_{T\text{point}}\right)=1-F_{T\text{point}}\left(C_{T\text{point}}\right)=1-\frac{1}{\sigma_{\text{median}}\sqrt{2\pi}}\int_{-\infty}^{C_{T\text{point}}}\exp\left[-\frac{1}{2}\left(\frac{x-\mu_{\text{median}}}{\sigma_{\text{median}}}\right)^{2}\right]\text{d}x \tag{14}$$

Based on the above analysis, in the research of performance degradation for tensile damage, the median value of correlation coefficient series, named C_{median}, would be choose as characteristic parameters of the material performance degradation. The result of the failure criteria of the material performance degradation obeyed the normal distribution N (μ_{median}, σ^2_{median}).

4. Conclusions

In this paper, the acoustic emission instrument was used for the performance degradation research of tensile damage of a high-strength aluminum alloy. Through data analysis and transformation of AE signals, the characteristics parameter of the material performance degradation C_{median} has been determined, the failure criteria of the material performance degradation has been developed and obeyed normal distribution N (μ_{median}, σ^2_{median}). This result will provide basic research for further life prediction research.

The material researched in this paper is used for the gearbox shell of high-speed train. This research will provide non-destructive and real-time performance degradation analysis as a new method and approach, and also provide a new application background of the acoustic emission instrument. The result of this work expected to be widely used in other material performance degradation.

Acknowledgments

The authors would like to acknowledge the financial support provided by the National Natural Science Foundation of China (Grant No. 61273205), the Fundamental Research Funds for the Central Universities of China (Grant No. FRF-SD-12-028A), the 111 Project (Grant No. B12012).

Author Contributions

Yibo Ai and Weidong Zhang conceived and designed the study. Yibo Ai and Chang Sun performed the experiments. Hongbo Que provided the basic performance analysis of the material. Yibo Ai made the data analysis. Yibo Ai and Chang Sun wrote the paper. All authors read and approved the manuscript.

Conflicts of Interest

The authors declare no conflict of interest.

References

1. Chen, C.Y. Innovation Strategy Research for Locomotive and Rolling Stock Industry of China. Master's Thesis, Beijing Jiaotong University, Beijing, China, 2007.
2. Li, Y. Cooperation Agreement of independent innovation Joint Action Plan for China high-speed train. *Technol. Ind. China* **2008**, *3*, 90.
3. Zhang, W.D.; Zhang, X.W.; Yang, B.; Ai, Y.B. Damage characterization and recognition of aluminum alloys based on acoustic emission signal. *J. Univ. Sci. Technol. Beijing* **2013**, *35*, 626–633.
4. Wang, B.M. *Overall System and Bogie of High-Speed Motor Train Units*; Southwest Jiaotong University Press: Chengdu, China, 2008.
5. Li, X.J. CRH bullet trains overview. *Railw. Tech. Superv.* **2007**, *9*, 26–28.
6. Ding, S.B.; Liu, F.J. *New Non-Destructive Testing Technology and Application*; Higher Education Press: Beijing, China, 2012.
7. Lugo, M.; Jordon, J.B.; Horstemeyer, M.F.; Tschopp, M.A.; Harris, J.; Gokhale, A.M. Quantification of damage evolution in a 7075 aluminum alloy using an acoustic emission technique. *Mater. Sci. Eng. A* **2011**, *528*, 6708–6714.
8. Behrens, B.A.; Santangelo, A.; Buse, C. Acousic emission technique for online monitoring during cold forging of steel components: A promising approach for online crack detection in metal forging processes. *Prod. Eng. Res. Dev.* **2013**, *7*, 423–432.
9. Han, Z.Y.; Luo, H.Y.; Zhang, Y.B.; Cao, J.W. Effects of micro-structure on fatigue crack propagation and acoustic emission behaviors in a micro-alloyed steel. *Mater. Sci. Eng. A* **2013**, *559*, 534–542.
10. Maillet, E.; Morscher, G.S. Waveform-based selection of acoustic emission events generated by damage in composite materials. *Mech. Syst. Signal Process.* **2015**, *52*, 217–227.
11. Roca, A.S.; Fals, H.C.; Fernández, J.B.; Macías, E.J.; Adán, F.S. New stability index for short circuit transfer mode in GMAW process using acoustic emission signals. *Sci. Technol. Weld. Join.* **2007**, *12*, 460–466.

12. Roca, A.S.; Fals, H.C.; Fernández, J.B.; Macías, E.J.; Parte, M.P. Artificial neural networks and acoustic emission applied to stability analysis in gas metal arc welding. *Sci. Technol. Weld. Join.* **2009**, *14*, 117–124.
13. Cholette, M.E.; Djurdjavonic, D. Degradation modeling and monitoring of machines using operation-specific hidden Markov models. *IIE Trans.* **2014**, *46*, 1107–1123.
14. Hong, S.; Wang, B.Q.; Li, G.Q.; Hong, Q. Performance degradation assessment for bearing based on ensemble empirical mode composition and Gaussian mixture model. *J. Vibr. Acoust.* **2014**, *136*, 1006–1014.
15. Savage, J.J.; Seecharan, T.S.; Son, Y.K. Probability-based prediction of degrading dynamic systems. *J. Mech. Des*. **2013**, *135*, 2–15.
16. Dobrica, M.B.; Fillon, M. Performance degradation in scratched journal bearings. *Tribol. Int.* **2012**, *51*, 1–10.
17. China Standardization Committee. *GB228–2002 Tensile Test for Metals Specimens*; China Standard Press: Beijing, China, 2002.
18. Yang, M.W. *Acoustic Emission Testing*; Machinery Industry Press: Beijing, China, 2005.

Twinning-Detwinning Behavior during Cyclic Deformation of Magnesium Alloy

Soo Yeol Lee [1,*], Huamiao Wang [2] and Michael A. Gharghouri [3]

[1] Department of Materials Science and Engineering, Chungnam National University, Daejeon 305-764, Korea

[2] Materials Science and Technology, Los Alamos National Laboratory, Los Alamos, NM 87544, USA; E-Mail: wanghm@lanl.gov

[3] Canadian Neutron Beam Centre, Canadian Nuclear Laboratories, Chalk River, ON K0J 1J0, Canada; E-Mail: Michael.Gharghouri@cnl.ca

* Author to whom correspondence should be addressed; E-Mail: sylee2012@cnu.ac.kr

Academic Editor: Klaus-Dieter Liss

Abstract: *In situ* neutron diffraction has been used to examine the deformation mechanisms of a precipitation-hardened and extruded Mg-8.5wt.%Al alloy subjected to (i) compression followed by reverse tension (texture T1) and (ii) tension followed by reverse compression (texture T2). Two starting textures are used: (1) as-extruded texture, T1, in which the basal pole of most grains is normal to the extrusion axis and a small portion of grains are oriented with the basal pole parallel to the extrusion axis; (2) a reoriented texture, T2, in which the basal pole of most grains is parallel to the extrusion axis. For texture T1, the onset of extension twinning corresponds well with the macroscopic elastic-plastic transition during the initial compression stage. The non-linear macroscopic stress/strain behavior during unloading after compression is more significant than during unloading after tension. For texture T2, little detwinning occurs after the initial tension stage, but almost all of the twinned volumes are detwinned during loading in reverse compression.

Keywords: magnesium alloy; deformation; twinning; detwinning; neutron diffraction

---

1. Introduction

Many investigations have been devoted to understanding the deformation behavior of magnesium and its alloys, because of their potential applications for lightweight materials in the aircraft and automotive industries, and for portable electronic devices [1,2]. Magnesium alloys have poor formability at room temperature, which can be attributed to the limited number of available slip systems [3,4]. <a> slip with a 1/3<11.0> Burgers vector on the close-packed (00.2) basal plane is the primary slip system in hexagonal close-packed (HCP) magnesium. Non-basal <a> slip on the {10.0} prismatic and {10.1} pyramidal plane has often been observed at higher stresses. While none of these slip modes can accommodate deformation along the c-axis, deformation twinning can provide limited deformation along the c-axis. {10.2}<10.1> extension twinning is commonly found during plastic deformation at room temperature in favorably oriented grains relative to the applied loading direction, leading to tension-compression yield asymmetry and strong plastic anisotropy in magnesium alloys [5–8]. When deformation twinning occurs, the lattice is reoriented approximately 86.3° relative to the parent lattice, resulting in pronounced changes in the crystallographic texture during deformation [9–12]. *In situ* neutron diffraction experiments have been employed extensively to study the plastic deformation behavior of magnesium alloys. The technique provides information on the distribution of internal stresses and strains among the various crystallographic orientations, as well as on bulk texture evolution caused by twinning [13–19].

In the current work, neutron diffraction is used to study the plastic deformation behavior of a precipitation-hardened and extruded Mg-8.5wt.%Al alloy. Lattice strains and diffraction peak intensities for several grain orientations are monitored *in situ* during deformation to examine the evolution of the stress state and the occurrence of twinning and detwinning in various grain orientations. The loading paths consist of (i) compression followed by reverse tension and (ii) tension followed by reverse compression.

2. Experimental Section

2.1. Materials

The extruded Mg-8.5wt.%Al alloy used in this study was solution treated and aged, resulting in equiaxed grains with an average size of ~60 μm. Details of the sample preparation are provided elsewhere [13]. The bulk crystallographic texture was measured using the E3 neutron diffractometer of the Canadian Neutron Beam Centre, located at the National Research Universal (NRU) reactor at Canadian Nuclear Laboratories, Canada. The orientation distribution function (ODF) for each sample was determined from the {10.0}, {00.2}, {10.1} and {10.2} pole figures. Two starting textures were used: (1) as-extruded texture, T1 (Figure 1a), in which the basal pole of most grains is normal to the extrusion axis and a very small portion of grains are oriented with the basal pole parallel to the extrusion axis; (2) a reoriented texture, T2 (Figure 1b), in which the basal pole of most grains is parallel to the extrusion axis. The reoriented texture, T2, was obtained by compressing the T1 texture along the extrusion direction to a strain of ~9%, followed by annealing at a suitable temperature [18]. Therefore, {10.2} extension twinning could be easily activated under compression along the extrusion direction for T1, and under tension along the extrusion direction for T2.

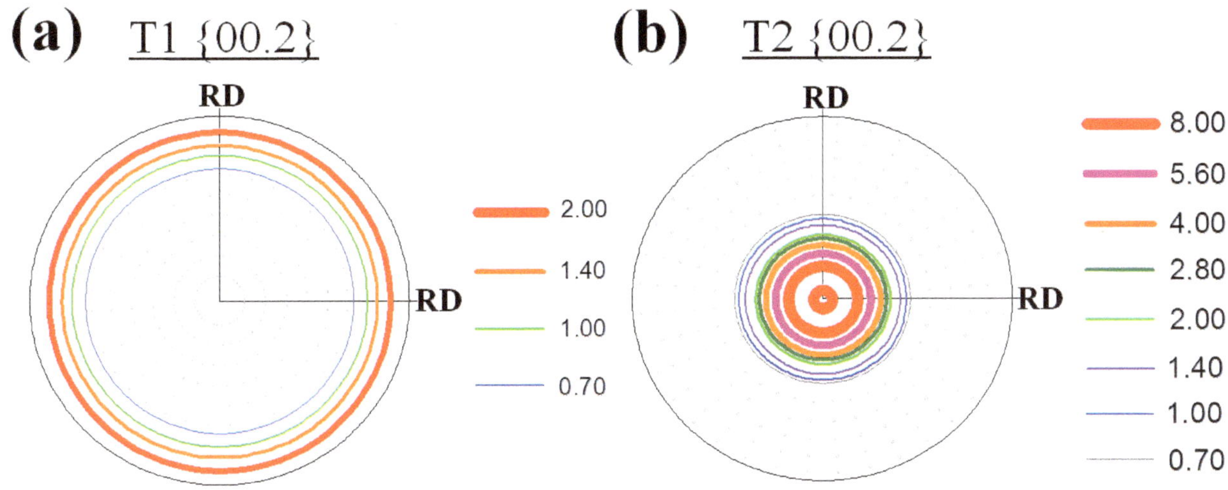

Figure 1. Initial textures of (**a**) T1 and (**b**) T2 samples determined by neutron diffraction. The center of each pole figure corresponds to the extrusion direction (ED), RD is the radial direction of the extruded bar.

2.2. Neutron Diffraction Strain Measurements

Neutron diffraction strain measurements were conducted on the L3 neutron diffractometer of the Canadian Neutron Beam Centre, Canadian Nuclear Laboratories. T1 and T2 samples were deformed in compression followed by reverse tension and in tension followed by reverse compression, respectively. Diffraction peaks for several grain families, characterized by the crystallographic plane which was normal to the loading direction, were acquired *in situ* during deformation, yielding interplanar spacings (*d*-spacings) parallel to the loading direction for each grain family as a function of applied load. The lattice strains for each family were calculated using the following equation:

$$\varepsilon_{\mathrm{hk.l}} = (d_{hk.l} - d_{hk.l}^{0})/d_{hk.l}^{0} \tag{1}$$

where $d_{hk.l}^{0}$ and $d_{hk.l}$ are the *d*-spacings of the $\{hk.l\}$ family of planes in the unloaded and loaded conditions, respectively.

3. Results and Discussion

The macroscopic stress-strain response of the T1 sample subjected to compression followed by reverse tension is shown in Figure 2.

The material yields at ~−100 MPa, at which the elastic-plastic transition is well underway, though plasticity is not fully developed. The non-linear macroscopic stress/strain behavior during unloading after compression is more significant than during unloading after tension.

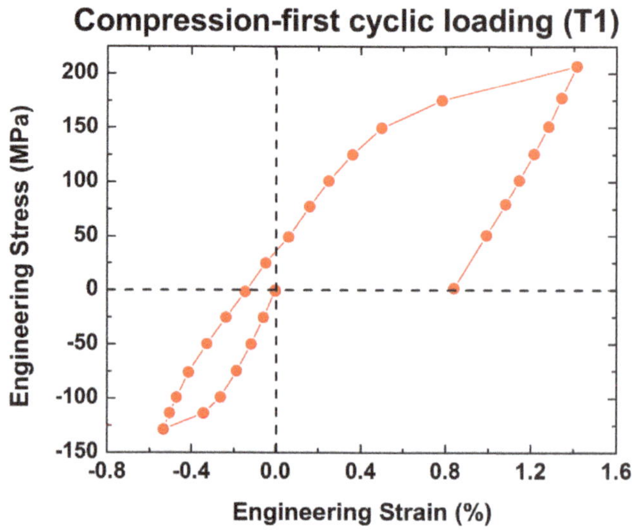

Figure 2. Macroscopic stress-strain response for the T1 sample. The symbols correspond to points in the loading history at which diffraction data were acquired.

Figure 3 shows lattice strain evolution for four different families as a function of applied stress for the T1 sample subjected to compression followed by reverse tension (Figure 2). The corresponding integrated intensity variations for all measured reflections are provided in Figure 4 [17]. A detailed discussion of the intensity variations in Figure 4 is provided in [17], based on which it is possible to identify the stress intervals over which twinning and detwinning occur. The twinning and detwinning intervals identified in Figure 3 are based on this previous analysis.

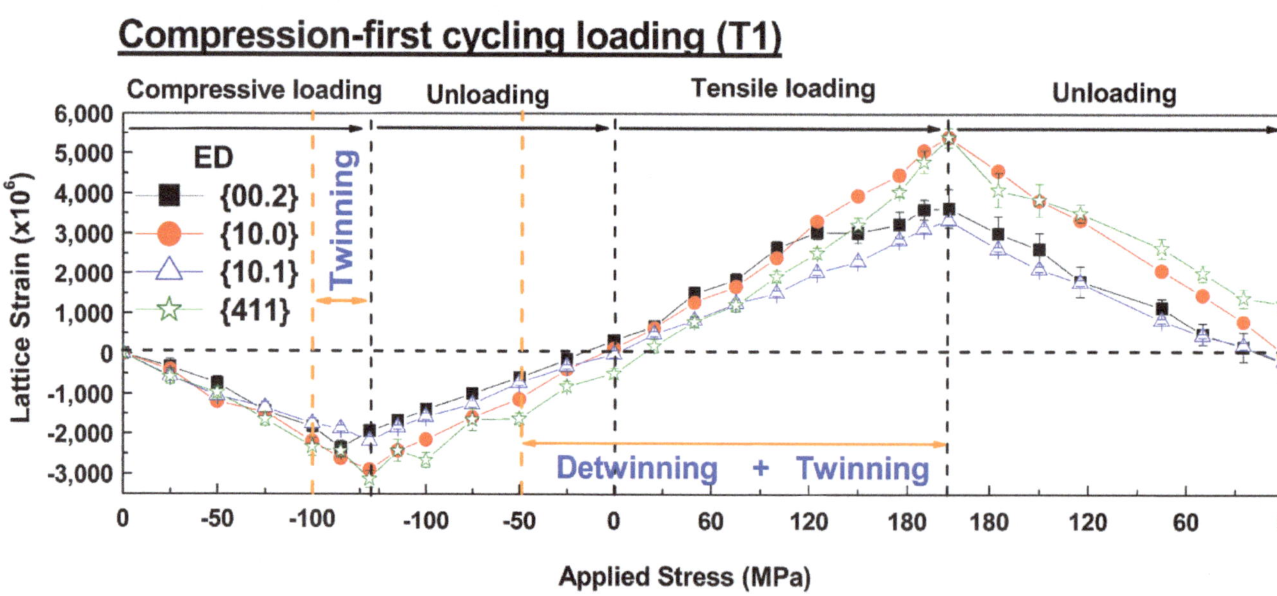

Figure 3. Lattice strain evolution for various grain families as a function of applied stress during the deformation shown in Figure 2. The *{hk.l}* plane normal is parallel to the applied loading direction, which is parallel to the extrusion direction (ED). The angles between the c-axis and the loading direction for the {00.2}, {10.0} and {10.1} families are 0°, 90° and 61.9°, respectively.

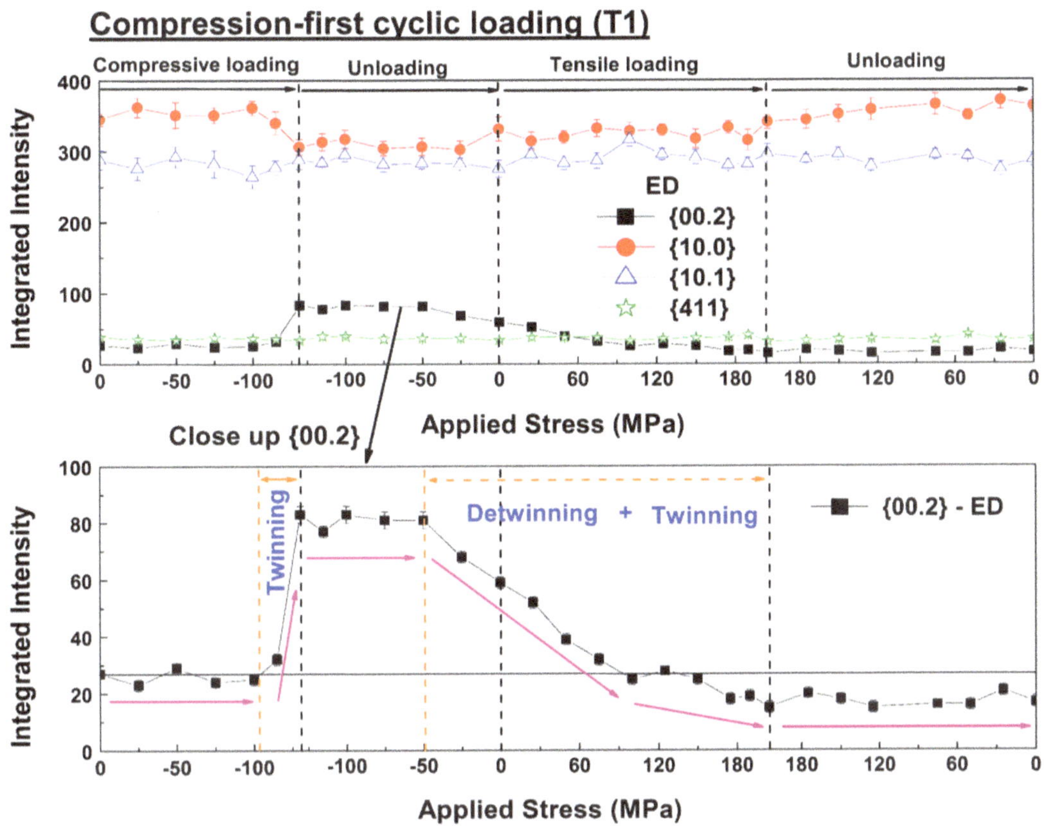

Figure 4. Intensity variations for various grain families as a function of applied stress during the deformation shown in Figure 2 [17].

In Figure 3, during the initial compressive loading step, the lattice strain increases linearly with applied stress for all four grain families up to ~−50 MPa. Beyond this stress, the lattice strain in the {10.1} family increases more slowly with applied stress compared with the other orientations, indicating that this family deforms plastically. The deformation mechanism is most likely basal <a> slip, which has the lowest critical resolved shear stress (CRSS) of all the commonly observed deformation modes, and for which the {10.1} family is favorably oriented. As the load increases beyond ~−100 MPa, the {00.2} and {10.0} lattice strains deviate from one another, corresponding to an increase in the {00.2} intensity and a concurrent decrease in the {10.0} intensity (see Figure 4 [17]). Despite being favorably oriented for {10.2}<10.1> extension twinning in compression, the strain in the {10.0} family increases linearly with the applied stress during the initial compressive loading step, resulting in almost zero residual strain upon unloading. On the other hand, the lattice strain in the {00.2} family, which is twin-related to the {10.0} grains, experiences relaxation during the final loading step, resulting in a slightly tensile residual strain at the end of unloading. The residual strain for the soft {10.1} family is slightly compressive, or zero. The strain in the β-$Mg_{17}Al_{12}$ precipitates ({411} reflection) increase linearly during loading, with no evidence of relaxation, resulting in relatively large compressive residual strains upon unloading. During unloading after compression, it was found that about 40% of the twinned volume detwins [17]. This detwinning behavior is thought to contribute significantly to the non-linear behavior observed in Figure 2.

During reverse loading in tension, the intensity data in Figure 4 [17] show that detwinning continues up to an applied stress of ~+100 MPa, after which the {00.2} minority grains undergo

{10.2} extension twinning, as revealed by a decrease in the {00.2} intensity beyond ~100 MPa. The {00.2} and {10.1} families, favorably oriented for extension twinning and basal slip respectively, show stress relaxation during reverse tensile loading, resulting in slightly compressive residual strains upon unloading. Conversely, the lattice strain in the {100} family, which is not favorably oriented for basal slip or extension twinning, varies linearly throughout the test, resulting in negligible residual strain at the end of the test. The {411} precipitates accumulate high levels of lattice strain during the reverse tensile loading, but, unlike the {10.0} family, the precipitates show large tensile residual strains upon final unloading.

Figure 5 shows the macroscopic stress-strain response for the T2 sample subjected to tension followed by reverse compression. The initial stress-strain response is linear up to an applied stress of ~75 MPa, after which the slope decreases as the material starts to yield. During unloading after tension and reloading in compression, the stress-strain response is clearly non-linear. The material undergoes general yielding in compression at ~−50 MPa. The unloading behavior at the end of the test is also clearly non-linear, but the effect is less significant than the non-linear behavior after compression for T1 (Figure 2).

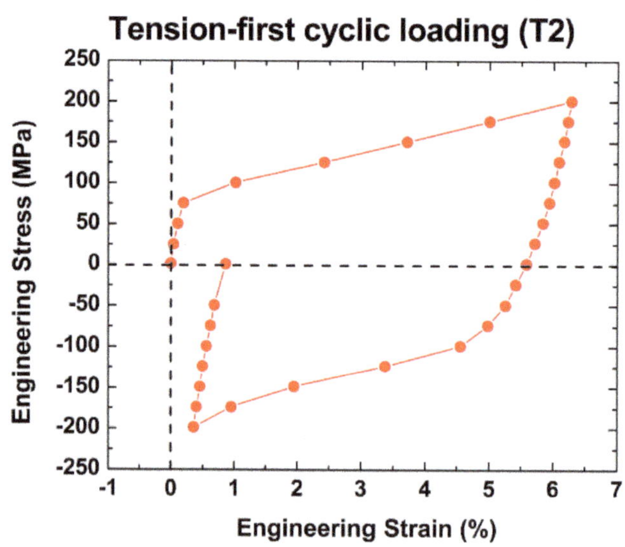

Figure 5. Macroscopic stress-strain response for the T2 sample. The symbols correspond to points in the loading history at which diffraction data were acquired.

Lattice strain and integrated intensity variations for four different grain orientations as a function of applied stress are shown in Figure 6 for the T2 sample subjected to tension followed by reverse compression.

During initial loading in tension, the lattice strain varies linearly with applied stress for all four grain orientations up to ~50 MPa showing the same slope, as expected based on the near elastic isotropy of magnesium. Beyond ~50 MPa, the {10.2} and {10.3} strains increase more slowly with stress, indicating that these grains undergo plastic deformation, most likely by basal <a> slip (Figure 6a), which is the easiest slip system to activate in Mg-Al alloys, and for which these grain families are favorably oriented. Beyond ~75 MPa, the lattice strain behavior of the {00.2} family deviates from its initial linear behavior, corresponding to a decrease in intensity for the {00.2} peak in Figure 6b, and macroscopic yielding in Figure 5. The {00.2} family, which represents a large fraction of the microstructure, has the basal pole aligned 0° relative to the applied loading direction. It is thus favorably oriented for {10.2} extension twinning in tension, which results in the observed decrease in intensity. Likewise, the changes in the {11.0} and {10.3} intensities for 75 < σ (MPa) < 200 are due to {10.2} extension twinning in the {10.3} families, as the two families are twin-related.

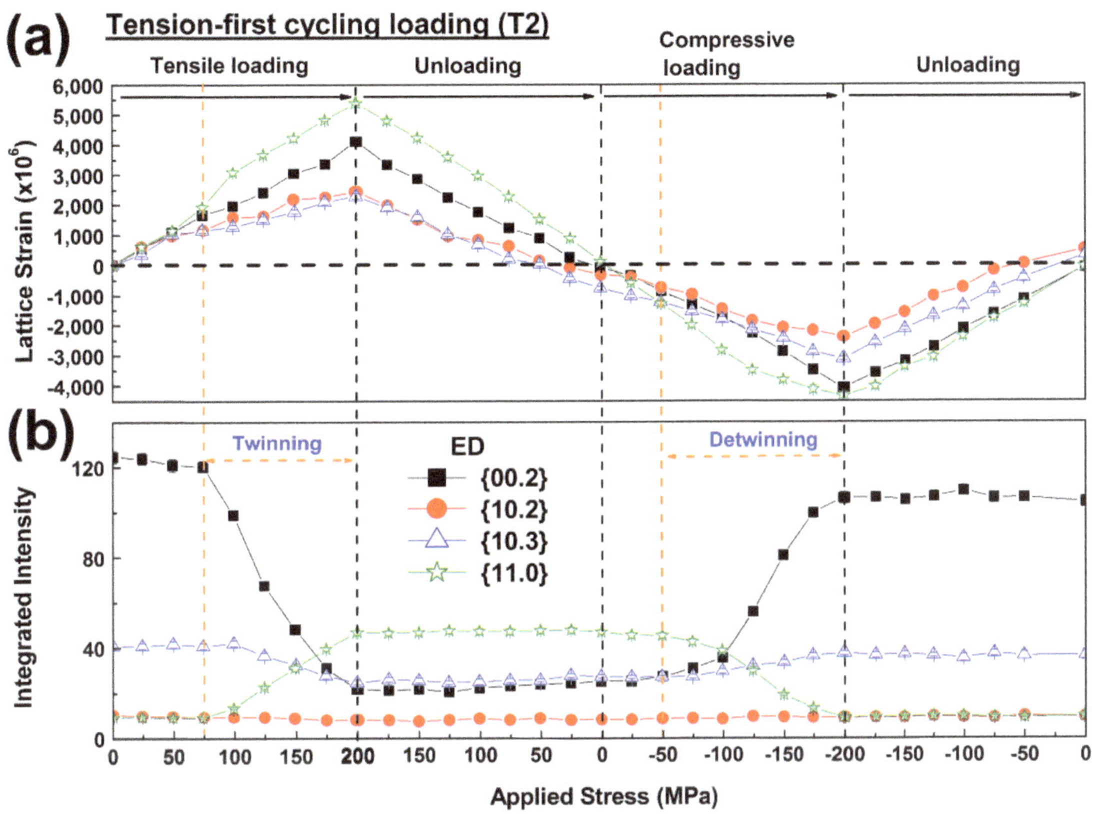

Figure 6. (**a**) Lattice strain evolution and (**b**) intensity variations for various grain families as functions of applied stress during the deformation shown in Figure 5. The {*hk.l*} plane normal is parallel to the applied loading direction, which is parallel to the extrusion direction (ED).

Figure 6a shows that the {10.2}, {10.3} and {00.2} families, which deform plastically by either basal slip or extension twinning during the initial tensile portion of the test, undergo stress relaxation, resulting in compressive residual strains upon unloading. On the other hand, the {11.0} family, which is unfavorably oriented for both extension twinning and basal slip, accumulates much higher lattice strains during tensile loading, resulting in slightly tensile residual strains at the end of unloading. In the compression stage of the test, the {10.2} and {10.3} families, which are favorably oriented for basal slip, accumulate relatively low levels of lattice strain during loading, resulting in small tensile residual strains at the end of the test. Conversely, the residual strain in the hard {00.2} family is slightly compressive, or zero. The {11.0} family presents an interesting case—despite being favorably oriented for extension twinning in compression, it still accumulates large lattice strains during loading, resulting in slightly negative (or zero) residual lattice strain at the end of the test. This lack of a significant relaxation effect associated with twinning is consistent with the behavior of the {10.0} family during the initial compressive loading step in T1, and is in marked contrast to the behavior of the {00.2} family in both textures, in which strong relaxation effects are associated with twinning. This difference in behavior between the {10.0}/{11.0} and {00.2} orientations may be due to the fact that extension twinning is activated by compression normal to the c-axis in the {10.0} and {11.0} family, but by tension parallel to the c-axis in the {00.2} family.

During unloading after tension, the intensities of all four reflections change little (the {00.2} and {10.3} intensities increase slightly), indicating that the twinned material generated during tension does not undergo significant detwinning. During reverse loading in compression, however, the twinned volume within the {00.2} family, which contributes to the {10.0} signal, is favorably oriented to detwin in compression. Thus, in reverse compression, detwinning starts at an applied stress of ~−50 MPa and continues up to an applied stress of −200 MPa, as revealed by the increase in the {00.2} intensity (Figure 6b). However, the {00.2} intensity at the start of the test is not fully recovered at −200 MPa. During unloading after the compressive portion of the test, the {00.2} peak intensity does not change, suggesting that the non-linear behavior observed in Figure 5 is not related to the behavior of deformation twins, but rather to dislocation phenomena. This behavior is likely due to the fact that almost all traces of twinning are gone when the highest compressive stress is reached, such that little material is favorably oriented for either twinning or detwinning when the final unloading starts.

4. Conclusions

In situ neutron diffraction has been used to investigate the plastic deformation behavior of Mg-8.5wt.%Al alloy subjected to compression followed by reverse tension (texture T1) and tension followed by reverse compression (texture T2). For texture T1, the onset of extension twinning corresponds well with the macroscopic elastic-plastic transition during the initial compression stage. The non-linear macroscopic stress/strain behavior during unloading after compression is more significant than during unloading after tension. For texture T2, little detwinning occurs after the initial tension stage, but almost all of the twinned volumes are recovered during loading in reverse compression. The development of residual strains in the various grain orientations during unloading after compression is strongly influenced by plastic anisotropy at the grain level. The residual strains observed in each grain family can be largely rationalized based on the ease of activation of basal slip and extension twinning.

Acknowledgments

This work was supported by the National Research Foundation of Korea (NRF) grant funded by the Korean government (MSIP) (No.2013R1A4A1069528, No.2013R1A1A1076023).

Author Contributions

S.Y.L. wrote the initial draft of the manuscript; S.Y.L., H.W., M.A.G. analyzed the data; All co-authors contributed to the interpretation of the data.

Conflicts of Interest

The authors declare no conflict of interest.

References

1. Avedesian, M.M.; Baker, H. *Magnesium and Magnesium Alloys*; ASM Specialty Handbook; ASM International: Materials Park, OH, USA, 1999.
2. Wang, C.; Han, P.; Zhang, L.; Zhang, C.; Yan, X.; Xu, B.S. The strengthening effect of Al atoms into Mg-Al alloy: A first-principles study. *J. Alloy. Comp.* **2009**, *482*, 540–543.
3. Roberts, C.S. *Magnesium and Its Alloys*; John Wiley & Sons, Inc.: New York, NY, USA, 1960.
4. Muránsky, O.; Carr, D.G.; Barnett, M.R.; Oliver, E.C.; Šittner, P. Investigation of deformation mechanisms involved in the plasticity of AZ31 Mg alloy: *In situ* neutron diffraction and EPSC modelling. *Mater. Sci. Eng. A* **2008**, *496*, 14–24.
5. Partridge, P.G. The crystallography and deformation modes of hexagonal close-packed metals. *Metall. Rev.* **1967**, *12*, 169–194.
6. Agnew, S.R.; Tomé, C.N.; Brown, D.W.; Holden, T.M.; Vogel, S.C. Study of slip mechanisms in a magnesium alloy by neutron diffraction and modeling. *Scripta Mater.* **2003**, *48*, 1003–1008.
7. Jain, J.; Poole, W.J.; Sinclair, C.W.; Gharghouri, M.A. Reducing the tension-compression yield asymmetry in a Mg-8Al-0.5Zn alloy via precipitation. *Scripta Mater.* **2010**, *62*, 301–304.
8. Hong, S.G.; Park, S.H.; Lee, C.S. Role of {10−12} twinning characteristics in the deformation behavior of a polycrystalline magnesium alloy. *Acta Mater.* **2010**, *58*, 5873–5885.
9. Clausen, B.; Tomé, C.N.; Brown, D.W.; Agnew, S.R. Reorientation and stress relaxation due to twinning: Modeling and experimental characterization for Mg. *Acta Mater.* **2008**, *56*, 2456–2468.
10. Proust, G.; Tomé, C.N.; Jain, A.; Agnew, S.R. Modeling the effect of twinning and detwinning during strain-path changes of magnesium alloy AZ31. *Int. J. Plasticity* **2009**, *25*, 861–880.
11. Muránsky, O.; Barnett, M.R.; Luzin, V.; Vogel, S. On the correlation between deformation twinning and Luders-like deformation in an extruded Mg alloy: *In situ* neutron diffraction and EPSC.4 modelling. *Mater. Sci. Eng. A* **2008**, *527*, 1383–1394.
12. Park, S.H.; Hong, S.G.; Lee, C.S. In-plane anisotropic deformation behavior of rolled Mg-3Al-1Zn alloy by initial {10−12} twins. *Mater. Sci. Eng. A* **2013**, *570*, 149–163.
13. Gharghouri, M.A.; Weatherly, G.C.; Embury, J.D.; Root, J. Study of the mechanical properties of Mg-7.7at.% Al by *in situ* neutron diffraction. *Phil. Mag. A* **1999**, *79*, 1671–1695.
14. Brown, D.W.; Agnew, S.R.; Bourke, M.A.M.; Holden, T.M.; Vogel, S.C.; Tomé, C.N. Internal strain and texture evolution during deformation twinning in magnesium. *Mater. Sci. Eng. A* **2005**, *399*, 1–12.
15. Wu, L.; Agnew, S.R.; Brown, D.W.; Stoica, G.M.; Clausen, B.; Jain, A.; Fielden, D.E.; Liaw, P.K. Internal stress relaxation and load redistribution during the twinning-detwinning-dominated cyclic deformation of a wrought magnesium alloy, ZK60A. *Acta Mater.* **2008**, *56*, 3699–3707.
16. Muránsky, O.; Barnett, M.R.; Carr, D.G.; Vogel, S.C.; Oliver, E.C. Investigation of deformation twinning in a fine-grained and coarse-grained ZM20 Mg alloy: Combined in situ neutron diffraction and acoustic emission. *Acta Mater.* **2010**, *58*, 1503–1517.
17. Lee, S.Y.; Gharghouri, M.A.; Root, J.H. Plastic deformation of magnesium alloy subjected to compression-first cyclic loading. In Proceeding of the 140th TMS Annual Meeting & Exhibition, San Diego, CA, USA, 27 February–3 March 2011; pp. 595–598.

18. Lee, S.Y.; Gharghouri, M.A. Pseudoelastic behavior of magnesium alloy during twinning-dominated cyclic deformation. *Mater. Sci. Eng. A* **2013**, *572*, 98–102.
19. Lee, S.Y.; Wang, H.; Gharghouri, M.A.; Nayyeri, G.; Woo, W.; Shin, E.; Wu, P.D.; Poole, W.J.; Wu, W.; An, K. Deformation behavior of solid-solution-strengthened Mg-9 wt.% Al alloy: *In situ* neutron diffraction and elastic-viscoplastic self-consistent modeling. *Acta Mater.* **2014**, *73*, 139–148.

13

A Mathematical Formulation to Estimate the Effect of Grain Refiners on the Ultimate Tensile Strength of Al-Zn-Mg-Cu Alloys

Halil Ibrahim Kurt [1,*], Murat Oduncuoglu [1,†] and Mehmet Kurt [2,†]

[1] Technical Sciences, University of Gaziantep, Gaziantep 27310, Turkey;
E-Mail: oduncuoglu@gmail.com

[2] Department of Metallurgical and Materials Engineering, Engineering Faculty, Fırat University, Elazığ 23119, Turkey; E-Mail: mehmetkurt_27@mynet.com

[†] These authors contributed equally to this work.

[*] Author to whom correspondence should be addressed; E-Mail: hiakurt@gmail.com

Academic Editor: Hugo F. Lopez

Abstract: In this study, the feed-forward (FF) neural networks (NNs) with back-propagation (BP) learning algorithm is used to estimate the ultimate tensile strength of unrefined Al-Zn-Mg-Cu alloys and refined the alloys by Al-5Ti-1B and Al-5Zr master alloys. The obtained mathematical formula is presented in great detail. The designed NN model shows good agreement with test results and can be used to predict the ultimate tensile strength of the alloys. Additionally, the effects of scandium (Sc) and carbon (C) rates are investigated by using the proposed equation. It was observed that the tensile properties of Al-Zn-Mg-Cu alloys improved with the addition of 0.5 Sc and 0.01 C wt.%.

Keywords: Al-Zn-Mg-Cu alloys; ultimate tensile strength; modeling; ANN

---

1. Introduction

Al-Zn-Mg-Cu alloys, using ultra-high strength Al-based alloys, are widely used in aviation, automotive, and aerospace industries due to of their low density, high specific strength, toughness, and fatigue durability [1–7]. The mechanical properties of the alloys can be improved by some important factors such as: minimizing inclusions, applying thermo-mechanical treatments and changing the composition [8]. One of the most important methods to increase the strength is the refining of the grain size, which is described by the well-known Hall-Petch Equation [9]:

$$\sigma_y = \sigma_0 + kd^{-1/2} \quad (1)$$

where σ_y is the yield stress, σ_0 a friction stress, d the grain size and k a constant.

The existing knowledge of grain refinement can be divided into three basic methods: mechanical, chemical, and thermal, whereas the addition of master alloys (chemical) is the most economical one. In the chemical method, the addition of master alloys promotes nucleation and hinders growth [10–14]. Grain refiners used in aluminum (Al) and its alloys are Al-Ti, Al-B, Al-Ti-B, Al-Ti-C, Al-Zr, Al-Sc, and Al-Sr [15–20]. A fine equiaxed grain structure leads to better mechanical properties. Moreover, grain refinement can also bring other advantages, such as enhanced machinability, good surface finish, resistance to hot tearing, high toughness and high yield strength [21,22]. The type and addition level of these master alloys also affect the mechanical properties of alloys. The relation among grain size, changing operation parameters and strength are quite complex and non-linear. Therefore, it is very important to select the correct master alloy to obtain the maximum strength. As an experiment, this phenomenon is costly and time consuming. Therefore, an attempt has been made to model the complex grain refinement phenomena in Al-Zn-Mg-Cu alloys.

Successful modelings of artificial neural networks (ANN) have been reported in materials science [23]. Reddy *et al.* [24] used the feed forward neural network (FFNN) with back-propagation (BP) learning algorithm to predict the grain size as a function of titanium (Ti) and boron (B) addition levels, and holding time during grain refinement of Al–7Si alloy. The authors stated that the grain size of Al–7Si alloy, with good learning precision and generalization, can be predicted with the FFNN model. Rashidi *et al.* [25] modeled the effect of the process parameters, namely current density, saccharin concentration, and bath temperature on the grain size of nano-crystalline nickel coating by using a feed-forwarded multilayer perceptron ANN framework. They reported that there is a remarkable agreement between the model prediction and the experimental observation. Tofigh *et al.* [26] used the FFNN in prediction of hardness, tensile and compressive yield stress, ultimate tensile strength, and elongation percentage of Al alloy reinforced alumina nano-particles. The authors notified that the prediction ability of the model is in good agreement with experimental data.

The aim of this work is to obtain an explicit formulation of the ultimate tensile strength as a function of complex master alloys and, theoretically, to investigate the effect of scandium (Sc) and carbon (C) rates on the ultimate tensile strength of the alloys by using the obtained equation.

2. Background of Neural Networks

ANNs are computationally in the context of artificial intelligence, which is imitating the neural behavior of human beings. ANNs can be called a computer model of the human brain's neural structure. The basic elements of the ANNs are "neurons", which are the processing elements of ANNs. The "network" is defined as the structure in which the neurons act simultaneously in a group. The network involves an input layer, hidden layer, and the output layer [27]. In an ANN, neurons are tightly interconnected in various layers, where the adaptive weights are, conceptually, connection strengths between neurons. The three main characteristics of a NN are weights (w_i), bias (b_i), and transfer function. Each neuron receives inputs, attached with a weight. Each input data is multiplied by the corresponding weight of the neuron connection. Next, a bias value is added to the summation of inputs and corresponding weights (u) according to the following equation:

$$u_i = \sum_{j=1}^{H} w_{ij} x_j + b_i \tag{2}$$

The summation u_i is converted as the output with an activation (transfer) function, $F(u_i)$ yielding a value called the unit's "activation", as the following formula:

$$O = f(u_i) \tag{3}$$

The activation or transfer function that converts a neuron's weighted input to its output activation play the substantial role in the overall performance of an ANN's implementation [28].

3. Neural Network Studies

An extensive literature survey has been performed for available experimental results, as shown Table 1. The addition of grain refiners promotes the formation of fine equiaxed grains by suppressing the growth of columnar and twin columnar grain. The finer grain size also decreases the size of defects, such as micro-pores and second phase particles, thereby contributing to improved mechanical properties [29]. It is clear from Table 1 that the grain refiners and the applied heat treatment result in an increase in the ultimate tensile strength of the alloys. Ebrahimi and Emamy [30] reported that Al-5Ti-1B and Al-5Zr master alloy decreases the grain size of Al-12Zn-3Mg-2.5Cu aluminum alloy, and Al-5Ti-1B master alloy is more effective than Al-5Zr master alloy in reducing the grain size of Al-12Zn-3Mg-2.5Cu aluminum alloy. They expressed that the mechanical properties of the alloy are improved with the addition of grain refiners and heat treatment, and that this improvement can be attributed the increase to nano-metric precipitates, the homogeneity of the microstructure, and the decrease in grain size.

Table 2 shows the change of the input elements, from minimum to maximum, in training and testing sets of NN. The input variables are silicon (Si), magnesium (Mg), zinc (Zn), cupper (Cu), nickel (Ni), iron (Fe), manganese (Mn), chromium (Cr), Ti, zirconium (Zr), Sc, B, C, Al (in wt.%), and heat treatment (HT). The output variable is the ultimate tensile strength (UTS) in MPa. In the program, the alloys with and without heat treatment is coded as "1" and "0", respectively. In ANN system, each input variable is scaled to the range of 0 to 1 by the following the formula:

$$x_N = \frac{x - x_{min.}}{x_{max.} - x_{min.}} \tag{4}$$

where x_N is the normalized value of variable x, and x_{max} and x_{min} are the maximum and minimum values of the variable, respectively.

Table 1. Data set for training and testing.

Ref.	Master Alloys		HT	AGS	UTS
	Type	wt.%		(μm)	(MPa)
Ebrahimi and Emamy [30]	-	-	No	310	328
	-	-	Yes	310	504
	Al-5Ti-1B	0.05	No	38	352
	Al-5Ti-1B	0.05	Yes	38	621
	Al-5Zr	0.3	No	150	328
	Al-5Zr	0.3	Yes	150	530
Fakhraei and Emamy [31]	-	-	No	305	168
	Al-8B	0.1	No	155	190
	Al-8B	0.3	No	130	222
	Al-8B	0.5	No	113	236
	Al-8B	1	No	111	197
	Al-15Zr	0.1	No	215	193
	Al-15Zr	0.3	No	168	226
	Al-15Zr	0.5	No	150	243
	Al-15Zr	1	No	152	202
Fakhraei and Emamy [32]	-	-	No	305	168
	Al-5Ti-1B	0.05	No	178	178
	Al-5Ti-1B	0.1	No	150	183
	Al-5Ti-1B	0.3	No	116	200
	Al-5Ti-1B	0.5	No	100	253
	Al-5Ti-1B	1	No	88	242
Wang *et al.* [33]	-	-	Yes	320	170
	Al-5Ti-1B	0.2	Yes	200	220
	Al-5Ti-1B	0.5	Yes	180	229
	Al-5Ti-1B	1	Yes	100	249
	Al-5Ti-1B	1.5	Yes	80	250
	Al-5Ti-1B	2	Yes	75	241
	Al-5Ti-1B	3	Yes	40	230
	Al-5Ti-1B	5	Yes	70	215
	Al-5Ti-0.25C	0.2	Yes	200	230
	Al-5Ti-0.25C	0.5	Yes	150	260
	Al-5Ti-0.25C	1	Yes	80	267
	Al-5Ti-0.25C	1.5	Yes	65	271
	Al-5Ti-0.25C	2	Yes	40	249
	Al-5Ti-0.25C	3	Yes	50	240
	Al-5Ti-0.25C	5	Yes	80	215
Shabani *et al.* [34]	-	-	Yes	560	260
	Al-5Ti-1B	0.05	Yes	300	283
	Al-5Zr	0.3	Yes	345	276

Table 1. *Cont.*

Ref.	Master Alloys		HT	AGS	UTS
	Type	wt.%		(μm)	(MPa)
Ravi *et al.* [29]	-	-	No	630	168
	-	-	No	630	177
	Al-5Ti-1B	2	No	425	189
	Al-1Ti-3B	2	No	215	201
	Al-1Ti-3B	0.5	No	180	227
Kamali *et al.* [35]	-	-	No	190	174
	Al-5Ti-1B	0.05	No	48	217
	Al-5Ti-1B	0.1	No	39	225
	Al-5Ti-1B	0.3	No	41	237
	Al-5Ti-1B	0.5	No	42	245
	Al-5Ti-1B	0.05	Yes	48	318
	Al-5Ti-1B	0.1	Yes	39	330
	Al-5Ti-1B	0.3	Yes	41	350
	Al-5Ti-1B	0.5	Yes	42	354
Wang *et al.* [36]	-	-	No	165	167
	Al-5Ti-1B	0.01	No	87	161
	Al-5Ti-1B	0.02	No	47	183
	Al-5Ti-1B	0.05	No	40	181
	Al-5Ti-1B	0.1	No	52	166
Liu *et al.* [37]	Al-0.04Ti	0.036	No	110	270
	Al-3Sc	0.2	No	172	372
	Al-3Zr+Al-3Sc	0.15+0.2	No	100	393
	Al-0.04Ti+Al-3Zr+Al-3Sc	0.036+0.15+0.2	No	61	395
He *et al.* [38]	-	-	Yes	370	260
	Al-3.8Zr	0.1	Yes	196	272
	Al-3.6Sc	0.2	Yes	370	296
	Al-3.6Sc	0.6	Yes	72	360
	Al-3.8Zr+Al-3.6Sc	0.1+0.2	Yes	42	398

- Any grain refiner was not added in the alloys.

Table 2. Input elements (wt.%).

Elements	Si	Mg	Zn	Cu	Ni	Fe	Mn	Cr	Ti	Zr	Sc	B	C	Al	HT
Min.	0.03	0.02	0	0.03	0	0.15	0	0	0	0	0	0	0	Rem.	No = 0
Max.	13.2	20.4	12.24	4.47	0.95	0.45	0.2	0.1	1	1	0.6	1	0.2		Yes = 1

Output values resulted from ANN are also in the range [0.1], and transformed to their equivalent values, based on a reverse method of the normalization technique [39]. The unnormalized method is as:

$$x = x_N\,(x_{max} - x_{min}) + x_{min} \tag{5}$$

The two main processing phases of NN include training and testing. The training process is the adjustment of weights and biases in order to obtain an output through applying a proper method. Therefore, the experimental results are divided into training and testing sets. The datasets for training and testing are randomly selected from among the experimental results where 61 sets are training set and

6 sets are testing set. It is well known that increasing the data used in the training process of NN enhances its learning ability. Matlab NN toolbox is employed for the network training. Back-propagation (BP) learning algorithm (Levenberg–Marquardt-Trainlm, Boston, MA, USA), the most popular, and an effective, supervised learning method, and sigmoid function, an act activation function that joins curvilinear, linear, and constant behavior, depending on the values of the input, are used for the training of NN [40,41]. Training of the networks starts with adjusting initial random values for weights and biases. After submitting the input vector, forward propagation of the intermediate results leads to the producing of the output vector. Then, the weights and biases would be modified in order to reduce the error. The network replies to an input, without any change in the structure, within the testing process [42,43]. After the network is trained, the testing dataset is used to verify the effectiveness of the network and to estimate the expected performance in the future.

4. Results and Discussion

The main aim is to obtain the explicit formulation of the ultimate tensile strength as a function of input variables. One of the most difficult duties in NN works is the determination of the number of hidden layers and the number of neurons in the hidden layers. It is well known that NNs are typically in layers, which are made up of a number of interconnected nodes that contain an activation function. The input layer, which communicates to one or more hidden layers, where the actual processing is made via a system of weighted connections. The hidden layers then link to an output layer, where the answer is the output [44]. There is no well-defined procedure to find the optimal parameter settings and the network architecture. The trial and error approach is used to determine the number of neurons in the hidden layer. The various neuron numbers in one hidden layer (5–20) are used in this study. It is observed that the optimal architecture of NN with logistic sigmoid transfer function is 15–17–1. The NN toolbox in Matlab is used to obtain the proposed equation. It should be noted that the developed explicit formulation is valid for the ranges of the training set. Additionally, the correlation coefficient of the proposed equation is also evaluated, and is found to be 0.8509. In other words, the average prediction accuracy of the established ANN model is 85.09%.

$$T = 460 \times \dot{T} + 161 \tag{6}$$

where T is ultimate tensile strength and where:

$$\dot{T} = \left(\frac{1}{1+e^{-w}}\right)$$

where:

$$\begin{aligned} w = {} & (0.94889) \times \left(\frac{1}{1+e^{-u1}}\right) + (2.13402) \times \left(\frac{1}{1+e^{-u2}}\right) + (0.45077) \times \left(\frac{1}{1+e^{-u3}}\right) + (-3.32533) \\ & \times \left(\frac{1}{1+e^{-u4}}\right) + (2.39273) \times \left(\frac{1}{1+e^{-u5}}\right) + (-2.47205) \times \left(\frac{1}{1+e^{-u6}}\right) + (0.17732) \\ & \times \left(\frac{1}{1+e^{-u7}}\right) + (1.59606) \times \left(\frac{1}{1+e^{-u8}}\right) + (1.95837) \times \left(\frac{1}{1+e^{-u9}}\right) + (-3.33818) \\ & \times \left(\frac{1}{1+e^{-u10}}\right) + (-0.36762) \times \left(\frac{1}{1+e^{-u11}}\right) + (-2.96723) \times \left(\frac{1}{1+e^{-u12}}\right) + (-2.99594) \\ & \times \left(\frac{1}{1+e^{-u13}}\right) + (-1.59159) \times \left(\frac{1}{1+e^{-u14}}\right) + (-5.78338) \times \left(\frac{1}{1+e^{-u15}}\right) + (-0.19006) \\ & \times \left(\frac{1}{1+e^{-u16}}\right) + (1.50506) * \left(\frac{1}{1+e^{-u17}}\right) + (0.15734) \end{aligned}$$

and

$$u1 = (-0.49993 \times X1) + (-0.00354 \times X2) + (-0.61144 \times X3) + (0.49105 \times X4) + (0.72856 \times X5) + (-0.06895 \times X6) + (-0.03897 \times X7) + (-0.66021 \times X8) + (-0.92799 \times X9) + (0.13060 \times X10) + (0.13452 \times X11) + (0.54220 \times X12) + (0.40258 \times X13) + (0.13213 \times X14) + (0.36274 \times X15) + (-0.23276)$$

$$u2 = (-0.35276 \times X1) + (0.18702 \times X2) + (-0.24048 \times X3) + (1.03259 \times X4) + (0.64191 \times X5) + (-0.32554 \times X6) + (-2.70935 \times X7) + (0.14497 \times X8) + (-1.54846 \times X9) + (1.09702 \times X10) + (-2.60705 \times X11) + (-2.10947 \times X12) + (-0.02828 \times X13) + (-0.29674 \times X14) + (0.09758 \times X15) + (-0.26159)$$

$$u3 = (1.69292 \times X1) + (-0.15616 \times X2) + (0.17488 \times X3) + (-2.36728 \times X4) + (-0.61096 \times X5) + (-2.04671 \times X6) + (0.02049 \times X7) + (0.59184 \times X8) + (0.43818 \times X9) + (-0.31379 \times X10) + (1.14791 \times X11) + (1.12623 \times X12) + (-0.77400 \times X13) + (0.25359 \times X14) + (-1.75379 \times X15) + (-0.15834)$$

$$u4 = (-1.19858 \times X1) + (-0.65841 \times X2) + (1.26693 \times X3) + (-0.12854 \times X4) + (-0.64606 \times X5) + (0.19024 \times X6) + (1.29375 \times X7) + (-0.54070 \times X8) + (-0.85351 \times X9) + (-3.19349 \times X10) + (-1.19332 \times X11) + (-0.36499 \times X12) + (-2.22971 \times X13) + (0.79672 \times X14) + (-0.28841 \times X15) + (-0.05969)$$

$$u5 = (0.31740 \times X1) + (0.23849 \times X2) + (-0.16464 \times X3) + (0.72392 \times X4) + (0.08065 \times X5) + (-0.22394 \times X6) + (-0.45745 \times X7) + (-0.59003 \times X8) + (-0.53258 \times X9) + (-0.09952 \times X10) + (-0.59419 \times X11) + (-0.29143 \times X12) + (-0.45241 \times X13) + (0.52024 \times X14) + (0.60874 \times X15) + (0.58822)$$

$$u6 = (-1.65634 \times X1) + (-0.28557 \times X2) + (0.74052 \times X3) + (-2.75057 \times X4) + (-0.18806 \times X5) + (1.22060 \times X6) + (3.32170 \times X7) + (0.08922 \times X8) + (0.75341 \times X9) + (0.77162 \times X10) + (0.61279 \times X11) + (0.99672 \times X12) + (1.04996 \times X13) + (-1.24403 \times X14) + (-0.36127 \times X15) + (-1.26639)$$

$$u7 = (0.18672 \times X1) + (-0.21317 \times X2) + (-0.06951 \times X3) + (0.23293 \times X4) + (0.18344 \times X5) + (0.86836 \times X6) + (0.15483 \times X7) + (-0.49808 \times X8) + (-0.39467 \times X9) + (0.70716 \times X10) + (0.52081 \times X11) + (-0.15581 \times X12) + (0.04996 \times X13) + (-0.47618 \times X14) + (0.49369 \times X15) + (0.13833)$$

$$u8 = (0.20311 \times X1) + (-0.18870 \times X2) + (0.71925 \times X3) + (0.59576 \times X4) + (0.04333 \times X5) + (-1.38544 \times X6) + (0.22941 \times X7) + (-0.92876 \times X8) + (-0.49392 \times X9) + (-0.35084 \times X10) + (-0.85939 \times X11) + (0.69356 \times X12) + (0.92923 \times X13) + (0.76245 \times X14) + (0.80903 \times X15) + (0.78233)$$

$$u9 = (0.45508 \times X1) + (0.45380 \times X2) + (-0.53263 \times X3) + (0.62014 \times X4) + (0.48778 \times X5) + (-0.42787 \times X6) + (-1.03911 \times X7) + (-0.16981 \times X8) + (0.13517 \times X9) + (2.24826 \times X10) + (-0.05111 \times X11) + (-1.26004 \times X12) + (0.30575 \times X13) + (1.73001 \times X14) + (-0.22815 \times X15) + (0.74025)$$

$$u10 = (-3.04675 \times X1) + (-1.47749 \times X2) + (0.52853 \times X3) + (1.76084 \times X4) + (-1.14597 \times X5) + (4.48201 \times X6) + (0.88897 \times X7) + (-2.78018 \times X8) + (-0.76322 \times X9) + (-1.89132 \times X10) + (-0.37789 \times X11) + (-7.27827 \times X12) + (-0.16189 \times X13) + (1.69861 \times X14) + (-3.22967 \times X15) + (-1.60096)$$

$$u11 = (0.38619 \times X1) + (0.00559 \times X2) + (-0.13668 \times X3) + (-0.13627 \times X4) + (-0.00132 \times X5) + (-0.42122 \times X6) + (0.12256 \times X7) + (0.12089 \times X8) + (0.30673 \times X9) + (-0.19105 \times X10) + (0.09498 \times X11) + (0.33387 \times X12) + (0.13216 \times X13) + (-0.34123 \times X14) + (-0.09914 \times X15) + (-0.47796)$$

$$u12 = (-0.88070 \times X1) + (-1.26162 \times X2) + (1.00008 \times X3) + (0.44580 \times X4) + (-0.69643 \times X5) + (0.54684 \times X6) + (1.47863 \times X7) + (0.26171 \times X8) + (-0.07183 \times X9) + (-2.17460 \times X10) + (1.83544 \times X11) + (-1.22216 \times X12) + (0.19267 \times X13) + (-1.94418 \times X14) + (0.16993 \times X15) + (-0.74403)$$

$$u13 = (-1.16712 \times X1) + (-1.05184 \times X2) + (0.73395 \times X3) + (0.04286 \times X4) + (-0.39426 \times X5) + (2.84218 \times X6) + (1.99864 \times X7) + (-0.22580 \times X8) + (1.15608 \times X9) + (-0.03548 \times X10) + (-1.92515 \times X11) + (-4.03396 \times X12) + (0.38855 \times X13) + (-0.37459 \times X14) + (1.85893 \times X15) + (-0.99138)$$

$$u14 = (-0.04539 \times X1) + (-0.07940 \times X2) + (0.32185 \times X3) + (-0.79070 \times X4) + (-0.20130 \times X5) + (1.03768 \times X6) + (1.17849 \times X7) + (0.17644 \times X8) + (1.38321 \times X9) + (-1.21757 \times X10) + (-1.73489 \times X11) + (-2.77648 \times X12) + (-0.46636 \times X13) + (0.18136 \times X14) + (0.82488 \times X15) + (-0.16546)$$

$$u15 = (-0.97802 \times X1) + (-1.35732 \times X2) + (0.46732 \times X3) + (-1.94090 \times X4) + (0.45079 \times X5) + (0.86831 \times X6) + (0.96056 \times X7) + (0.86766 \times X8) + (-0.27002 \times X9) + (2.17786 \times X10) + (-3.34955 \times X11) + (1.57693 \times X12) + (-0.51299 \times X13) + (1.69850 \times X14) + (-0.16939 \times X15) + (-1.56104)$$

$$u16 = (-0.37956 \times X1) + (-0.13332 \times X2) + (0.05849 \times X3) + (-0.34892 \times X4) + (0.22692 \times X5) + (-0.26460 \times X6) + (-0.97223 \times X7) + (0.30502 \times X8) + (0.22059 \times X9) + (-0.10065 \times X10) + (-0.65479 \times X11) + (0.50434 \times X12) + (-0.54470 \times X13) + (-0.16661 \times X14) + (-0.27592 \times X15) + (-0.04856)$$

$$u17 = (0.18078 \times X1) + (0.16781 \times X2) + (-0.13242 \times X3) + (1.59311 \times X4) + (0.28523 \times X5) + (-0.66845 \times X6) + (-0.07740 \times X7) + (0.03953 \times X8) + (-0.20656 \times X9) + (-0.12734 \times X10) + (-0.86300 \times X11) + (0.42924 \times X12) + (0.24858 \times X13) + (-0.62146 \times X14) + (0.95215 \times X15) + (0.52058)$$

where X1, X2, X3, X4, X5, X6, X7, X8, X9, X10, X11 X12, X13, X14, and X15 are normalized input data of Al, Si, Mg, Zn, Cu, Ni, Fe, Mn, Cr, Ti, Zr, Sc, B, C wt.%, and HT.

Table 3 shows statistical parameters of training and test data sets. The performance of NN is evaluated by the correlation coefficient (R). Mean absolute error (MAE) and mean square error (MSE) are also used as error evaluation criteria in order to facilitate the comparison between predicted values and desired values.

Table 3. Statistical parameters.

Datasets	R	MSE	MAE
Training set	0.985	0.141	2.606
Testing set	0.946	0.528	6.313

The correlation coefficients of training and testing sets are 0.985 and 0.946, respectively, which means that the performance of the NN model is quite high and acceptable. MAE and MSE values for the training set are 2.606 and 0.141, and are 6.313 and 0.528 for testing set. If the MSE reaches zero, the performance of the model is regarded as excellent [45]. It can be said that the proposed NN model is in good agreement with the experimental data, and that all the errors are within acceptable ranges.

Figure 1 illustrates the correlation of NN and test data for the training and testing sets. The "test" means the determined experimental value of UTS and "NN" means the predicted value of UTS by NN, as shown in Figure 1. R^2 values of training and testing are 0.9711 and 0.8946, respectively. R^2 value compares the accuracy of the model to the accuracy of a trivial assessment model. A high R^2 (1) value tells that all points lie exactly on the curve with no scatter, and the results are perfect. It is clear that all the values are higher than 0.89. It can be said that the proposed NN model can predict the ultimate tensile strength of Al-Zn-Mg-Cu alloys with high accuracy and reliability. A few minor deviations between the experimental and theoretical results are observed in the training and testing sets of NN, which can be attributed to the variation in experimental conditions.

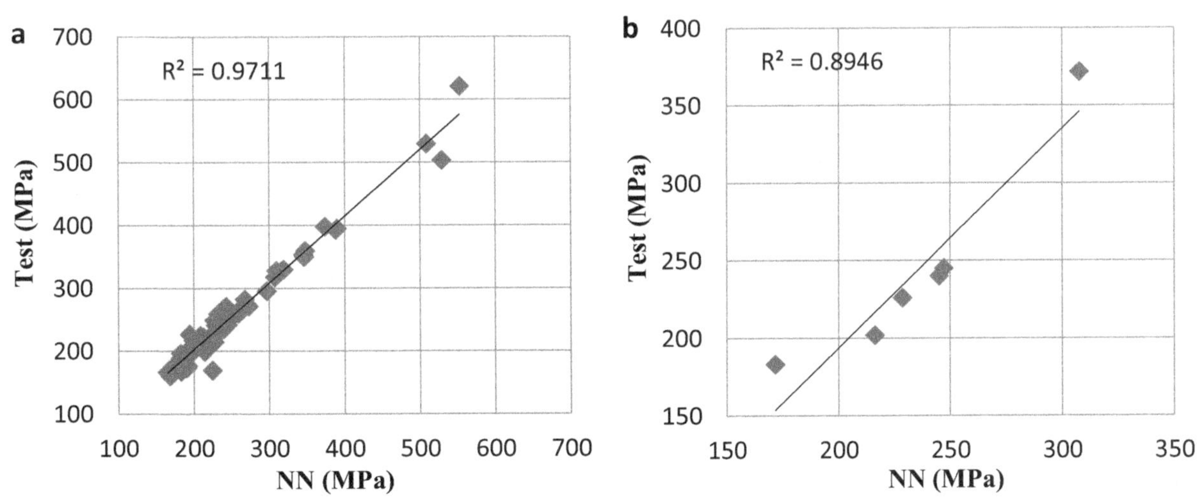

Figure 1. Correlation of NN and Test (**a**) Training set (**b**) Testing set.

One of the most effective elements promoting grain refinement in aluminum alloys is scandium. Aluminum and Al_3Sc have the same lattice structure and the misfit between lattice parameters is about 1.3%. The grain refinement mechanism of primary aluminum with scandium is due to heterogeneous nucleation, and the addition of scandium has been shown to have a beneficial effect on the strength of a cast Al-Mg alloy [46,47]. Figure 2 demonstrates the effect of Sc rates on the ultimate tensile strength of unrefined Al-Zn-Mg-Cu alloys and refined the alloys by Al-5Ti-1B and Al-5Zr using the proposed formulation above (Equation 6). In the case where there is no applied heat treatment, the ultimate tensile strength of Al-Zn-Mg-Cu alloys clearly increases with the addition of Sc. The main reason for this improvement can be attributed to the reduced grain size and the precipitation of more primary complex particles from the melt [48].

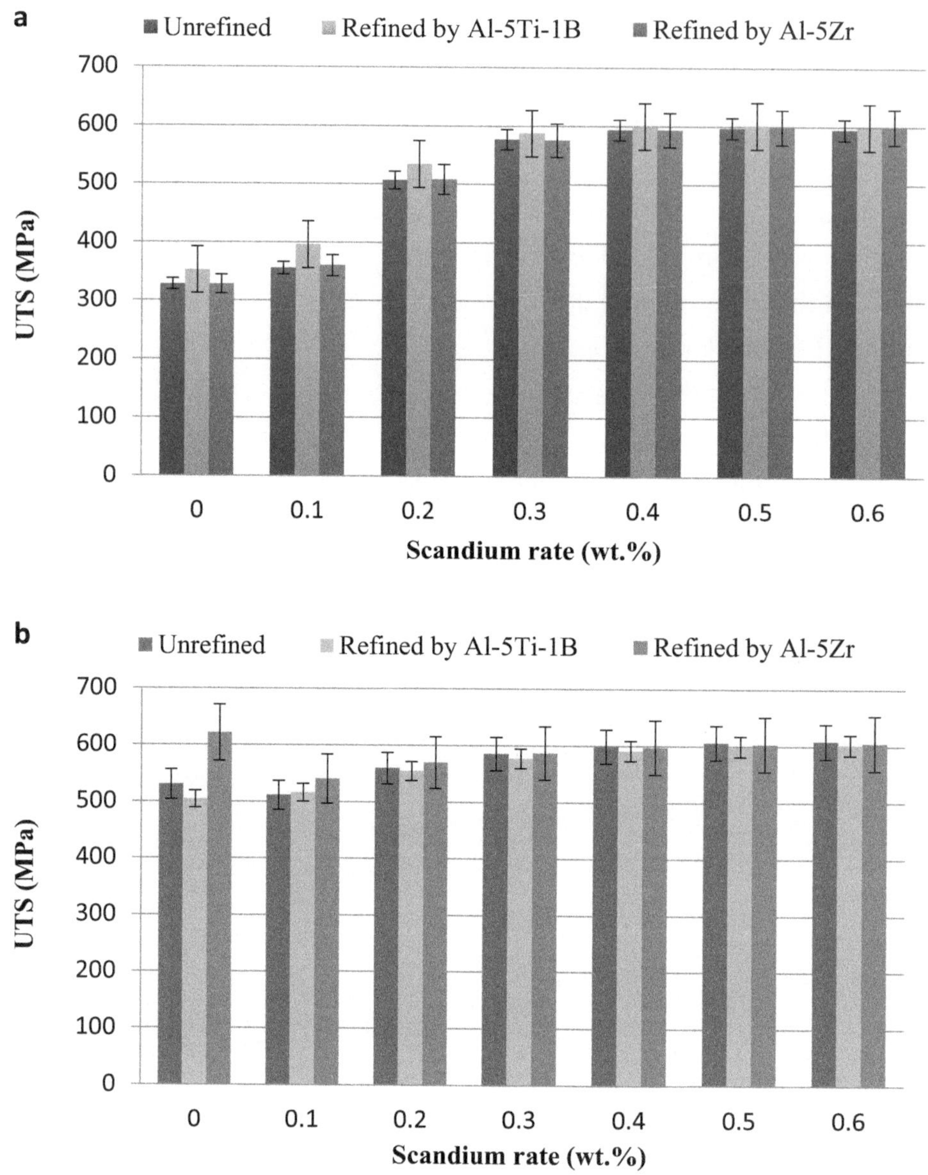

Figure 2. Effect of Sc rate on UTS of Al-Zn-Mg-Cu alloys: (**a**) non heat treatment (**b**) heat treatment.

It is clearly seen that, with the tensile properties of the alloys, first, the applied heat treatment reduces, then increases, the addition of Sc. The comparison of the UTS values revealed that the addition of 0.5% Sc is the most effective method to enhance tensile properties. It is concluded that the ultimate tensile strength of Al-Zn-Mg-Cu alloys, refined by Al-5Ti-1B or Al-5Zr, can be improved with the addition of 0.5% Sc without the need for heat treatment. Figure 3 shows the effect of the rate of C on the ultimate tensile strength of Al-Zn-Mg-Cu alloys containing 0.5 wt.% Sc. It is clear that the ultimate tensile strength of Al-Zn-Mg-Cu alloys, refined by Al-5Zr, enhances with the addition of C to a degree, but that the ultimate tensile strength of unrefined the alloys and refined the alloys by Al-5Ti-1B reduces with the addition of C. The optimum amount of is observed to be 0.01 wt.%.C. It is well known that Al-Ti-C master alloys, which contain $TiAl_3$ and TiC particles, are good refiners for aluminum alloys. These particles in Al-Ti-C master alloys remain stable in the melt and act the active nucleant substrates. These particles very fine and act quickly, but rapidly dissolve in molten aluminum. The morphology, dimension, and distribution of these particles have major impacts on the grain refining efficiency. The homogeneous distribution of these particles increases the mechanical properties of aluminum alloys [49,50].

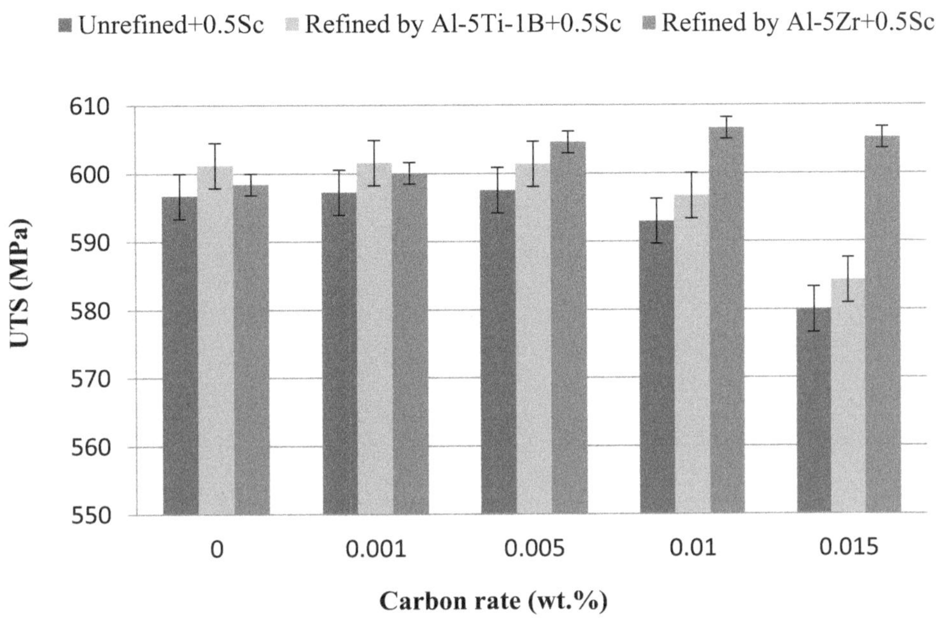

Figure 3. Effect of C rate to UTS of Al-Zn-Mg-Cu alloys.

5. Conclusions

This work proposes an approach for the prediction of ultimate tensile strength of Al-Zn-Mg-Cu alloys refined by Al-5Ti-1B and Al-5Zr. Sixty-seven data are used in this model. Feed-forward NN with back propagation is used for the training process, and the proposed NN model shows good agreement with experimental results. Therefore, the mathematical function is derived in an explicit form by using ANN outputs. The effects of Sc and C on the ultimate tensile strength of Al-Zn-Mg-Cu alloys are predicted with a high success rate. It is suggested that the strength of the alloys, refined by Al-5Ti-1B or Al-5Zr, can be increased with the addition of 0.5% Sc without heat treatment. All R and R^2 values for training, testing, and formula are larger than 0.94 and 89, respectively. This notes that the composed prediction model has a high reliability rate. The mean absolute error for the predicted values does not exceed 6.4%. Hence, it can be concluded that considerable savings, in terms of cost and time, could be obtained by using the advanced neural network model.

Author Contributions

All authors contributed equally to this work. All authors analyzed and interpreted the data, discussed the conclusions, and prepared the manuscript.

Conflicts of Interest

The authors declare no conflicts of interest.

References

1. Fan, X.; Jiang, D.; Meng, Q.; Zhong, L. The microstructural evolution of an Al–Zn–Mg–Cu alloy during homogenization. *Mater. Lett.* **2006**, *60*, 1475–1479.
2. Chen, J.-Z.; Zhen, L.; Yang, S.-J.; Dai, S.-L. Effects of precipitates on fatigue crack growth rate of aa 7055 aluminum alloy. *Trans. Nonferrous Met. Soc. China* **2010**, *20*, 2209–2214.
3. Wang, S.C.; Starink, M.J.; Gao, N. Precipitation hardening in Al–Cu–Mg alloys revisited. *Scr. Mater.* **2006**, *54*, 287–291.
4. Miller, W.S.; Zhuang, L.; Bottema, J.; Wittebrood, A.J.; de Smet, P.; Haszler, A.; Vieregge, A. Recent development in aluminium alloys for the automotive industry. *Mater. Sci. Eng. A* **2000**, *280*, 37–49.
5. Cui, J.; Roven, H.J. Recycling of automotive aluminum. *Trans. Nonferrous Met. Soc. China* **2010**, *20*, 2057–2063.
6. Rendigs, K. Aluminium structures used in aerospace-status and prospects. *Mater. Sci. Forum* **1997**, *242*, 11–24.
7. Williams, J.C.; Starke, E.A., Jr. Progress in structural materials for aerospace systems1. *Acta Mater.* **2003**, *51*, 5775–5799.
8. Srivatsan, T.S.; Sriram, S.; Veeraraghavan, D.; Vasudevan, V.K. Microstructure, tensile deformation and fracture behaviour of aluminium alloy 7055. *J. Mater. Sci.* **1997**, *32*, 2883–2894.
9. Hall, E. The deformation and ageing of mild steel: III discussion of results. *Proc. Phys. Soc. Sect. B* **1951**, *64*, 747.
10. McCartney, D. Grain refining of aluminium and its alloys using inoculants. *Int. Mater. Rev.* **1989**, *34*, 247–260.
11. Campbell, J. Effects of vibration during solidification. *Int. Met. Rev.* **1981**, *26*, 71–108.
12. Chalmers, B. *Principles of Solidification*; John Wiley and Sons, Inc.: New York, NY, USA, 1964.
13. Easton, M.; StJohn, D. Grain refinement of aluminum alloys: Part I. The nucleant and solute paradigms—A review of the literature. *Metall. Mat. Trans. A* **1999**, *30*, 1613–1623.
14. Jiang, K.; Liu, X. The effect of melting temperature and time on the TiC particles. *J. Alloy. Compd.* **2009**, *484*, 95–101.
15. Li, P.; Kandalova, E.G.; Nikitin, V.I. Grain refining performance of Al–Ti master alloys with different microstructures. *Mater. Lett.* **2005**, *59*, 723–727.
16. Wang, T.; Chen, Z.; Fu, H.; Gao, L.; Li, T. Grain refinement mechanism of pure aluminum by inoculation with Al–B master alloys. *Mater. Sci. Eng.: A* **2012**, *549*, 136–143.
17. Wang, T.; Fu, H.; Chen, Z.; Xu, J.; Zhu, J.; Cao, F.; Li, T. A novel fading-resistant Al–3Ti–3B grain refiner for Al–Si alloys. *J. Alloy. Compd.* **2012**, *511*, 45–49.
18. Liu, X.; Wang, Z.; Zang, Z.; Bian, X. The relationship between microstructures and refining performances of Al–Ti–C master alloys. *Mater. Sci. Eng.: A* **2002**, *332*, 70–74.

19. Li, B.; Pan, Q.; Huang, X.; Yin, Z. Microstructures and properties of Al–Zn–Mg–Mn alloy with trace amounts of sc and zr. *Mater. Sci. Eng.: A* **2014**, *616*, 219–228.
20. Yang, M.-B.; Pan, F.-S.; Cheng, R.-J.; Tang, A.-T. Effects of Al-10Sr master alloys on grain refinement of AZ31 magnesium alloy. *Trans. Nonferrous Met. Soc. China* **2008**, *18*, 52–58.
21. Birol, Y. Production of Al–Ti–B master alloys from ti sponge and kbf4. *J. Alloy. Compd.* **2007**, *440*, 108–112.
22. Wróbel, T. Review of inoculation methods of pure aluminium primary structure. *Arch. Mater. Sci. Eng.* **2011**, *50*, 110–119.
23. Kurt, H.I.; Oduncuoglu, M. Formulation of the effect of different alloying elements on the tensile strength of the *in situ* Al-Mg_2si composites. *Metals* **2015**, *5*, 371–382.
24. Reddy, N.S.; Rao, A.K.P.; Chakraborty, M.; Murty, B.S. Prediction of grain size of Al–7Si alloy by neural networks. *Mater. Sci. Eng. A* **2005**, *391*, 131–140.
25. Rashidi, A.M.; Eivani, A.R.; Amadeh, A. Application of artificial neural networks to predict the grain size of nano-crystalline nickel coatings. *Comput. Mater. Sci.* **2009**, *45*, 499–504.
26. Tofigh, A.A.; Rahimipour, M.R.; Shabani, M.O.; Alizadeh, M.; Heydari, F.; Mazahery, A.; Razavi, M. Optimized processing power and trainability of neural network in numerical modeling of al matrix nano composites. *J. Manuf. Process.* **2013**, *15*, 518–523.
27. Yurdakul, M.; Akdas, H. Modeling uniaxial compressive strength of building stones using non-destructive test results as neural networks input parameters. *Constr. Build. Mater.* **2013**, *47*, 1010–1019.
28. Jaiswal, S.; Benson, E.R.; Bernard, J.C.; van Wicklen, G.L. Neural network modelling and sensitivity analysis of a mechanical poultry catching system. *Biosyst. Eng.* **2005**, *92*, 59–68.
29. Ravi, K.R.; Manivannan, S.; Phanikumar, G.; Murty, B.S.; Sundarraj, S. Influence of Mg on grain refinement of near eutectic Al-Si alloys. *Metall. Mat. Trans. A* **2011**, *42*, 2028–2039.
30. Seyed Ebrahimi, S.H.; Emamy, M. Effects of Al–5Ti–1B and Al–5Zr master alloys on the structure, hardness and tensile properties of a highly alloyed aluminum alloy. *Mater. Des.* **2010**, *31*, 200–209.
31. Fakhraei, O.; Emamy, M. Effects of Zr and B on the structure and tensile properties of Al–20% Mg alloy. *Mater. Des.* **2014**, *56*, 557–564.
32. Fakhraei, O.; Emamy, M.; Farhangi, H. The effect of Al–5Ti–1B grain refiner on the structure and tensile properties of Al–20%Mg alloy. *Mater. Sci. Eng. A* **2013**, *560*, 148–153.
33. Wang, E.; Gao, T.; Nie, J.; Liu, X. Grain refinement limit and mechanical properties of 6063 alloy inoculated by Al–Ti–C (B) master alloys. *J. Alloy. Compd.* **2014**, *594*, 7–11.
34. Shabani, M.J.; Emamy, M.; Nemati, N. Effect of grain refinement on the microstructure and tensile properties of thin 319 Al castings. *Mater. Des.* **2011**, *32*, 1542–1547.
35. Kamali, H.; Emamy, M.; Razaghian, A. The influence of Ti on the microstructure and tensile properties of cast Al–4.5Cu–0.3Mg alloy. *Mater. Sci. Eng.: A* **2014**, *590*, 161–167.
36. Wang, D.; de Cicco, M.P.; Li, X. Using diluted master nanocomposites to achieve grain refinement and mechanical property enhancement in as-cast Al–9Mg. *Mater. Sci. Eng.: A* **2012**, *532*, 396–400.
37. Liu, Z.; Li, Z.; Wang, M.; Weng, Y. Effect of complex alloying of Sc, Zr and Ti on the microstructure and mechanical properties of Al–5Mg alloys. *Mater. Sci. Eng.: A* **2008**, *483–484*, 120–122.

38. He, Y.-D.; Zhang, X.-M.; You, J.-H. Effect of minor Sc and Zr on microstructure and mechanical properties of Al-Zn-Mg-Cu alloy. *Trans. Nonferrous Met. Soc. China* **2006**, *16*, 1228–1235.
39. Hassan, A.M.; Alrashdan, A.; Hayajneh, M.T.; Mayyas, A.T. Prediction of density, porosity and hardness in aluminum–copper-based composite materials using artificial neural network. *J. Mater. Process. Technol.* **2009**, *209*, 894–899.
40. Cevik, A.; Kutuk, M.A.; Erklig, A.; Guzelbey, I.H. Neural network modeling of arc spot welding. *J. Mater. Process. Technol.* **2008**, *202*, 137–144.
41. Rumelhart, D.E.; Hinton, G.E.; Williams, R.J. Learning internal representations by error propagation. In *Parallel Distributed Processing: Explorations in the Microstructure of Cognition*; David, E.R., James, L.M., Group, C.P.R., Eds.; MIT Press: Cambridge, MA, USA, 1986; Volume 1, pp. 318–362.
42. Prasad, B.K.R.; Eskandari, H.; Reddy, B.V.V. Prediction of compressive strength of scc and hpc with high volume fly ash using ann. *Constr. Build. Mater.* **2009**, *23*, 117–128.
43. Shafabakhsh, G.H.; Ani, O.J.; Talebsafa, M. Artificial neural network modeling (ANN) for predicting rutting performance of nano-modified hot-mix asphalt mixtures containing steel slag aggregates. *Constr. Build. Mater.* **2015**, *85*, 136–143.
44. Pandit, S.; Mandal, C.; Patra, A. *Nano-Scale Cmos Analog Circuits: Models and Cad Techniques for High-Level Design*; CRC Press: Boca Raton, FL, USA, 2014.
45. Sun, Y.; Zeng, W.D.; Zhang, X.M.; Zhao, Y.Q.; Ma, X.; Han, Y.F. Prediction of tensile property of hydrogenated Ti600 titanium alloy using artificial neural network. *J. Mater. Eng. Perform.* **2011**, *20*, 335–340.
46. Davydov, V.G.; Rostova, T.D.; Zakharov, V.V.; Filatov, Y.A.; Yelagin, V.I. Scientific principles of making an alloying addition of scandium to aluminium alloys. *Mater. Sci. Eng. A* **2000**, *280*, 30–36.
47. Lathabai, S.; Lloyd, P.G. The effect of scandium on the microstructure, mechanical properties and weldability of a cast Al–Mg alloy. *Acta Mater.* **2002**, *50*, 4275–4292.
48. Norman, A.F.; Hyde, K.; Costello, F.; Thompson, S.; Birley, S.; Prangnell, P.B. Examination of the effect of Sc on 2000 and 7000 series aluminium alloy castings: For improvements in fusion welding. *Mater. Sci. Eng. A* **2003**, *354*, 188–198.
49. Ding, H.; Li, H.; Liu, X. Different elements-induced destabilisation of TiC and its application on the grain refinement of mg–al alloys. *J. Alloy. Compd.* **2009**, *485*, 285–289.
50. Quested, T.E.; Greer, A.L. The effect of the size distribution of inoculant particles on as-cast grain size in aluminium alloys. *Acta Mater.* **2004**, *52*, 3859–3868.

14

Nanoreinforced Cast Al-Si Alloys with Al_2O_3, TiO_2 and ZrO_2 Nanoparticles

Iman S. El-Mahallawi [1,*], Ahmed Yehia Shash [2,*] and Amer Eid Amer [3]

[1] Mining Petroleum and Metallurgical Engineering Department, Cairo University, Giza 12613, Egypt
[2] Mechanical Design and Production Department, Cairo University, Giza 12613, Egypt
[3] Mechanical Design and Production Engineering Department, Faculty of Engineering, Beni-Suef University, Beni-Suef 62111, Egypt; E-Mail: aeid958@yahoo.com

* Authors to whom correspondence should be addressed;
E-Mails: ielmahallawi@bue.edu.eg (I.S.E.-M.); ahmed.shash@cu.edu.eg (A.Y.S.)

Academic Editor: Anders E. W. Jarfors

Abstract: This study presents a new concept of refining and enhancing the properties of cast aluminum alloys by adding nanoparticles. In this work, the effect of adding alumina (Al_2O_3), titanium dioxide (TiO_2) and zirconia (ZrO_2) nano-particles (40 nm) to the aluminum cast alloy A356 as a base metal matrix was investigated. Alumina, titanium dioxide and zirconia nano-powders were stirred in the A356 matrix with different fraction ratios ranging from (0%–5%) by weight at variable stirring speeds ranging from (270, 800, 1500, 2150 rpm) in both the semisolid (600 °C) and liquid (700 °C) state using a constant stirring time of one minute. The cast microstructure exhibited change of grains from dendritic to spherical shape with increasing stirring speed. The fracture surface showed the presence of nanoparticles at the interdendritic spacing of the fracture surface and was confirmed with EDX analysis of these particles. The results of the study showed that the mechanical properties (strength, elongation and hardness) for the nanoreinforced castings using Al_2O_3, TiO_2 and ZrO_2 were enhanced for the castings made in the semi-solid state (600 °C) with 2 weight% Al_2O_3 and 3 weight% TiO_2 or ZrO_2 at 1500 rpm stirring speed.

Keywords: nanoreinforced castings; semisolid casting; hypoeutectic aluminum alloys

1. Introduction

Al-Si alloys are used for several automotive applications and various aerospace industries due to their outstanding properties such as high castability, abrasion resistance and excellent strength to weight ratio. The Al-Si alloys in particular are greatly known for their rapidly advancing activities in aircrafts, aerospace and automotive industries. Intensive efforts have been made to understand the modification of these alloys by adding Na, Sr or Sb [1–4]. The addition of certain elements, such as calcium, sodium, strontium, and antimony, to hypoeutectic aluminum-silicon alloys results in a finer lamellar or fibrous eutectic network [4]. Gruzleski *et al.* [2] discussed the techniques commonly used in the foundry, and showed that the addition of P to the melt results in modified hypereutectic Al-Si alloys, with both coarse and unmodified eutectic silicon surrounding the refined primary silicon. Though Na and Sr are used to achieve refinement of the eutectic silicon, a combined effect does not happen at the same time with P due to chemical incompatibility of phosphors with other modifying chemicals such as strontium and sodium. This has been explained [2] to result from the formation of strontium phosphide or sodium phosphide upon the addition of strontium or sodium to phosphorus pre-refined alloys. Other investigations have shown that adding rare earth elements would result in modification and refinement of the primary and eutectic silicon particles [5,6].

The mechanical properties of the A356 alloy, not only depend on the dendritic structures, but also on the sizes and morphologies of the eutectic Si particles [7,8]. The eutectic Si of untreated A356 presents a coarse plate-like structure, which will deteriorate the mechanical properties (especially the ductility) of the alloy. The mechanical properties of the hypereutectic Al-Si alloys are highly affected by the morphology, size, and distribution of both primary Si particles (PSPs) and eutectic silicon. The morphology of silicon particles is dependent on the solidification rate, as under normal casting conditions, PSPs are very coarse and show star-like and other irregular shapes. Therefore, to improve the mechanical properties of the hypo and hypereutectic Al-Si alloys, size, distribution, and the morphology of PSPs and eutectic silicon should be controlled, as well as refining of the aluminum dendrites [2,3].

The use of ceramic particles to improve the yield and ultimate strength (UTS) of cast Al alloys has been considered by many researchers [9–15]. Prospects of using nanoparticles as refining and reinforcement agents to gain improved performance of Al-Si cast alloys by adding Al_2O_3 and TiO_2 particles have gained significant interest recently [16–22].

Rohatgi *et al.* [11] have predicted the significant role of producing Al-Si ceramic composites for bearings, pistons, cylinder liners, *etc.* leading to savings in materials and energy. Previous trials to modify the structure and enhance the mechanical properties of A356 via reinforcement with ceramic micro-sized particles, such as Al_2O_3 [12], SiC [13,14] and TiC [15] did not result in a breakthrough in casting modification, though a few engineering cast products entered the market [23].

It has been shown that enhanced as-cast properties of Al-Si alloys are obtained by treating the melt with nano-sized particles [16–22]. Introducing Al_2O_3 nanoparticles to the (A356) cast alloy in the semi-solid state with mechanical stirring has a beneficial effect on improving the strength–ductility relationship in these alloys [16,17]. This is attributed to the modification of the dendritic columnar structure into a smaller and equiaxed globular grain arrangement, resulting from semi-solid casting conditions. Moreover, the enhanced viscosity of the semi-solid processing would serve to improve the ceramic particle/melt wettability and entrap or capture the reinforcement material physically. The addition of ceramic particles to aluminum alloys raises the viscosity very quickly [24], providing suitable conditions for particle capture; moreover, the Al_2O_3 nanoparticles possess appropriate properties that are compatible with the Al alloy relatively high thermal conductivity and thermal expansion coefficients that affects its role as a nanodispersion for reinforcement in the Al alloy matrix [16–19].

For most applications, a homogeneous distribution of the particles is desirable in order to maximize the mechanical properties. In order to achieve a good homogeneous distribution of the particles in the matrix, the process parameters related with the stir casting method must be considered, the stirring speed and stirring time are key parameters. There is some debate in literature about the appropriate stirring speeds, while one study [25] has investigated stirring speeds within the range of 500, 600 and 700 rpm and the stirring times in the range of 5, 10 and 15 min, another study [26] has used higher stirring speeds in the range 1000–1500 rpm.

Haizhi Ye [23] has presented an overview of Al-Si based alloys for engine applications where it was shown that fatigue failure and wear caused by surface delamination are the most important causes for failure or end of life of engine parts. The main causes according to the analysis presented by Haizhi Ye [23] are the size and shapes (morphology) of the Primary Si Particles (PSPs). Also, these materials have low specific gravity that makes their properties particularly superior in strength and modulus to many traditional engineering materials.

Therefore, the aim of this work is to compare the effect of adding three different nanoparticles as reinforcements (Al_2O_3, TiO_2 and ZrO_2) on the microstructure, the mechanical properties and the wear behavior of A356 aluminum alloy cast from the semi-solid state. The ultimate goal of the work is to discuss the options for producing nanocomposites via the casting route, in order to evaluate the options for using nanoparticles dispersion for the production of semi-solid cast A356 alloy with improved properties.

2. Statistical Analysis

The first step in this work was to statistically-analyze the results obtained through previous work by the authors (represented by nine samples for each condition), for A356 alumina (Al_2O_3) nanoreinforced cast alloys [16–18], on the relation between the tensile strength and the amount of nanoparticles added (in weight%) and the addition temperature. The ANOVA technique was applied on the responses to find fisher factors (f) and compare it with the tabulated values at the levels of confidence (90%, 95%, and 99%), using response surface methodology (RSM) technique. Table 1 shows the orthogonal array L4 that was used to apply the ANOVA technique. The obtained results showed that the effect of nano-reinforcement is significant at 90%, and the effect of the temperature is significant at 85%. The analysis results were used to predict the measured values. Accordingly, the following experimental plan was made for this work.

Table 1. Orthogonal L4 array for ANOVA analysis of previous results.

Trial number	Factor	Nano%	Stirring Temperature	Predicted Tensile Strength	Average Value of Measured Tensile Strength	Error	Error%
1		0%	600–610	157.254	155	2.254	1.45419355
2		0%	700	138.756	141	2.244	1.59148936
3		2%	600–610	192.754	195	2.246	1.15179487
4		2%	700	174.256	172	2.256	1.31162791

3. Experimental Work

3.1. Material Preparation

Titanium dioxide (TiO_2) and zirconia (ZrO_2) ceramic particles of 40 nm particle size were added as reinforcement materials to A356 alloy with the chemical composition illustrated in Table 2. A charge of 1 kg of aluminum alloy A356 was placed in a crucible furnace, and then heated to the required temperature (640 °C). The addition of the nanoparticles was made either at 710 °C or 640 °C in the liquid or semisolid state, respectively, by direct immersion of small packages of alumina nanoparticles wrapped in aluminum foil with simultaneous mechanical stirring of the melt. The melting and casting conditions are fully presented in previous work [16–18]. In the case of liquid state, the melt was brought to 700 °C when the addition was made, then stirred and poured. While in the case of the semisolid state, the temperature of the melt was brought down to 600 °C before pouring. The nanoparticles were added, while the melt was stirred mechanically at varied stirring speeds of (270, 800, 1500, 2150 rpm), and using the melting unit illustrated in Figure 1. The stirring was carried out mechanically using four blades impeller. The Al_2O_3, TiO_2 and ZrO_2 nanoparticles were preheated to 700 °C, 400 °C and 500 °C, respectively, to avoid entering the sintering stage. After completion of stirring and mixing, the alloys were poured in preheated steel molds at 300 °C. Table 3 summarizes the different casting conditions used in this work.

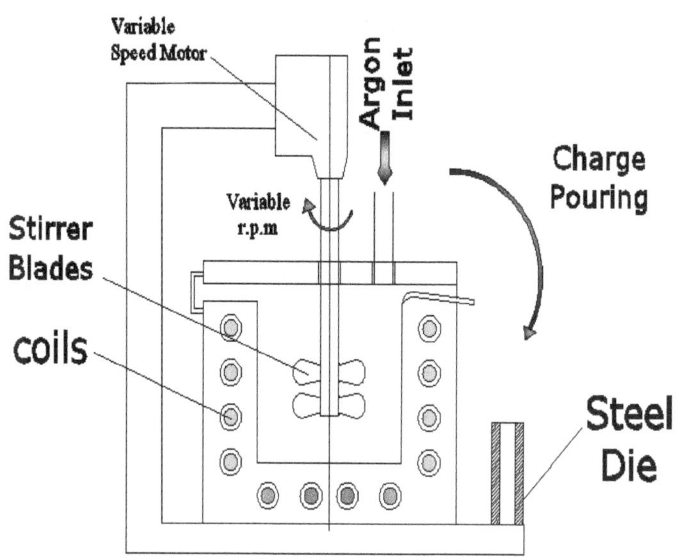

Figure 1. Furnace used for preparing the castings.

Table 2. Chemical Composition (in weight%) of A356 cast Al-Si.

Alloy	Chemical Composition (weight%)							
	Al	Si	Mg	Fe	Cu	Pb	Zn	Mn
A356	Bal.	7.44	0.3	0.27	0.02	0.022	0.01	Nil

Table 3. Casting conditions used in this work.

No.	Al_2O_3 Weight%	TiO_2 Weight%	ZrO_2 Weight%	Temperature (°C)	Stirring Speed (rpm)
1	0	0	0	600	1500
2	1	1	1	600	1500
3	2	2	2	600	1500
4	4	3	3	600	1500
5	--	5	5	600	1500
6	2	3	3	600	270
7	2	3	3	600	800
8	2	3	--	600	2150
9	2	3	--	700	1500

--: Condition not relevant to this addition.

3.2. Mechanical Testing

The mechanical properties mainly tensile strength, ductility, and hardness were determined in the as-cast conditions for the investigated material. The tensile tests were conducted on round tension test specimens of diameter 5.02 mm and gage length 25.2 mm using a universal testing machine according to DIN 50125. The reported result for each case is the highest value among three samples. The hardness tests were conducted on Rockwell hardness testing machine using a (1/16)″ diameter hardened steel ball and a 62.5 kg applied load.

3.3. Microstructure and SEM Analysis

The microstructure examination was carried out using OLYMPUS DP12 optical metallurgical microscope (New York Microscope Company, Beni Suef, Egypt) equipped with a high-resolution digital camera for investigating the microstructure. The surface topography and fracture characteristics were studied using Scanning Electron Microscope (JSM-5400 SEM, JEOL Company, Centre for Metallurgical Research CMRDI Cairo, Egypt) to understand the fracture and wear mechanisms. The favorable sites for Al2O3 particle incorporation were identified by Scanning Electron Microscope (JSM-7000 FEG-SEM, JEOL Company, Gemeinshaftslabor fuer elektronmikroskopie GFE, Aachen, Germany), lateral resolution 1.2 nm at 30 kV combined with EDX/EBSD-System EDAX Pegasus for chemical and crystallographic analysis. The ZrO2 agglomerates were detected by using JSM-5410 Scanning Electron Microscope (JEOL Company, Beni Suef, Egypt) with a high- resolution of 3.5 nm.

3.4. Wear Test for the Al_2O_3 Nanoreinforced Alloy

PLINT TE 79 Multi Axis Tribometer Machine (Phoenix Tribology Limited Company, Cairo University-Giza, Giza, Egypt) was used for measuring the friction force; friction coefficient and wear rate for the manufactured materials. A standard specimen with diameter of 8 mm and 20 mm length is prepared; then a computerized pin on disc machine, used for friction and wear testing of materials, is loaded vertically downwards onto the horizontal Disc. The wear tests were then performed for the A356 cast material with the following parameters: velocity = 0.8 m/s, time = 1200 s and load = 10 N. The differences in the weight of the samples were taken as an indication of the wear resistance of the material.

4. Results and Discussion

4.1. Mechanical Properties

4.1.1. Effect of Nanoparticles Addition

Figures 2–4 present the Mechanical properties (ultimate tensile strength (UTS), elongation% and hardness) of the produced castings with Al_2O_3, TiO_2 and ZrO_2 nanoparticles reinforcement.

It can be seen from Figure 2 that as the weight% of Al_2O_3 nanoparticles increases, the UTS increases; the maximum UTS was reached (195 MPa) at 2 weight% Al_2O_3 (increase of about 26%). Beyond this limit, the UTS decreases as the weight% increases. It can also be seen that the ductility increases when Al_2O_3 exceeds 1 weight%. The ductility increases by about 40% with 2 weight% Al_2O_3 nanoparticles added. The ductility reaches a minimum value with 4 weight% Al_2O_3 nanoparticles added. It can also be noted that the hardness increases by about 30% with 4 weight% Al_2O_3 nanoparticles added.

It is shown from Figure 3 that as the weight% of the TiO_2 nanoparticles increases up to 3%, the UTS increases until reaching 185 MPa. Beyond this weight%, the UTS decreases with increase in the weight% of the TiO_2 nanoparticles. The ductility reaches its maximum with 3 weight% TiO_2 nanoparticles added, and then drops to a minimum value with 5 weight% TiO_2 nanoparticles added. The hardness reaches its minimum value of 49 HRB with the addition of 5 weight% TiO_2 nanoparticles, which means that the material becomes softer and this is supported by the drop in the strength.

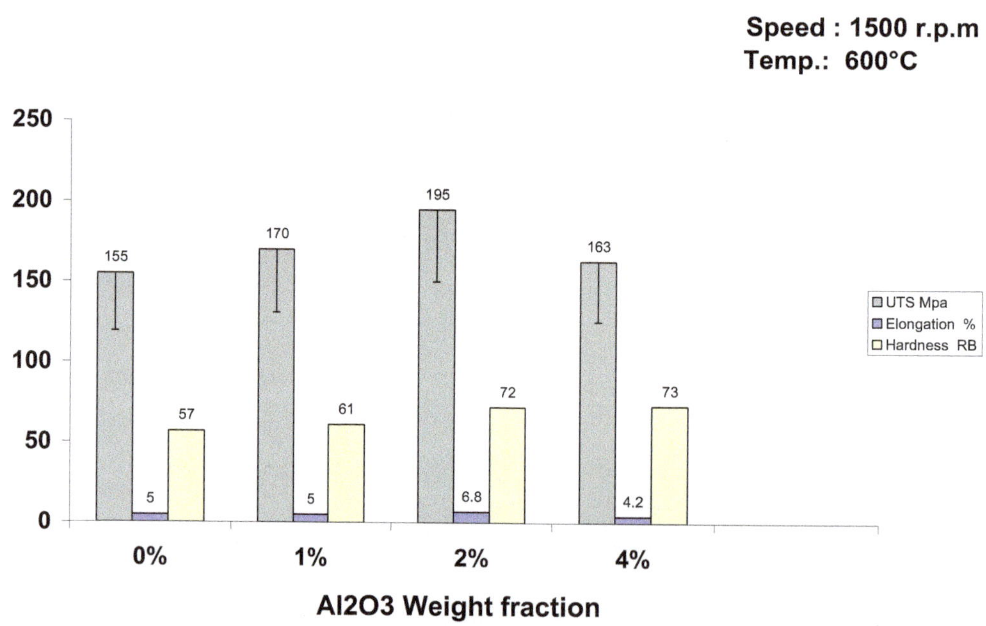

Figure 2. The effect of Al_2O_3 nanoparticles% on the ultimate tensile strength (UTS), hardness and ductility using 1500 rpm stirring speed at semi-solid state (600 °C).

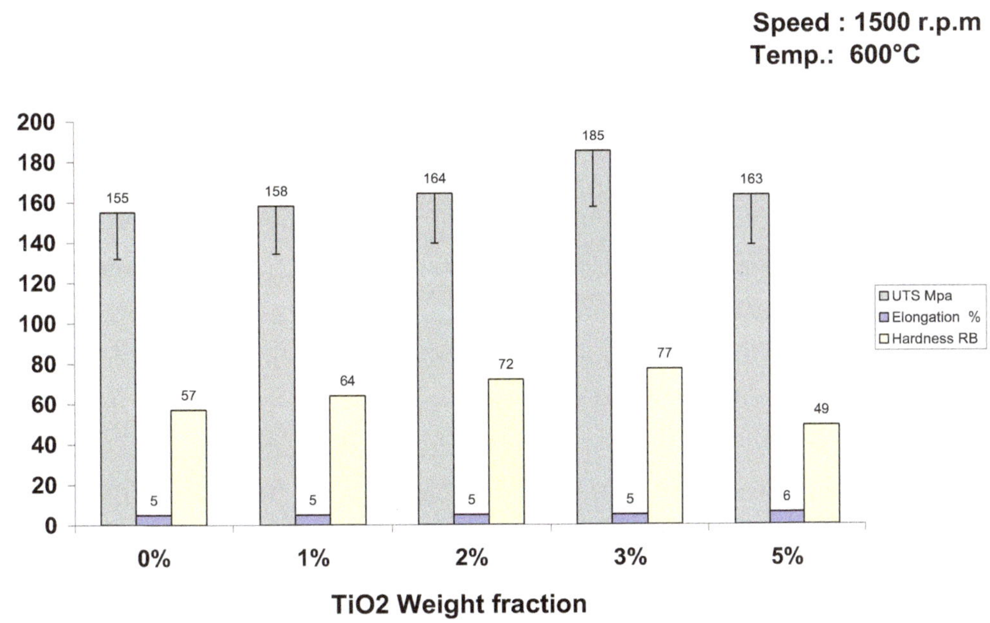

Figure 3. The effect of TiO_2 nanoparticles% on the UTS, hardness and ductility using 1500 rpm stirring speed at semi-solid state (600 °C).

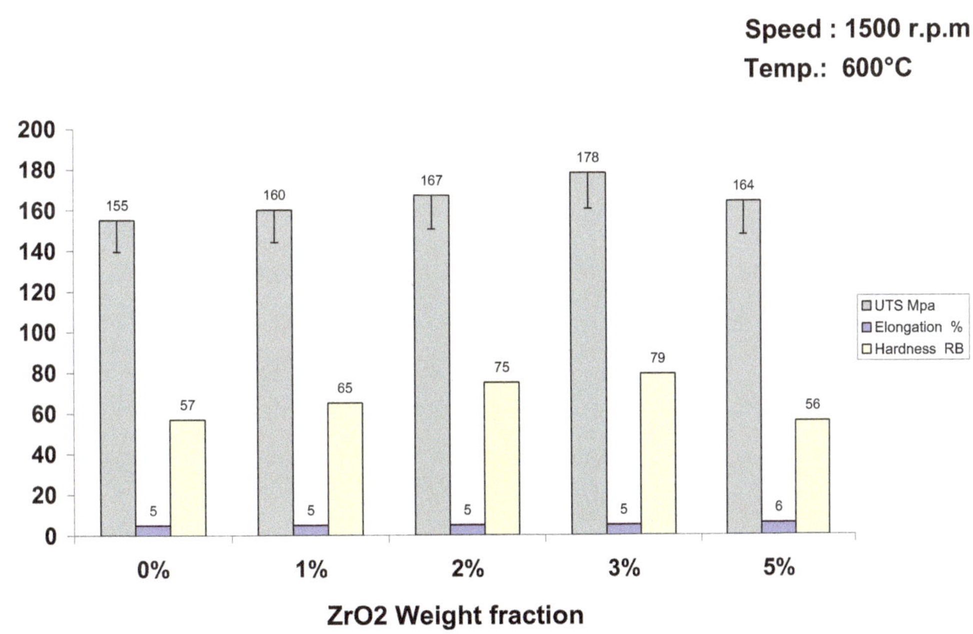

Figure 4. The effect of ZrO_2 nanoparticles% on the UTS, hardness and ductility using 1500 rpm stirring speed at semi-solid state (600 °C).

Figure 4 shows the effect of ZrO_2 nanoparticles% on the UTS using 1500 rpm stirring speed in the semi-solid state (600 °C). It can be seen that the addition of 3 weight% ZrO_2 nanoparticles results an increase in the tensile strength and the hardness. Whereas, the increase in ductility was only observed at 5 weight% ZrO_2 coupled with a decrease in both hardness and strength.

The previous results show that the inclusion of 2–3 weight% nanoparticles increases the tensile strength of the nanoreinforced A356 alloy, without loss in ductility and even an increase in ductility, in some cases. Beyond this limit all properties drop. The observed increase in the strength of the A356 nanoreinforced alloys suggests a misfit effect in the lattice parameter of the matrix phase. This would be a result of cooling induced changes caused by the difference between the coefficients of thermal expansion between the primary phase (Al or Si) and the nanoparticles resulting an increase in the hardness of the first. The coefficients of thermal expansion are: (22.2×10^{-6} m/m K for Al, 3×10^{-6} m/m K for Si, 5.4×10^{-6} m/m K for Al_2O_3 nanoparticles, 9×10^{-6} m/m K for TiO_2 nanoparticles, and 6.53×10^{-6} m/m K for ZrO_2 nanoparticles) [27]. Additionally, enhanced dislocation generation and reduced sub grain size owing to the presence of the nanoparticles may be contributing to the increased hardness of the aluminum phase. The increase in the hardness is also due not only to the mismatch in thermal expansion, but also mismatch in elastic moduli of the nanoparticles. The addition of the nanoparticles also results in a constraint in the plastic flow of matrix. The matrix could flow only with the movement of the nanoparticles or over the particles during plastic deformation, the matrix gets constrained considerably to the plastic deformation because of smaller inter-particle distance and this results in higher degree of improvement in the flow stress.

Rapid quenching from the melt and solid state reactions have been identified as processing routes for obtaining nanostructured materials leading to two phase nanostructures, however, it is highly desirable to obtain such nanostructures directly in bulk form, e.g., through casting [28]. The current work focuses on the properties of nanodispersed A356 alloy obtained by rheocasting, which is a challenging production method for producing nanodispersed alloys. The potential for nanodispersed materials for being high-tech materials has been explored [29] and it has been shown that the addition of nano-oxides results improvement in the hardness, the tensile strength and the ductility of stir cast aluminum. This has been explained by the tendency of the oxide nanoparticles to agglomerate on the grain boundaries in columnar structures, whereas these oxide nanoparticles are more likely to disperse uniformly along the equaxied structures resulting from stirring.

It is shown from Table 4, that alumina and zirconia have close coefficients of thermal expansion, whereas titanium oxide has a relatively higher coefficient. It is also shown that alumina and titanium oxide have higher elastic moduli, whereas zirconium oxide has the least elastic modulus among the three oxides. This may explain the difference in the obtained strength results for the three castings as the alloy reinforced with alumina shows higher strength compared to that dispersed with zirconia.

Table 4. Properties of Al_2O_3, TiO_2 and ZrO_2 Reinforcement Nanoparticles.

Reinforcement	**γ-Al_2O_3 ***	**TiO_2 ***	**ZrO_2 ***
Density (g/cm^3)	3.60	4.23	5.89
Crystal Structure	FCC	Tetragonal	monoclinic
Appearance	White Solid	White solid	White powder
Young's Modulus (GPa)	380	244	171
Average Size (nm)	50	50	50
Melting point °C	2054	1870	2715
Thermal conductivity, at 800 °C, W/m °K	29	8	5.5
Coefficient of thermal expansion (10^{-6} m/m °K)	5.4	9	6.53

* CRC Materials Science and Engineering Handbook.

4.1.2. Effect of Stirring Speed

It can be seen, from Figure 5 that as the stirring speed increases up to 1500 rpm, the UTS increases to reach 195 MPa. Beyond this stirring speed, the UTS decreases as the stirring speed increases possibly due to the increase in internal porosity [18] and as well the semi-solid cast cooling rapidly below the ranges permitting sound castings to be obtained. Increasing the speed of stirring beyond 850 rpm increases the ductility. The ductility increases by about 45% at a stirring speed of 1500 rpm. At 2150 rpm stirring speed, the ductility reaches its minimum values; possibly due to the presence of high porosity content [18]. This could also explain that the hardness at this stirring speed decreases after reaching a maximum value at 1500 rpm.

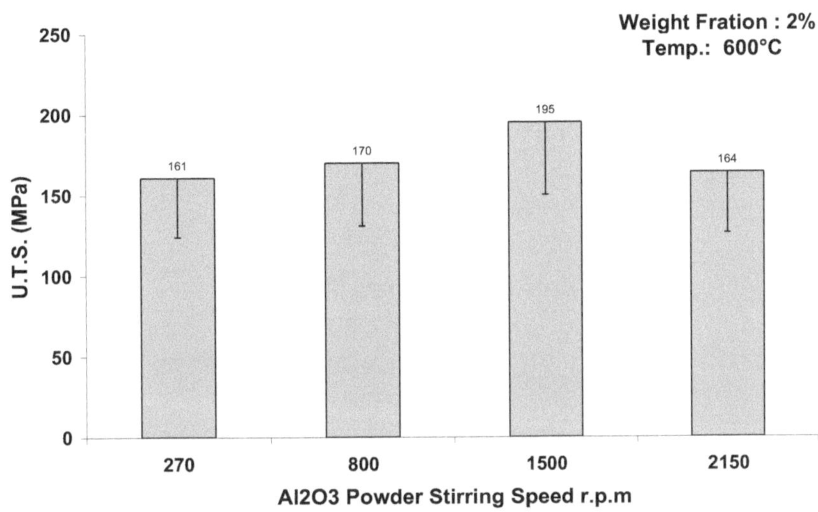

Figure 5. The effect of stirring speed on the UTS of 2 weight% Al_2O_3 nanoreinforced alloy at 600 °C.

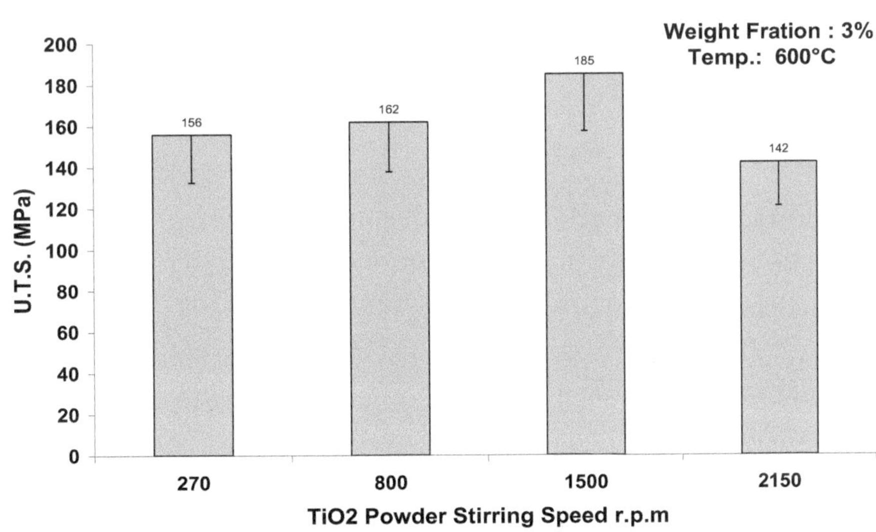

Figure 6. The effect of stirring speed on the UTS of 3 weight% TiO_2 nanoreinforced alloy at 600 °C.

It is shown in Figure 6 that as the stirring speed of nano TiO_2 reinforced casting increases to 1500 rpm, the UTS increases to reach 185 MPa. Beyond this stirring speed, the UTS decreases as the stirring speed increases, possibly due to increase in internal porosity [18] and cooling of the semi-solid slurry, as explained before. When the speed of stirring increases beyond 850 rpm, the ductility starts to increase. The ductility increases slightly at the stirring speed of 1500 rpm. At 2150 rpm stirring speed, the ductility reaches its minimum value; possibly due to the high porosity [18] content created in the composite. Again, the hardness at this stirring speed decreases after reaching its maximum value at 1500 rpm.

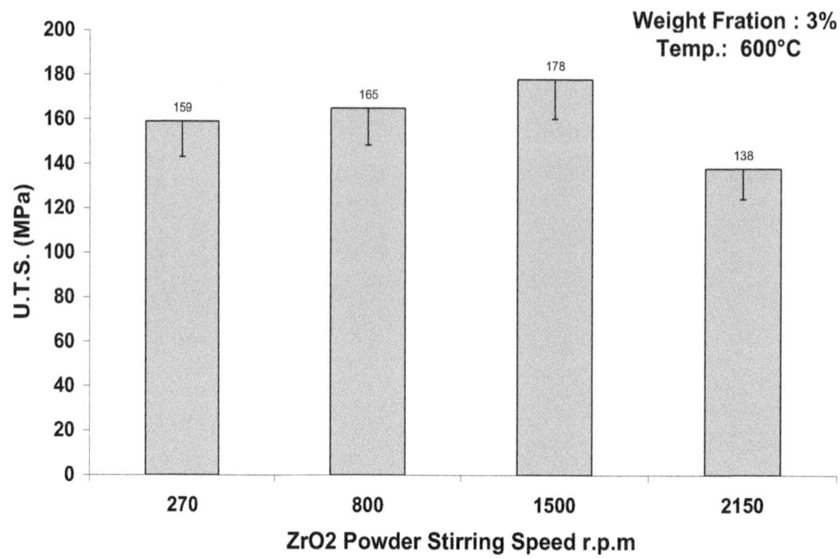

Figure 7. The effect of stirring speed on the UTS of 3 weight% ZrO_2 nanoreinforced alloy at 600 °C.

Figure 7 shows that as the stirring speed of the ZrO_2 nanoreinforced material increases to 1500 rpm, the UTS increases to reach 192 MPa. Beyond this stirring speed, the UTS decreases as the stirring speed increases, possibly due to internal porosity [18] and faster cooling of the slurry. At 2150 rpm stirring speed, the ductility reaches its minimum values, possibly due to the high porosity content [18] created in the casting, where its hardness at this stirring speed decreases after reaching its maximum value at 1500 rpm.

4.1.3. Effect of Casting Temperature

Figure 8 shows that the max. UTS is achieved for all conditions for the casting poured from 600 °C with stirring. This could be explained by the beneficial effect of the semi-solid state (mushy zone) on the distribution of the nanoparticles through the A356 matrix, which leads to enhanced reinforcement of the matrix. The good distribution of the particles leads to higher UTS and higher ductility with noticeable change in the hardness values. The improved strength and ductility exhibited by the nanoreinforced alloy cast from the semi-solid state compared to that cast from the liquid state (700 °C) is attributed to the high effective viscosity of the metal slurry that prevents particles from settling, floating, or agglomerating.

The known models addressing the incorporation of particles into a solidifying matrix have been identified [30] to be: (i) the kinetic models based on the interaction between the velocity of the solid/liquid interface and the velocity of the particle; (ii) the thermodynamic models which are closely related to classical heterogeneous nucleation theory; and (iii) the models based on the ratio of the thermophysical properties of the particles and the melt. A good correlation among all three models [31] has integrated many of the effective parameters, and has suggested that there are three possible pathways for nanoparticle capture: viscous capture, Brownian capture and spontaneous capture. Viscous capture occurs under conditions of extremely high cooling rates or increased viscosity of the melt. Brownian capture is likely to occur when the nanoparticles have slightly lower Hamaker constants than the melt. Spontaneous capture was proposed as the most favorable for nanoparticle capture during solidification of metal melt, the key being to select or design/fabricate suitable nanoparticles that offer a higher Hamaker constant than that of the liquid metal [30]. However, the problem with these models is that they consider capturing of the nanoparticles, but ignore strengthening mechanisms for the proposed matrix/reinforcement combinations.

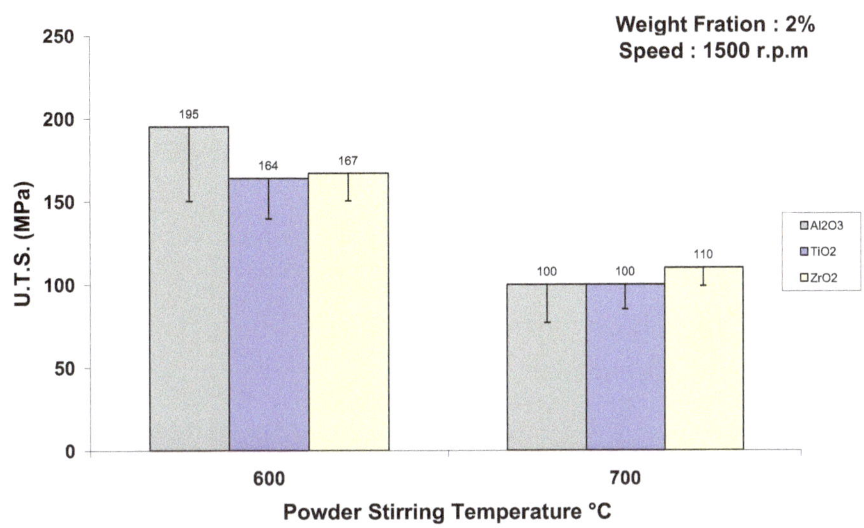

Figure 8. The effect of stirring temperature on the UTS at 1500 rpm stirring speed.

This work as well as others have shown that adding nanoparticles to metal matrix of A356 (and A390 [19]) has significant potential of enhancing both strength and ductility. It has been shown [24] that in eutectic and hypereutectic matrix alloys, such as A1-Si, particle capture take place for all growth conditions, whereas, in hypoeutectic Al alloy systems, where a non-planar solid-liquid interface is typically present, particle pushing takes place. The classical theory for strengthening in dispersed structures forming composites [24] is based on the dislocation motion being constrained by the precipitates and dispersoid particles. These particles form coherency strains and dislocation networks around them, leading to a misfit between the particles and the matrix, which causes the hardness and strength to increase. The usual drop in ductility associated with increase in strength is explained by the fact that dislocations move on specific slip planes under the action of a shear stress. If the slip plane has obstacles penetrating it, e.g., precipitates or dispersoids, a dislocation moving on this slip plane must interact with these obstacles. The high volume fraction of interfacial regions offered by the nanodispersion cause the improvement in ductility where grain boundary sliding becomes the major deformation mechanism [28].

4.2. Microstructural Evolution

Figure 9a–c shows the optical microstructure of the base matrix A356 and the alloys reinforced with 2 weight% Al_2O_3 and 3 weight% TiO_2, using stir casting (rheocast), respectively. The typical microstructure for this alloy consists of primary aluminum matrix (white phase) and Al-Si eutectic (dark lamellar structure). The as-cast microstructures of the three castings show that the phases are uniformly distributed. The figures clearly show a morphological change in the microstructure. In the base matrix sample, the microstructure is dendritic, whereas in the other rheocast samples, the primary dendrites are fragmented due to mechanical stirring. However, with the continued stirring, the plastic strains within the fragmented grains would be considerably less and the process of coarsening will start (ripening effect) [13]. Since the coarsening is driven by interfacial energy, the process will lead to a reduction in the surface area and eventually spheroidal morphology is obtained. It was observed that not only the primary Al grains (white portions) but also eutectic Si particles (dark portions) are globular in rheocast samples in comparison to the base matrix, without stirring, cast samples. As the structure contains good amount of eutectic phases it should give a range of mechanical properties with mechanical stirring during processing [13]. The microstructure observations well support the results of the mechanical properties of the alloys reinforced with nanoparticles using mechanical stirring. The observed refinement of the Si eutectic has been also observed and recorded before [22] and has been attributed to the improved nucleation of Si particles and depletion of Si element near the Si particles [22]. However, the authors [22] have attributed this effect to heterogeneous nucleation initiated by the γ-Al_2O_3 nanoparticles. The effect on modifying the eutectic Si has been attributed to the result from the effect of the pushed γ-Al_2O_3, during solidification into the remaining melt, on restricting growth of the particles.

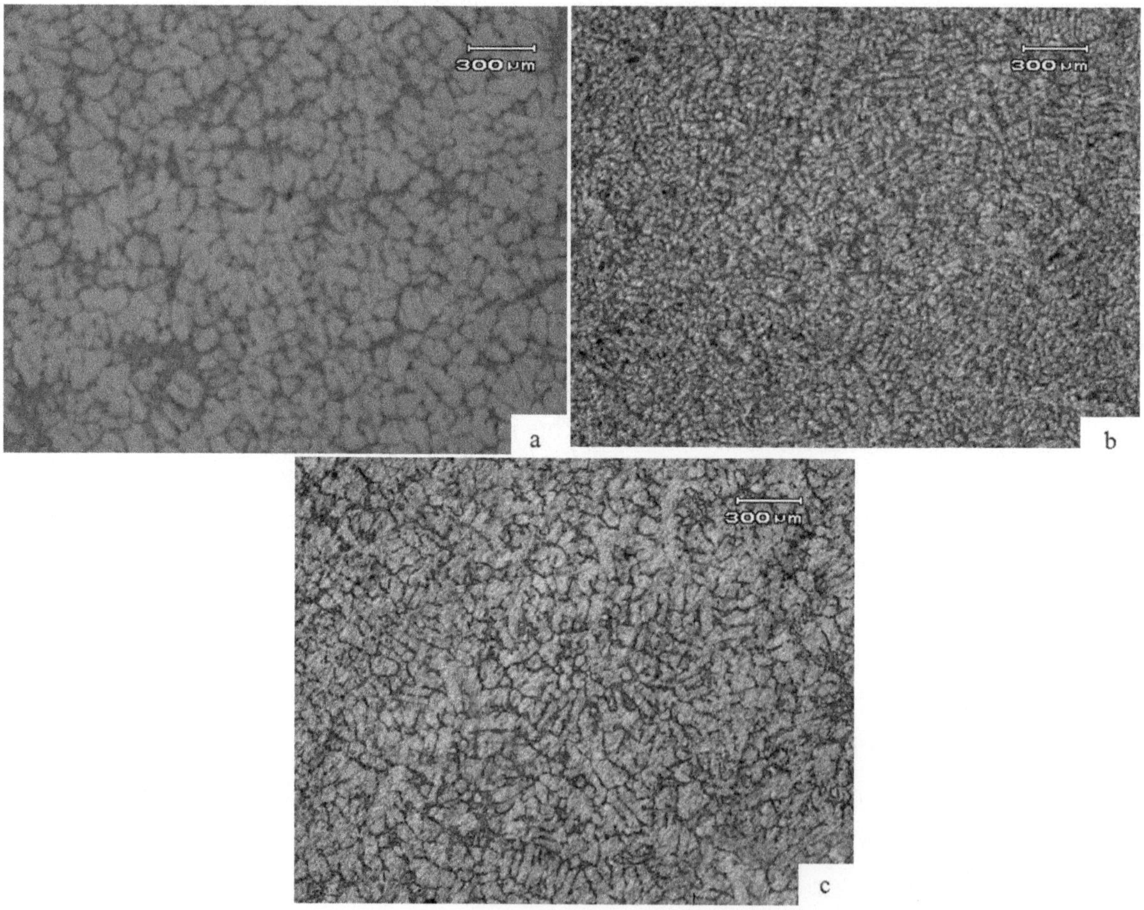

Figure 9. (**a**) Microstructure of the A356 matrix alloy with no stirring. (**b**) Microstructure of A356 nanoreinforced with Al_2O_3. (**c**) Microstructure of A356 nanoreinforced with TiO_2.

Stirring as an external applied force is believed to bring thermal homogeneity within the semi-solid slurry establishing shallow temperature gradients that alleviate nucleation barrier within the bulk liquid. In addition, stirring may assist in disintegration of the secondary and tertiary dendrite arms. Such phenomenon is the main recipe for equiaxed grain growth in SSM cast billets with distinct deformation and flow characteristics [14].

Figure 10 represents the microstructure of the A356 matrix alloy reinforced with 2 weight% Al_2O_3 at 600 °C with different stirring speeds. Figure 10a illustrates the presence of large grains of primary α-Al caused by low stirring at 270 rpm, which causes poor mechanical properties. When the stirring speed increases, this results in finer grains, which lead to an improvement in the mechanical properties of the castings represented by microstructures in Figure 10b,c.

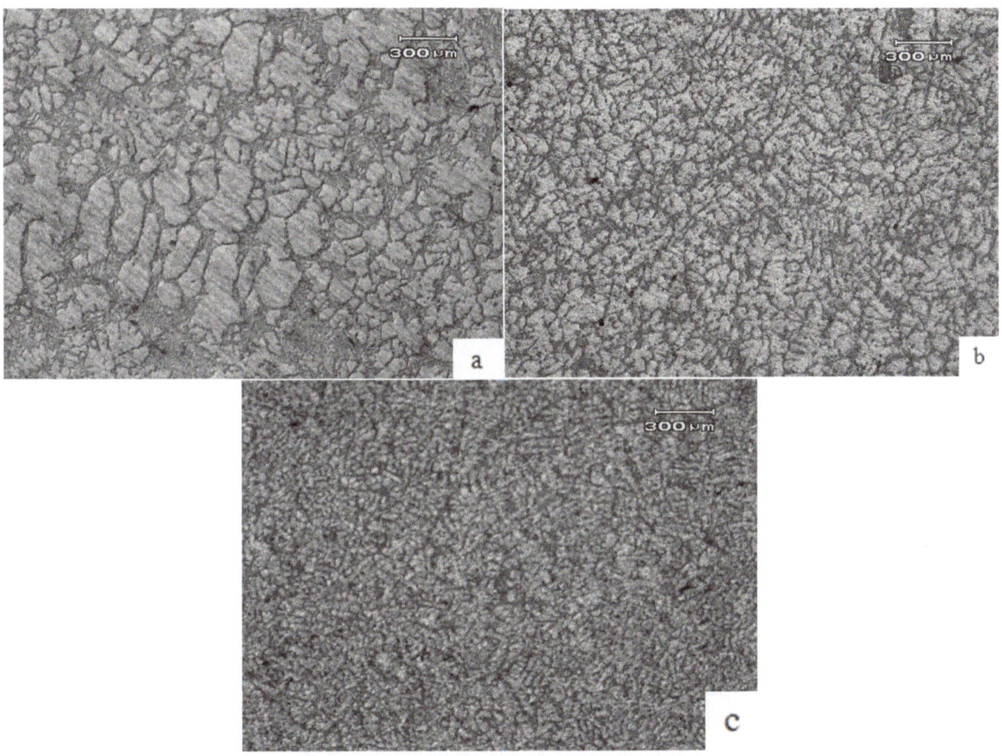

Figure 10. The microstructure obtained for different stirring speeds for the casting reinforced with 2 weight% Al_2O_3 at 600 °C: (**a**) 270 rpm; (**b**) 800 rpm; and (**c**) 1500 rpm.

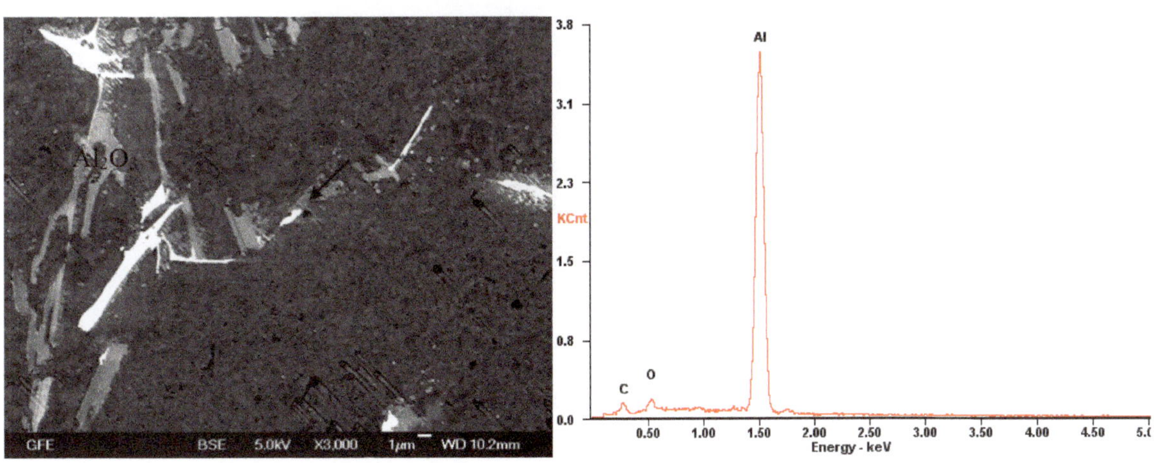

Figure 11. The SEM (**left**) and EDX (**right**) for the Al-Al_2O_3 nano-reinforced matrix.

Figure 11 shows the SEM and EDX for the Al-Al_2O_3 nano-reinforced matrix, which clearly shows the existence of the Al_2O_3 dispersed nano-particles in the Al-Matrix.

Figure 12 illustrates the microstructure of the A356 matrix alloy reinforced with 3 weight% ZrO_2 at 600 °C with 270 rpm stirring speed. Figure 12 shows the presence of large grains of primary α-Al resulting from the low stirring speed which causes pour mechanical properties.

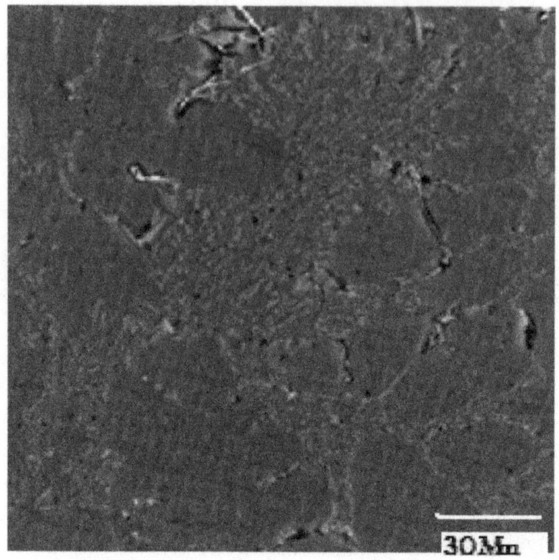

Figure 12. The microstructure of A356 reinforced with 3 weight% ZrO_2 at 270 rpm stirring speed.

Figure 13 shows the SEM fracture surface of the A356 alloy containing Al_2O_3, on which the agglomerated particles are detected. Agglomeration of nano-particles occurs during the melting stage and results in non-homogenous distribution inside the matrix. The Al_2O_3 agglomerated particles have a size of about 3 μm and are located in the interdendritic spaces, attached to the matrix. The EDXS analysis confirmed that these are Al_2O_3 particles, though a strong reflection from the matrix was inevitable.

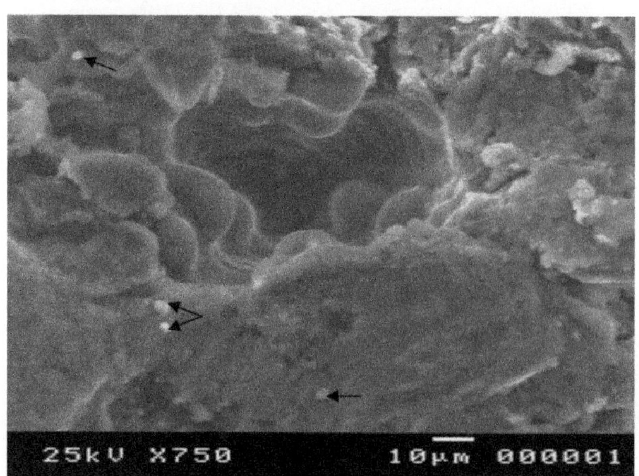

Figure 13. SEM for A356 Alloy containing Al_2O_3 agglomerated particles.

Figure 14 illustrates the brittle fracture surface of the specimen of the A356 alloy reinforced with 3 weight% TiO_2, which showed the optimum mechanical properties. It was difficult to detect TiO_2 nanoparticles; possibly due to the absence of agglomerated particles, which could appear clearly with the Al_2O_3 agglomerated particles.

Figure 15 shows the SEM of the A356 alloy containing ZrO_2 nanoparticles that have agglomerated during the melting stage, leading to a drop in the mechanical properties.

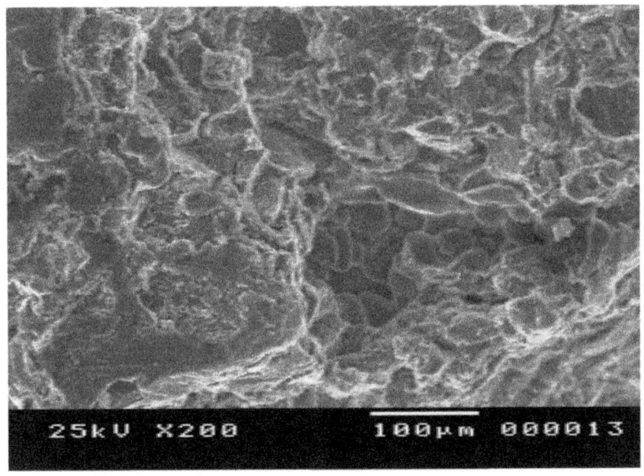

Figure 14. The (SEM) of the fracture surface specimen reinforced with 3 weight% TiO_2 nanoparticles.

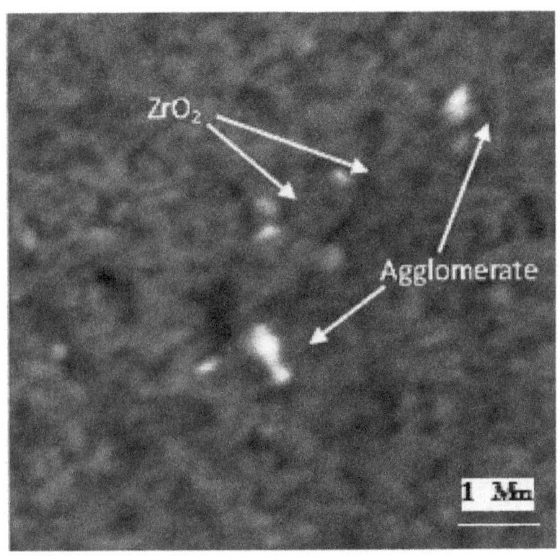

Figure 15. SEM for A356 Alloy containing ZrO_2 agglomerated particles.

4.3. Wear Test Results

The average-wear results of A356 samples reinforced with 0, 1, 2 and 4 weight% Al_2O_3 nanoparticles are shown in Table 5. It can be seen from Table 5 that the addition of 1 weight% nanoparticles resulted in a significant drop in friction coefficient, unaccompanied by any improvement in terms of weight loss. Increasing nanoparticles up to 4% resulted deterioration in wear resistance in terms of friction coefficient and weight loss.

Table 5. The average wear results of A356 samples reinforced with 0, 1, 2 and 4 weight% Al_2O_3 nanopowder.

Sample No.	Additions	Weight Loss (mg)	Friction Coefficient
1	A356	3.9	0.4
2	A356 + 1% Al_2O_3	4.0	0.361
3	A356 + 2% Al_2O_3	4.5	0.385
4	A356 + 4% Al_2O_3	5.5	0.430

The results show that the addition of 1 weight% nanoparticles did not produce a significant change in the wear resistance of the hypo-eutectic alloy A356, though there was a reduction in the friction coefficient. The fact that the wear resistance deteriorated after adding 2 and 4 weight% nanoparticles (though the hardness and strength increased) may be attributed to microstructural effects as well as the high load used in the test. Previous work on A390 alloy [19] has shown that the wear resistance of nanoreinforced A390 was enhanced compared to monolithic A390 alloy, which has been attributed to higher hardness resulting from refinement in microstructural constituents. This behavior is also attributed to microstructural effects in both alloys, where nano additions promote Al phase formation for the hypoeutectic alloys, whereas their refining effect is more pronounced for the hypereutectic alloys. The main wear mechanism observed [19] was delamination and tearing. Other work [32] has reported enhancement in the wear resistance of A356 rheocast alloy compared to those obtained by conventional casting, which has been attributed to the equiaxed structure obtained by rheocasting. The main wear mechanism was found to be plowing at small stresses and mixed plowing and delamination at high stresses. However, no data has been found to evaluate the wear performance of the rheocast nanodispersed A356 alloy.

The samples subjected to wear tests have been examined using SEM. Figure 16 show clearly the Al_2O_3 nanoparticles agglomerates distributed on the surface of the samples with 1 and 2 weight% Al_2O_3 addition subjected to wear, respectively. The figures also show that the dominant wear mechanism was a combination of adhesion and delamination mechanisms, similar to findings in a previous work for hypereutectic Al-Si alloy A390 [19].

In the base matrix sample without stirring, the microstructure is dendritic, whereas in the other rheocast samples, the primary dendrites are fragmented due to mechanical stirring which explains the improvement in the mechanical properties. Analysis using both scanning electron microscope (SEM) and high magnification shows evidence for the possibility of incorporating and entrapping nanosized particles within the interdendritic interface developing during the solidification of the dispersed alloys.

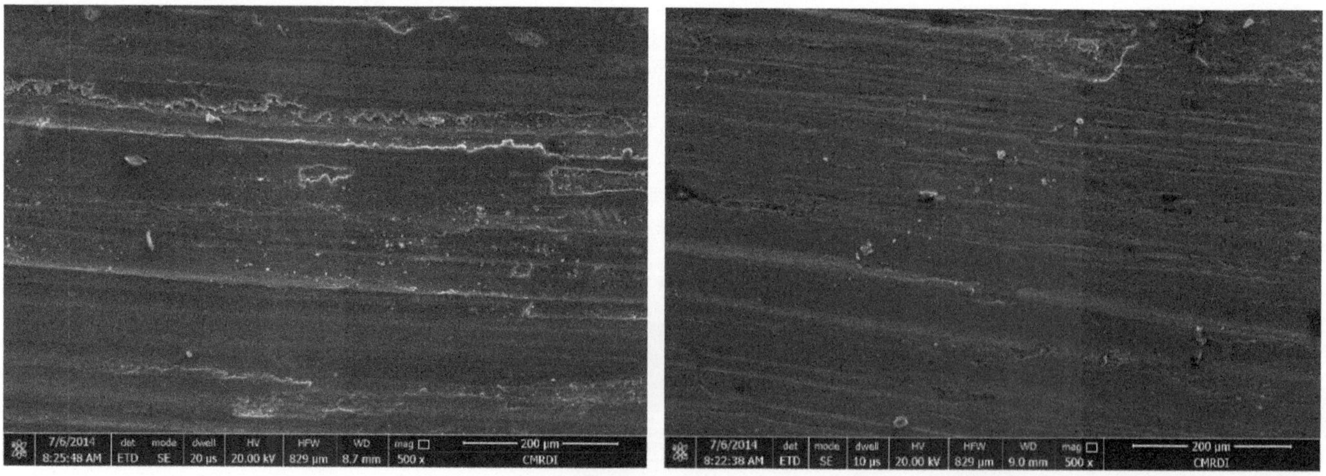

(a) (b)

Figure 16. SEM for wear surface of (**a**) 1 weight% Al_2O_3 and (**b**) 2 weight% Al_2O_3 nanoparticles.

The difference between the wear performance of the nanodispersed hyper- [19] and hypo-eutectic alloys (reported in this work) may be explained by the effect of the nanoparticles additions on both alloys. The addition of nanoparticles promotes the formation of α-Al phase and has been attributed to the improved nucleation of Si particles and depletion of Si element near the Si particles [19,33]. This would produce a deterioration effect on the wear resistance of the hypo-eutectic alloys, whereas the effect would be less significant on the hyper-eutectic alloys, since the morphology of the Si particles and the shape of the α-Al, as well as the morphology of the eutectic Si have a major contribution. It has been stated [19,33] that the cracks easily initiate inside the brittle primary Si or eutectic Si, and then propagate through boundaries with α-Al phase, thus the change in the Si morphology (according to restricted nucleation and growth theory of modification), and the crystalline appearance of the α-Al phase would improve the wear resistance for the hyper-eutectic alloys [33]. Also, the addition of nanoparticles to the hypereutectic alloys has been shown to produce a refining effect on the primary silicon particles, which has been attributed to increasing the nucleation sites [19] or inducing a thermal sub-cooling effect.

It has been shown [18,19] that nanodispersed aluminum alloys have refined microstructural constituents (Al dendrites, interlamellar spacing and primary Si particles) compared to their monolithic counterparts. The mechanical strength of crystalline metals or alloys is largely controlled by the grain size d. The well-known empirical Hall–Petch equation relates the yield strength σy to the average grain size d according to $\sigma_y = \sigma_0 + k(d)^{-1/2}$. Where σ_0 is the friction stress and k is a constant. A similar relation exists between the hardness and the grain size. Consequently, reducing d to the nanometer regime increases the strength considerably. However, the limits of the conventional description of yielding and of new mechanisms that may come into play at these small dimensions need to be explored and studied in much more detail [28].

Although the mechanisms operative in the nanoscale grain size regime differ from those for coarse-grained materials, strengthening mechanisms that are well documented for coarse-grained metals and alloys can also operate in nanocrystalline metals and alloys, although modified by the nano-scale grain size and non-equilibrium microstructures [34]. It is suggested that the smaller sizes of nanoparticles are expected to contribute to Orowan hardening; while the very large sizes (or agglomerates) would likely produce rule-of-mixtures behavior. Nanodispersed cast structures carry both natures of composites and refined structures. It is important at this stage to understand the active strengthening mechanisms in these materials, as this will be the key to developing new structures.

A new model has been recently suggested considering that the high mechanical resistance of nanoreinforced composites is the result of the incremental summation of several strengthening mechanism contributions, namely: load transfer effect, Hall-Petch strengthening, Orowan strengthening, coefficient of thermal expansion (CTE) and elastic modulus (EM) mismatch [35]. The new proposed model was shown to be consistent with experimental data for the system Al/2weight% Al_2O_3 [35].

5. Conclusions

- The castings made by adding nano-sized dispersoids using semi-solid route exhibited higher strength and ductility when compared with those prepared by liquid metallurgy rout.
- The stirring speed has a significant effect on the mechanical properties of the nano-dispersed castings. Increasing stirring speed more than 1500 rpm causes reduction in the tensile strength. The alloy stirred with 1500 rpm exhibits the highest tensile strength and elongation%.

- The A356 matrix alloy reinforced with 2% weight of Al_2O_3 showed the highest strength properties when compared with the one reinforced with 3% ZrO_2 weight and 3% weight of TiO_2 nano particles at the conditions of 1500 rpm stirring speed at semi solid state temperature 600 °C. In all cases this was not combined with loss in ductility.
- Analysis using scanning electron microscope (SEM) at high magnification shows evidence for the possibility of incorporating and entrapping nano-sized particles within the interdendritic interface developing during the solidification of the dispersed alloys.
- The introduction of varying amounts of nanosized particles to the A356 alloy did not produce a significant change on the wear resistance of the tested hypo-eutectic alloy A356 material with 1% nanoparticles and resulted deterioration after adding 2% and 4% nanoparticles, though a drop in the friction coefficient occurred at 1% addition.
- The options for making nanodispersed cast alloys still meet the challenge of obtaining homogeneous distribution of the nanoparticles.
- Another challenge is quantifying the amount of nanoparticles entering the slurry.

Acknowledgments

The authors would like to acknowledge the role of late -Ing. Y. Shash, the head of Mechanical Design and Production Dept., Faculty of Engineering, Cairo University who advised and supported this work at its early stages.

Author Contributions

Iman El-Mahallawi has led the research group, designed and performed the experiments for casting preparation, and microstructural and mechanical properties investigation for the Al oxide nanoreinforced material and wrote the introduction and discussion parts of the paper. Ahmed Shash has designed and performed the experiments of casting preparation, and microstructural and mechanical properties investigation for the Ti oxide nanoreinforced material and wrote the relevant parts. Amer Eid Amer has designed and performed the experiments of casting preparation, and the microstructural investigations for the Zr oxide nanoreinforced material and wrote the relevant parts.

Conflicts of Interest

The authors declare no conflict of interest.

References

1. Kashyap, K.T.; Muralli, S.; Raman, K.S.; Murthy, K.S.S. Casting and heat treatment variables of Al–7Si–Mg alloy. *Mater. Sci. Technol.* **1993**, *9*, 189–203.
2. Gruzleski, J.E.; Closset, B.M. *The Treatment of Liquid Alumnuim-Silicon Alloys*; AFS: Des Plaines, IL, USA, 1990; pp. 107–126.
3. Cacers, C.H. Microstructure effects on the strength-ductility relationship of AI-7Si-Mg casting alloys. *Mater. Sci. Forum* **2000**, *331*, 223–228.

4. Rooy, E.L. Aluminum and aluminum alloys. In *ASM Metals Handbook*; Aluminum Company of America: New York, NY, USA, 1992; Volume 15.

5. Shi, W.; Gao, B.; Tu, G.; Li, S.; Hao, Y.; Yu, F. Effect of neodymium on primary silicon and mechanical properties of hypereutectic Al-15%Si alloy. *J. Rare Earths* **2010**, *28*, 367–370.

6. Chen, C.; Liu, Z.; Ren, B.; Wang, M.; Weng, Y.; Liu, Z. Influences of complex modification of P and RE on microstructure and mechanical properties of hypereutectic Al-20Si alloy. *Trans. Nonferrous Met. Soc. China* **2007**, *17*, 301–306.

7. Dahle, A.K.; Nogita, K.; McDonald, S.D.; Dinnis, C.; Lu, L. Eutectic modification and microstructure development in Al–Si Alloys. *Mater. Sci. Eng. A* **2005**, *413–414*, 243–248.

8. Cruz, K.S.; Meza, E.S.; Fernandes, F.A.P.; Quaresma, J.M.V.; Casteletti, L.C.; Garcia, A. Dendritic arm spacing affecting mechanical properties and wear behavior of Al–Sn and Al–Si alloys directionally solidified under unsteady-state conditions. *Metall. Mater. Trans. A* **2010**, *41*, 972–984.

9. Mehrabian, R.; Riek, R.G.; Flemings, M.C. Preparation and casting of Metal-Particulate Non-Metal Composites. *Metall. Trans.* **1974**, *5*, 1899–1905.

10. Eliasson, J.; Sandstorm, R. *Applications of Aluminium Matrix Composites, Part 1*; Newaz, G.M., Neber-Aeschbacherand, H., Wohlbier, F.H., Eds.; Trans. Tech. Publications: Pfaffikon, Switzerland, 1995; pp. 3–36.

11. Rohatgi, P.K.; Asthana, R.; Das, S. Solidification, Structure and Properties of Cast Metal-Ceramic Particle Composites. *Int. Met. Rev.* **1986**, *31*, 115–139.

12. Singh, J.; Alpas, A.T. Elevated temperature wear of Al6061 and Al6061 20% Al_2O_3. *Scr. Metall. Mater.* **1995**, *32*, 1099–1105.

13. Seyed Reihani, S.M. Processing of squeeze cast Al6061–30vol% SiC composites and their characterization. *Mater. Des.* **2006**, *27*, 216–222.

14. Amirkhanlou, S.; Niroumand, B. Development of Al356/SiCp cast composites by injection of SiCp containing composite powders. *Mater. Des.* **2011**, *32*, 1895–1902.

15. Jerome, S.; Ravisankar, B.; Kumar Mahato, P.; Natarajan, S. Synthesis and evaluation of mechanical and high temperature tribological properties of in-situ Al–TiC composites. *Tribol. Int.* **2010**, *43*, 2029–2036.

16. El Mahallawi, I.S.; Egenfeld, K.; Kouta, F.H.; Hussein, A.; Mahmoud, T.S.; Rashad, R.M.; Shash, A.Y.; Abou-AL-Hassan, W. Synthesis and Characterization of New Cast A356/(Al_2O_3) p Metal Matrix Nano- Composites. In Proceedings of the 2nd Multifunctional Nanocomposites & Nanomaterials: International Conference& Exhibition MN2008, Cairo, Egypt, 11–13 January 2008.

17. El-Mahallawi, I.; Shash, Y.; Eigenfeld, K.; Mahmoud, T.; Rashad, R.; Shash, A.; El Saeed, M. Influence of Nano-dispersions on Strength Ductility Properties of Semi-solid Cast A356 Al Alloy. *Mater. Sci. Technol.* **2010**, *26*, 1226–1231.

18. El-Mahallawi, I.; Abdelkader, H.; Yousef, L.; Amer, A.; Mayer, J.; Schwedt, A. Influence of Al_2O_3 nanodispersions on microstructure features and mechanical properties of cast and T6 heat-treated Al Si hypoeutectic alloys. *Mater. Sci. Eng. A* **2012**, *556*, 76–87.

19. El Mahallawi, I.; Shash, Y.; Rashad, R.M.; Abdelaziz, M.H.; Mayer, J.; Schwedt, A. Hardness and Wear Behavior of Semi-Solid Cast A390 Alloy Reinforced with Al_2O_3 and TiO_2 Nanoparticles. *Arab. J. Sci. Eng.* **2014**, *39*, 5171–5184.

20. Mazahery, A.; Baharvandi, H.R.; Abdizadeh, H. Development of high performance A356-nano Al_2O_3 composites. *Mater. Sci. Eng. A* **2009**, *518*, 23–27.
21. Sajjadi, S.A.; Ezatpour, H.R.; Beygi, H. Comparison of microstructure and mechanical properties of A356 aluminum alloy/Al_2O_3 composites fabricated by stir and compo-casting processes. *Mater. Sci. Eng. A* **2011**, *528*, 8765–8771.
22. Choi, H.; Konishi, H.; Li, X. Al_2O_3 nanoparticles induced simultaneous refinement and modification of primary and eutectic Si particles in hypereutectic Al–20Si alloy. *Mater. Sci. Eng. A* **2012**, *541*, 159–165.
23. Haizhi, Y. An overview of the development of Al-Si-alloy based material for engine applications. *J. Mater. Eng. Perform. JMEPEG* **2003**, *12*, 288–297.
24. Chawla, K.K. *Metal Matrix Composites*; Springer: Berlin, Germany, 2006.
25. Prabu, S.B. Influence of stirring speed and stirring time on distribution of particles in cast MMC. *J. Mater. Process. Technol.* **2006**, *171*, 268–273.
26. Shash, A.Y.; Sc, M. Effect of Processing Parameters on the Mechanical Characteristics of A356/(Al_2O_3)p Cast Metal Matrix Nano-Composites (MMNCs). Master's Thesis, Cairo University, Cairo, Egypt, 2007.
27. Shackelford, James F.; Alexander, W. *"Frontmatter" Materials Science and Engineering Handbook*; Shackelford, J.F., Alexander, W., Eds.; CRC Press LLC: Boca Raton, FL, USA, 2001.
28. Koch, C.C. *Nanostructured Materials, Processing Properties, and Potential Applications*; CRC Press: Boca Raton, FL, USA, 2002; pp. 423–526.
29. Kim, G.; Hong, S.; Lee, M.; Kim, S.; Ioka, I.; Kim, B.; Kim, I. Effect of oxide dispersion on dendritic grain growth characteristics of cast aluminum alloy. *Mater. Trans.* **2010**, *51*, 1951–1957.
30. Boostani, A.F.; Tahamtan, S.; Jiang, Z.Y.; Wei, D.; Yazdani, S.; Khosroshahi, R.A.; Mousavian, R.T.; Xu, J.; Zhang, X.; Gong, D. Enhanced tensile properties of aluminium matrix composites reinforced with graphene encapsulated SiC nanoparticles. *Compos. Part A* **2015**, *68*, 155–163.
31. Xu, J.Q.; Chen, L.Y.; Choi, H.; Li, X.C. Theoretical study and pathways for nanoparticle capture during solidification of metal melt. *J. Phys. Condens. Matter* **2012**, *24*, 255304.
32. Dey, A.K.; Poddar, P.; Singh, K.K.; Sahoo, K.L. Mechanical and Wear Properties of Rheocast and Conventional Gravity Die Cast A356 Alloy. *Mater. Sci. Eng. A* **2006**, *435–436*, 521–529.
33. El Mahallawi, I.; Othman, O.A.; Abdelaziz, M.H.; Raed, H.; El-Fatah, T.A.; Ali, S. Understanding the role of nanodispersions on the properties of A390 Hyper-Eutectic Al-Si cast alloy. *TMS Light Met.* **2014**, doi:10.1002/9781118888438.ch227.
34. Scattergood, R.O.; Koch, C.C.; Murty, K.L.; Brenner, D. Strengthening mechanisms in nanocrystalline alloys. *Mater. Sci. Eng. A* **2008**, *493*, 3–11.
35. Casati, R.; Vedani, M. Metal Matrix Composites Reinforced by Nano-Particles—A Review. *Metals* **2014**, *4*, 65–83.

15

Effect of Indium Content on the Microstructure, Mechanical Properties and Corrosion Behavior of Titanium Alloys

Mi-Kyung Han, Jae-Bong Im, Moon-Jin Hwang, Bong-Jun Kim, Hae-Young Kim and Yeong-Joon Park *

Department of Dental Materials and Medical Research Center for Biomineralization Disorders, School of Dentistry, Chonnam National University, Gwangju 500-757, Korea;
E-Mails: mikihan2@naver.com (M.-K.H.); jabim@daum.net (J.-B.I.); mjhwang@jnu.ac.kr (M.-J.H.); deckerd@jnu.ac.kr (B.-J.K.); khenm@hanmail.net (H.-Y.K.)

* Author to whom correspondence should be addressed; E-Mail: yjpark@jnu.ac.kr

Academic Editor: Mark T. Whittaker

Abstract: Ti-*x*In (x = 0, 5, 10, 15 and 20 wt%) alloys were prepared to investigate the effect of indium on the microstructure, mechanical properties, and corrosion behavior of titanium with the aim of understanding the relationship between phase/microstructure and various properties of Ti-*x*In alloys. The Ti-*x*In alloys exhibited a lamellar α-Ti structure at an indium content of up to 20 wt%. High-resolution TEM images of the Ti-*x*In alloys revealed that all the systems contained a fine, acicular martensitic phase, which showed compositional fluctuations at the nanoscopic level. The mechanical properties and corrosion behavior of Ti-*x*In alloys were sensitive to the indium content. The Vickers hardness increased as the In content increased because of solid solution strengthening. The Ti-*x*In alloys exhibited superior oxidation resistance compared to commercially pure Ti (cp-Ti). Electrochemical results showed that the Ti-*x*In alloys exhibited a similar corrosion resistance to cp-Ti. Among the alloys tested, Ti-10In showed a potential for use as a dental material.

Keywords: Ti-*x*In alloys; mechanical properties; corrosion resistance; dental materials

---

1. Introduction

Titanium (Ti) and Ti-based alloys have received widespread attention for orthopedic and dental implant applications because of their favorable mechanical properties, including high specific strength, corrosion resistance, and good biocompatibility after implantation [1,2]. However, Ti has several drawbacks, such as low deformability, low wear resistance, and high reactivity with surrounding impurities, such as oxygen and nitrogen, at elevated temperatures [3,4]. Enhanced properties, such as the lower modulus of elasticity, increased corrosion resistance, and improved machinability, are possible through alloying Ti with a variety of elements.

Several Ti–*X* binary alloys have been developed, such as Ti–Nb [5–7], Ti–Ag [8,9], Ti–Au [10], Ti–Hf [11], Ti–Mn [12], Ti–Cr [13,14], Ti–Mo [15], Ti–Sn [16], Ti–Zr [17–19], Ti–Ta [20,21], Ti–Co [22], Ti–Pd [23], Ti–Ge [24] and Ti–Cu [25] alloys. The mechanical properties of Ti alloys are influenced by their microstructure, which, in turn, depends on chemical composition and synthetic processing. In fact, the key to this research resides in the judicious choice of the alloying elements. Therefore, in order to improve the mechanical properties of Ti, it is necessary to understand the effect of the alloying elements on the microstructure and the mechanical properties of the alloys.

There are several studies on developing Ti alloys containing indium. It was found that indium-containing Ti alloys were biocompatible. It was recently found that the addition of indium to Ti improves clinical performance of dental implant, such as mechanical properties, corrosion resistance, and biocompatibility [26,27]. Other than this, little experimental investigation has been attempted regarding Ti-In alloys. Sufficient information about the relationship between mechanical properties and phases is not available in the literature. Therefore, it is important to pursue a comprehensive structural analysis of this phase to provide an in-depth understanding of the mechanical properties.

In this study, the goal of which was the development of a dental titanium alloy with better mechanical properties than cp-Ti, the effect of alloying indium on the microstructure, mechanical properties, and corrosion behavior of Ti-In binary alloys was investigated by adding 5, 10, 15, and 20 wt% In. Henceforth, in this work, "Ti-*x*In" will stand for "Ti-*x* wt% In".

2. Experimental Section

2.1. Preparation of Ti-xIn Alloys

Experimental Ti-*x*In alloys (0, 5, 10, 15, and 20 wt% In) were prepared by arc-melting stoichiometric quantities of the elements in a water-cooled copper hearth using a tungsten electrode in a high-purity argon atmosphere. The starting materials (Ti sponge, Alfa Aesar, Ward Hill, MA, USA, 99.95%; In ingot, LS-Nikko, Seoul, Korea, 99.95%) were used without purification. During the arc-melting procedure, a titanium getter was heated prior to melting the reactant mixture to further purify the argon atmosphere. The samples were remelted seven times to promote sample homogeneity. Subsequently, the samples were heat treated for 4 h in an argon atmosphere using a tube furnace at temperatures below 150 °C to the respective solidus temperatures by cooling in a furnace at a rate of 10 °C/min down to 600 °C and then air-cooling to room temperature. These heat treatment conditions were chosen in accordance with the binary Ti–In phase diagrams [28]. Figure 1 shows a Ti-rich portion of the Ti–In phase diagram exhibiting the four alloys investigated and the temperature range of heat-treatment.

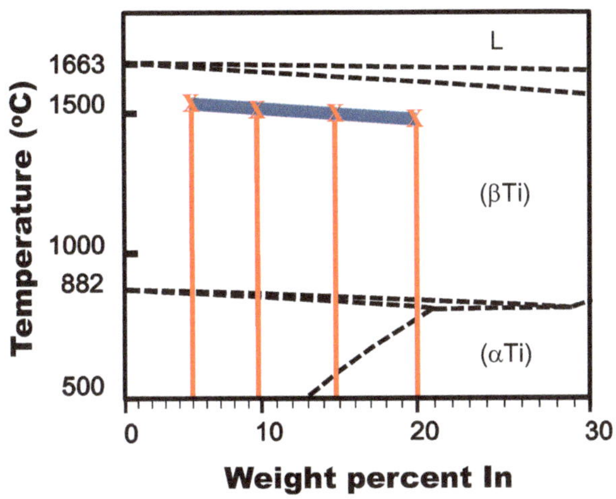

Figure 1. Ti-rich portion of the Ti–In system according to the published Ti-In equilibrium phase diagram [28]. The four alloys investigated and the temperature ranges of heat-treatment are marked.

2.2. Material Characterization

Phase analysis and structural characterization were performed using X-ray diffraction (XRD, PANalytical B.V., Almelo, The Netherlands). The XRD diffraction patterns were collected with an X'Pert PRO Multi-Purpose X-Ray Diffractometer (40 kV and 40 mA) using Cu K_{α} radiation (λ = 1.54056 Å). The scanning speed was 2°/min, and the scanning angle, 2θ, ranged from 20° to 80°. The lattice parameters were obtained by least squares refinement of the data with the aid of a Rietveld refinement program [29]. The microstructure of the samples was examined using a metallurgical microscope (Epiphot FX-35WA, Nikon, Tokyo, Japan), high-resolution transmission electron microscopy (HRTEM, Technai-F20, Philips, Amsterdam, The Netherlands) and selected area electron diffraction (SAED).

2.3. Measurement of Mechanical Properties

The microhardness of the polished alloys was measured using a Vickers microhardness tester (Zwick, Postfach 4350, Ulm, Germany) with a 500 g load maintained for 30 s. Elastic modulus testing was performed using a Nanoindenter XP (MTS Co., Dubuque, IA, USA) with a maximum indentation depth of 2 μm. The oxidation behavior of Ti with different In contents was tested using thermogravimetric analysis (TGA, SDTA 851e, Mettler-Toledo, Columbus, OH, USA), which enabled measurement of the change in mass due to oxidation. Samples measuring 4.5 × 4.2 × 1.0 mm^3 in size were heated to 795 °C and 1000 °C at a heating rate of 10 °C/min and air flow rate of 50 mL/min. After fracture by an Instron 4302 testing machine at a crosshead speed of 1.5 mm/min, the fracture surface morphology was investigated using a scanning electron microscope (SEM, S-3000N, Hitachi, Tokyo, Japan). The surface roughness of the ground surface (0.4 × 0.3 mm^2 in size) was measured using a Nanosurface 3D optical Profiler (NV-E1000, NanoSystem, Daejeon, Korea).

2.4. Electrochemical Analysis

To observe the corrosion behavior of the Ti-*x*In alloys, potentiodynamic anodic polarization tests were conducted at a scan rate of 5 mV/s from −1.5 V to +1.5 V (saturated calomel electrode: SCE) using a potentiostat (WAT100, WonA Tech Co., Ltd., Seoul, Korea) in a 0.9% NaCl solution at 37 ± 1 °C. Three samples were tested to confirm the experimental results through repetition. The surfaces of the samples 10 mm in diameter were mechanically polished using SiC paper up to 2000 grit. Electrochemical measurements were recorded using the three electrode technique, which involves the use of a working electrode (test samples), counter electrode (high density graphite), and reference electrode (SCE) [30]. Ar gas was bubbled into the electrolyte at 150 mL/min for over 20 min to eliminate residual oxygen in the electrolyte. The used electrolyte was replaced with a fresh one before each measurement. The exposed surface area of samples in the electrolyte was 0.283 cm^2. Potentiodynamic polarization curves were plotted using an automatic data acquisition system. Both the corrosion current density and potential were estimated with Tafel plots using both the anodic and cathodic branches.

3. Results and Discussion

3.1. Phase and Microstructure

Figure 2 shows the XRD patterns as a function of x for the Ti-*x*In (*x* = 0, 5, 10, 15, and 20 wt%) samples. There was no obvious change in the diffraction patterns and peak positions as the In content changed. The patterns of Ti-*x*In alloys can be indexed as a hexagonal α-Ti type structure (the space group of $P6_3/mmc$). In all of the Ti-*x*In alloys, no peak splitting was observed, which indicated that the structural transformation from α-Ti to β-Ti did not occur.

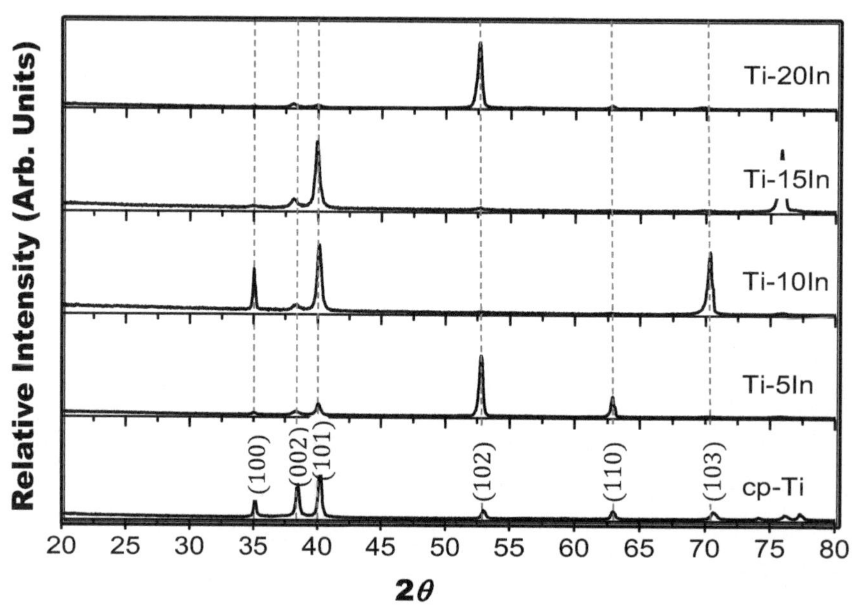

Figure 2. X-ray diffraction (XRD) patterns of commercially pure Ti (cp-Ti) and the series of binary Ti-*x*In alloys. The vertical dotted line in the figure is a guide to the eye, and it connects the peak positions of the α-Ti phase.

Rietveld refinements were done to investigate the effect of In on the lattice parameters, and the results are shown in Figure 3. The bigger atomic radius of In (approximately 1.67 Å) compared to Ti (approximately 1.47 Å) [31] meant the addition of In atoms caused the lattice parameters to increase. This caused the XRD peaks to shift toward lower angles. The lattice parameters a and c increased linearly with increasing In content in the Ti-xIn alloys, but the c axis increased much more rapidly than the a axis. As a result, the c/a ratio increased linearly as the In content increased, which is in agreement with previous findings [28]. The unit cell constants of cp-Ti were a = 2.951 (1) Å and c = 4.683 (1) Å (c/a ratio = 1.581), corresponding with those in the literature (JCPDS card No. 44-1294) [32], whereas the unit cell parameters of Ti-20In were a = 2.964 (1) Å and c = 4.740 (1) Å (c/a ratio = 1.599). Based on the linear relationship of the lattice parameters in the Ti-xIn alloys, we estimated that up to 20 wt% In dissolved to form a solid solution, in agreement with the binary Ti-In phase diagram [28].

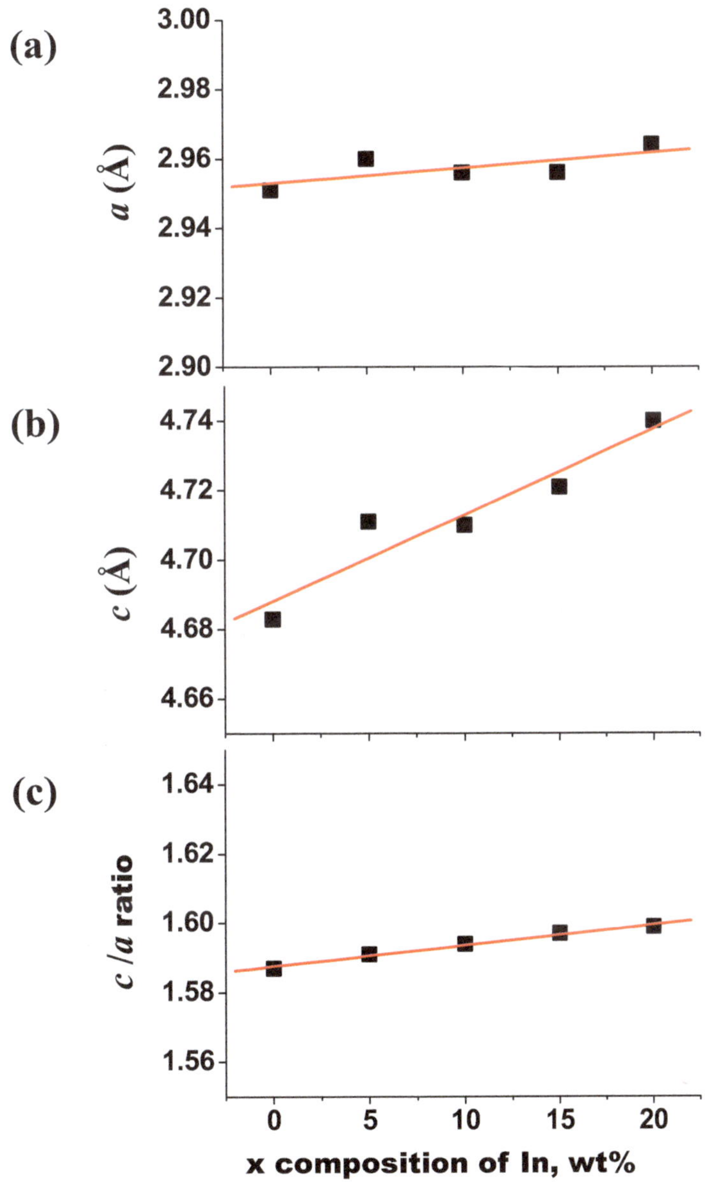

Figure 3. (**a**, **b**) Lattice parameters (**a** and **c**) of cp-Ti and Ti-xIn alloys and (**c**) the variation in the ratio (c/a) of lattice parameters.

Figure 4 shows the representative optical micrographs of the Ti-*x*In alloys. All the Ti-*x*In alloys exhibited lamellar pattern microstructure. Detailed microstructure analyses were performed using TEM. As shown in Figure 5, the fine, acicular martensitic phase was observed, especially in Ti-10In alloy. The SAED patterns of the Ti-10In alloy consisted of a characteristic single α-Ti phase (Figure 5c) in the matrix which showed a perfect atomic arrangement and a highly faulted sub-microstructure (Figure 5d) due to Ti and In solid solution formation.

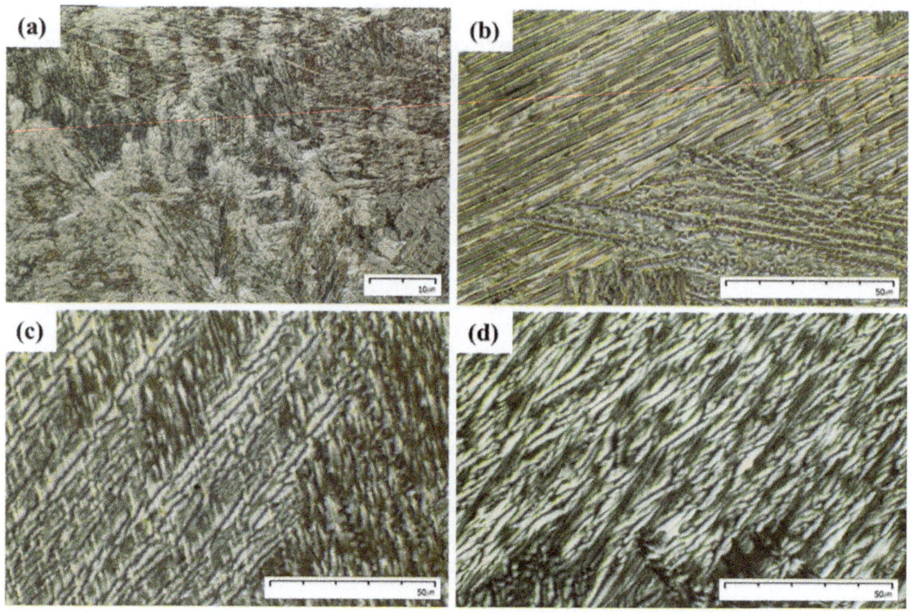

Figure 4. Optical micrographs of Ti-*x*In alloys: (**a**) Ti-5In, (**b**) Ti-10In, (**c**) Ti15In, and (**d**) Ti-20In.

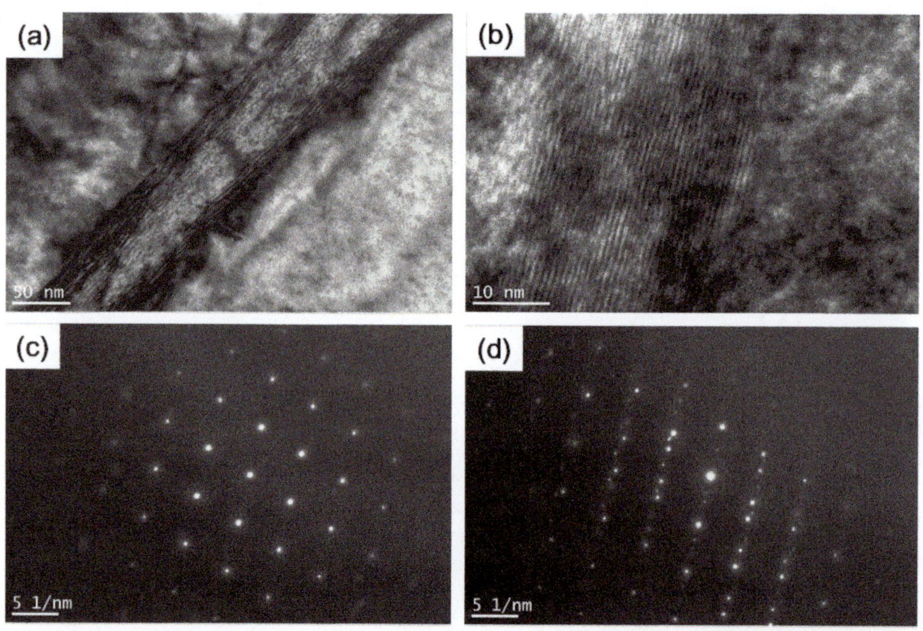

Figure 5. Transmission electron microscopy (TEM) and selected area electron diffraction (SAED) patterns of Ti-10In alloy. (**a**) TEM image, (**b**) magnified TEM image, and (**c**,**d**) representative SAED patterns.

3.2. Mechanical Properties

Table 1 shows the Vickers hardness and elastic modulus of Ti-*x*In alloys compared to those of cp-Ti ($n = 5$). All the Ti-*x*In alloys had significantly higher Vickers hardness values ($p < 0.05$) than that of cp-Ti. Of the Ti-In alloys, the alloy with 10 wt% In content showed the highest hardness value (522 HV), which was about three times larger than that of cp-Ti (165 HV). The abnormal hardness of the Ti-10In alloy was probably caused by the combined effect of the solid-solution strengthening of the α-phase and the hardening of the fine, acicular martensitic phase. This was in agreement with previous results by Ho *et al* [15]. They reported a sharp increase in hardness caused by the presence of a martensitic phase in the Ti-Mo system. On the other hand, no consistent trend in the elastic modulus related to changes in In concentration in the Ti-*x*In alloys was observed. The elastic modulus for cp-Ti was 132 GPa. The elastic moduli of the Ti-5In and Ti-15In alloys were slightly higher than that of cp-Ti, whereas the elastic moduli of the Ti-10In and Ti-20In alloys were lower ($p < 0.05$). The values for Ti-10In and Ti-20In alloys were about 124 GPa. Therefore, the Ti-10In and Ti-20In alloys are more suitable for dental applications because of their low modulus and high strength. The appearance of metal chips resulting from metal grinding at 10,000 rpm was examined using an SEM and is shown in Figure 6. Although no quantitative evaluation was performed, chips of Ti-*x*In alloys appeared larger than those of cp-Ti. As the In concentration increased, the chips became longer.

Table 1. Vickers hardness values of Ti-*x*In alloys compared with cp-Ti.

Alloy code	Hardness (VHN)	Elastic modulus (GPa)
cp-Ti	164.54 (3.54) ^{*,a}	132.35 (12.22) ^{a}
Ti-5In	402.60 (128.12) ^{b}	159.80 (2.57) ^{c}
Ti-10In	522.40 (61.03) ^{b}	124.10 (4.61) ^{a}
Ti-15In	303.20 (89.38) ^{c}	141.10 (4.01) ^{b}
Ti-20In	332.00 (46.80) ^{c}	124.43 (4.69) ^{a}

* Within the same column, the different alphabet in superscript indicated statistically different mean values at a 5% level ($p < 0.05$) according to Duncan's multiple range test.

Figure 7 shows profiles of the ground surfaces of the Ti-*x*In alloys, which were ground at 10,000 rpm using a SiC. Grinding marks were observed on all of the Ti-*x*In alloys. However, there was no pronounced difference in the appearances of the ground surfaces of all the Ti-*x*In samples. A surface roughness parameter (Ra) was selected to express the surface roughness. The cp-Ti showed a mean Ra value of 1.4 (3) μm. The Ra values for Ti-*x*In alloys were in the range of 1.2 (2)–1.8 (2) μm. The Ti-5In showed the highest mean Ra value of 1.8 (2) μm. The mean Ra value decreased as the In concentration increased in Ti-*x*In alloys. The oxidation behavior of Ti-*x*In alloys was assessed using thermogravimetric analysis (TGA).

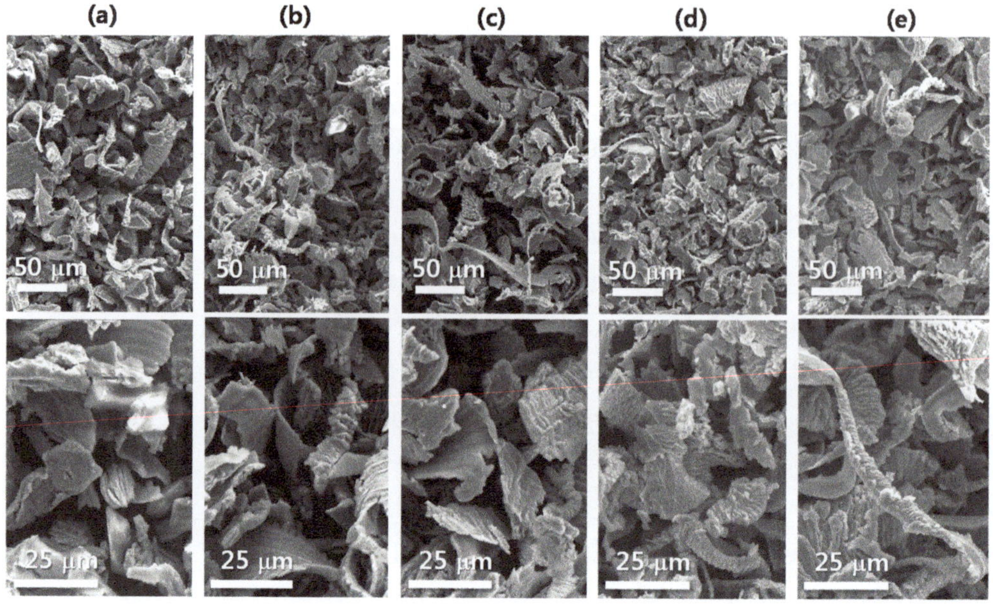

Figure 6. Metal chips resulting from grinding at 10,000 rpm (two magnifications): (**a**) cp-Ti, (**b**) Ti-5In, (**c**) Ti-10In, (**d**) Ti-15In, and (**e**) Ti-20In.

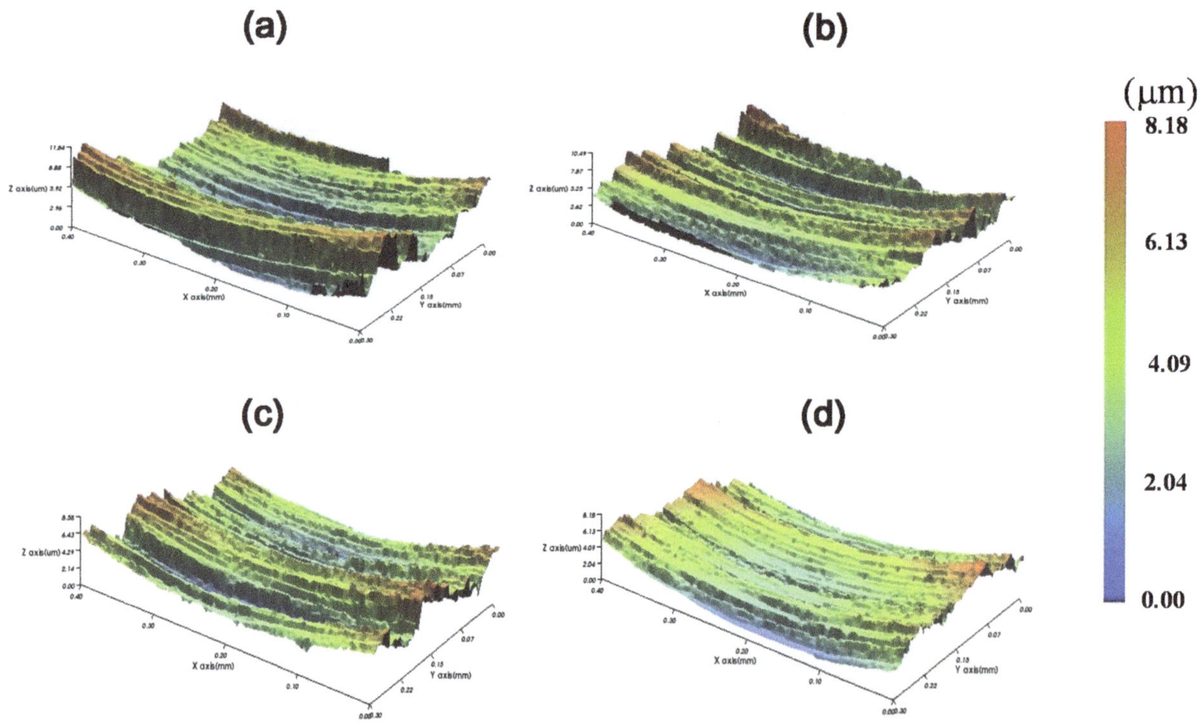

Figure 7. Profiles of ground surfaces of Ti-*x*In alloys at a grinding speed of 10,000 rpm: (**a**) Ti-5In, (**b**) Ti-10In, (**c**) Ti-15In, and (**d**) Ti-20In.

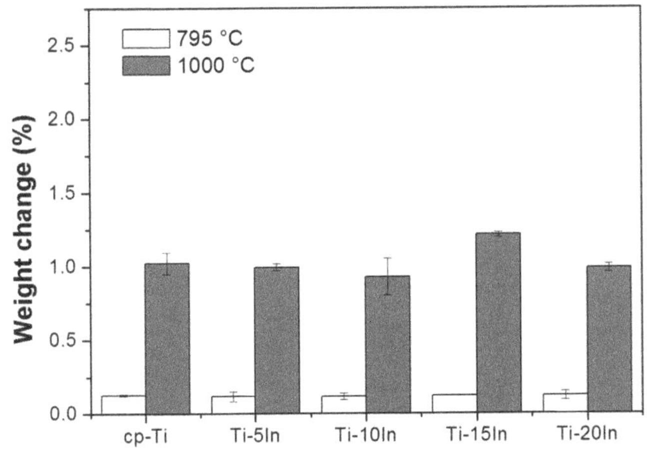

Figure 8. Thermogravimetric analysis (TGA) of cp-Ti and Ti-*x*In alloys showing various degrees of weight gain (%) after heating in air up to 795 °C and 1000 °C.

Figure 8 shows the result of TGA when the cp-Ti and Ti-*x*In alloys were heated up to 795 °C and 1000 °C at a heating rate of 10 °C/min in air ($n = 2$). At temperatures higher than 500 °C, the Ti-*x*In alloys were rapidly oxidized, resulting in a significant increase in mass. All of the In-containing samples showed weight gain of about 0.1% at a temperature of 795 °C, whereas at 1000 °C, a significant change in mass of 0.8%–1.2% was observed. Among the tested Ti-*x*In alloys, the final mass change in the Ti-10In alloy was significantly less than that after the oxidation of cp-Ti, indicating that the addition of In to cp-Ti could restrain the oxidation rate of the alloy, and Ti-10In alloys had stronger oxidation resistance.

3.3. Corrosion Behavior

The corrosion behavior of the Ti-*x*In alloys were evaluated using potentiodynamic polarization and the results are shown in Figure 9. The anodic polarization curves reveal typical anodic reactions below −0.08 V, initial passive oxide formation at −0.05–0.25 V. The cp-Ti, Ti-10In, and Ti-15In displayed passive behavior above 0.6 V, while the Ti-5In and Ti-20In revealed transpassive behavior above 0.6 V. Ti-15In showed a rapid current increase at approximately −0.08 V, which may be attributed to localized corrosion or pitting corrosion on the surface defects. The critical potentials of cp-Ti, Ti-5In, and Ti-10In were observed at −0.08, 0.04, and −0.05 V, respectively. Ti-20In showed no critical potential in the −0.08–0.16 V range. The initial passive current densities were in the following order: Ti-5In (2.6×10^{-6} μA/cm^2) > Ti-10In (1.9×10^{-6} μA/cm^2) > cp-Ti (1.7×10^{-6} μA/cm^2) > Ti-15In (1.2×10^{-6} μA/cm^2) > Ti-20In (1.1×10^{-6} μA/cm^2). Ti-20In and Ti-5In showed pseudopassive behavior. The current densities at 1.0 V were in the following order: Ti-5In (123×10^{-6} μA/cm^2) > Ti-20In (89×10^{-6} μA/cm^2) > Ti-10In (76×10^{-6} μA/cm^2) > Ti-15In (63×10^{-6} μA/cm^2) > cp-Ti (66×10^{-6} μA/cm^2). Using the Tafel extrapolation method, we calculated the corrosion parameters of cp-Ti and Ti-*x*In alloys from both the anodic and cathodic branches of the potentiodynamic polarization curves. These results are summarized in Table 2. A cathodic shift in E_{corr} compared to cp-Ti was observed in the Ti-5In, Ti-10In, and Ti-20In alloys. All investigated Ti-*x*In samples exhibited similar I_{corr} (0.35–0.73 μA/cm^2) compared to that (0.61 μA/cm^2) of cp-Ti.

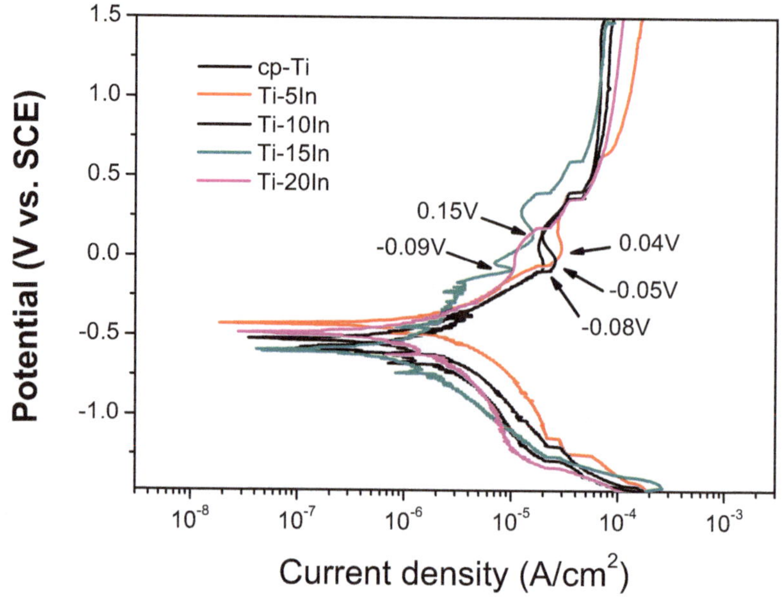

Figure 9. Representative potentiodynamic polarization curves for cp-Ti and Ti-*x*In alloys. The arrows indicate the critical potential.

Table 2. Corrosion potential (E_{corr}) and corrosion current density (I_{corr}) of cp-Ti and Ti-*x*In alloys.

Alloy code	E_{corr} (±SD) (V)	I_{corr} (±SD) (μA/cm²)
cp-Ti	−0.58 (0.04) ^{*,a}	0.61 (0.17) ^{a,b}
Ti-5In	−0.49 (0.05) ^{a}	0.73 (0.36) ^{b}
Ti-10In	−0.59 (0.07) ^{a}	0.42 (0.09) ^{a,b}
Ti-15In	−0.58 (0.12) ^{a}	0.35 (0.02) ^{a}
Ti-20In	−0.54 (0.05) ^{a}	0.44 (0.05) ^{a,b}

* Within the same column, the different alphabet in superscript indicated statistically different mean values at a 5% level ($p < 0.05$) according to Duncan's multiple range test.

4. Conclusions

This study investigated the influence of In on the microstructure, mechanical properties, and corrosion behavior of commercially pure titanium (cp-Ti). All the Ti-*x*In alloys with up to 20 wt% In content exhibited a hcp α structure. Based on the linear relationship of the lattice parameters with the In content, Ti alloys with up to 20 wt% In concentration formed solid solutions. The present study demonstrated that the mechanical properties and corrosion behavior of Ti-*x*In alloys are sensitive to the In content. The Vickers hardness increased as the In content increased because of solid solution strengthening. Ti-10In alloy with a fine, acicular martensitic phase had the best Vickers hardness and oxidation resistance among all of the tested samples. Electrochemical results showed that the Ti-*x*In alloys exhibited a similar corrosion resistance to cp-Ti. Considering the mechanical properties and corrosion behavior, Ti-*x*In alloy with an In content of 10 wt% is a good candidate for dental casting alloys.

Acknowledgments

This study was financially supported by the National Research Foundation of Korea (NRF) grant funded by the Korea government (MSIP) (No. 2011-0030121), and by the Basic Science Research Program through the NRF funded by the Ministry of Education, Science, and Technology (2011-0002706).

Author Contributions

The microstructure analysis of Ti-*x*In alloys was undertaken by Mi-Kyung Han. The mechanical analysis was performed by Jae-Bong Im. The electrochemical analysis was provided by Moon-Jin Hwang. The preparation of test samples was supported by Bong-Jun Kim and Hae-Young Kim. The research work outlined in this paper was designed and supervised by Yeong-Joon Park.

Conflicts of Interest

The authors declare no conflict of interest.

References

1. Leyens, C.; Peters, M. *Titanium and Titanium Aloys*; Wiley Online Library: Weinheim, Germany, 2003.
2. Lloyd, C.; Scrimgeour, S.; Brown, D.; Clarke, R.; Curtis, R.; Hatton, P.; Ireland, A.; McCabe, J.; Nicholson, J.; Setcos, J. Dental materials: 1995 literature review. *J. Dent.* **1997**, *25*, 173–208.
3. Ezugwu, E.; Wang, Z. Titanium alloys and their machinability—A review. *J. Mater. Process. Technol.* **1997**, *68*, 262–274.
4. Niinomi, M. Mechanical properties of biomedical titanium alloys. *Mater. Sci. Eng. A* **1998**, *243*, 231–236.
5. Lee, C.; Ju, C.-P.; Chern Lin, J. Structure—Property relationship of cast Ti-Nb alloys. *J. Oral Rehabil.* **2002**, *29*, 314–322.
6. Kikuchi, M.; Takahashi, M.; Okuno, O. Mechanical properties and grindability of dental cast Ti-Nb alloys. *Dent. Mater. J.* **2003**, *22*, 328–342.
7. Zhou, F.; Wang, B.; Qiu, K.; Lin, W.; Li, L.; Wang, Y.; Nie, F.; Zheng, Y. Microstructure, corrosion behavior and cytotoxicity of Zr–Nb alloys for biomedical application. *Mater. Sci. Eng. C* **2012**, *32*, 851–857.
8. Oh, K.-T.; Shim, H.-M.; Kim, K.-N. Properties of titanium-silver alloys for dental application. *J. Biomed. Mater. Res. Part B Appl. Biomater.* **2005**, *74*, 649–658.
9. Zhang, B.; Qiu, K.; Wang, B.; Li, L.; Zheng, Y. Surface characterization and cell response of binary ti-ag alloys with cp ti as material control. *J. Mater. Sci. Technol.* **2012**, *28*, 779–784.
10. Takahashi, M.; Kikuchi, M.; Okuno, O. Mechanical properties and grindability of experimental Ti-Au alloys. *Dent. Mater. J.* **2004**, *23*, 203–210.
11. Kikuchi, M.; Takahashi, M.; Sato, H.; Okuno, O.; Nunn, M.E.; Okabe, T. Grindability of cast Ti–Hf alloys. *J. Biomed. Mater. Res. Part B Appl. Biomater.* **2006**, *77*, 34–38.

12. Zhang, F.; Weidmann, A.; Nebe, B.; Burkel, E. Preparation of TiMn alloy by mechanical alloying and spark plasma sintering for biomedical applications. *J. Phys. Conf. Ser.* **2009**, 012007.
13. Hsu, H.-C.; Wu, S.-C.; Chiang, T.-Y.; Ho, W.-F. Structure and grindability of dental Ti–Cr alloys. *J. Alloys Compd.* **2009**, *476*, 817–825.
14. Ho, W.-F.; Chiang, T.-Y.; Wu, S.-C.; Hsu, H.-C. Mechanical properties and deformation behavior of cast binary Ti–Cr alloys. *J. Alloys Compd.* **2009**, *468*, 533–538.
15. Ho, W.; Ju, C.; Chern Lin, J. Structure and properties of cast binary Ti–Mo alloys. *Biomaterials* **1999**, *20*, 2115–2122.
16. Hsu, H.-C.; Lin, H.-C.; Wu, S.-C.; Hong, Y.-S.; Ho, W.-F. Microstructure and grindability of as-cast Ti–Sn alloys. *J. Mater. Sci.* **2010**, *45*, 1830–1836.
17. Ho, W.-F.; Chen, W.-K.; Wu, S.-C.; Hsu, H.-C. Structure, mechanical properties, and grindability of dental Ti–Zr alloys. *J. Mater. Sci. Mater. Med.* **2008**, *19*, 3179–3186.
18. Hsu, H.-C.; Wu, S.-C.; Hsu, S.-K.; Sung, Y.-C.; Ho, W.-F. Effects of heat treatments on the structure and mechanical properties of Zr–30Ti alloys. *Mater. Charact.* **2011**, *62*, 157–163.
19. Hsu, H.-C.; Wu, S.-C.; Sung, Y.-C.; Ho, W.-F. The structure and mechanical properties of as-cast Zr-Ti alloys. *J. Alloys Compd.* **2009**, *488*, 279–283.
20. Mareci, D.; Chelariu, R.; Gordin, D.-M.; Ungureanu, G.; Gloriant, T. Comparative corrosion study of Ti–Ta alloys for dental applications. *Acta Biomater.* **2009**, *5*, 3625–3639.
21. Zhou, Y.-L.; Niinomi, M. Ti–25Ta alloy with the best mechanical compatibility in Ti–Ta alloys for biomedical applications. *Mater. Sci. Eng. C* **2009**, *29*, 1061–1065.
22. Wang, R.; Welsch, G. Evaluation of an experimental Ti-Co alloy for dental restorations. *J. Biomed. Mater. Res. Part B Appl. Biomater.* **2013**, *101*, 1419–1427.
23. Moser, J.; Lin, J.C.; Taira, M.; Greener, E. Development of dental Pd−Ti alloys. *Dent. Mater.* **1985**, *1*, 37–40.
24. Lin, W.-J.; Wang, B.-L.; Qiu, K.-J.; Zhou, F.-Y.; Li, L.; Lin, J.-P.; Wang, Y.-B.; Zheng, Y.-F. Ti–Ge binary alloy system developed as potential dental materials. *J. Biomed. Mater. Res. Part B Appl. Biomater.* **2012**, *100*, 2239–2250.
25. Okabe, T.; Kikuchi, M.; Ohkubo, C.; Koike, M.; Okuno, O.; Oda, Y. The grindability and wear of Ti-Cu alloys for dental applications. *JOM* **2004**, *56*, 46–48.
26. Wang, Q.; Wang, Y.; Lin, J.; Zheng, Y. Development and properties of Ti-In binary alloys as dental biomaterials. *Mater. Sci. Eng. C* **2012**, *33*, 1601–1606.
27. Kim, T.-I.; Han, J.-H.; Lee, I.-S.; Lee, K.-H.; Shin, M.-C.; Choi, B.-B. New titanium alloys for biomaterials: A study of mechanical and corrosion properties and cytotoxicity. *Bio-Med. Mater. Eng.* **1997**, *7*, 253–263.
28. Gulay, L.; Schuster, J. Investigation of the titanium–indium system. *J. Alloys Compd.* **2003**, *360*, 137–142.
29. Hunter, B.A.; Howard, C.J. Australian Nuclear Science and Technology Organization. Available online: ftp://ftp.ansto.gov.au/pub/physics/neutron/rietveld/Rietica_LHPM95/MANUAL.PDF (accessed on 18 May 2015).
30. ASTM International. *Standard Reference Test Method for Making Potentiostatic and Potentiodynamic Anodic Polarization Measurements*; ASTM G5-94; ASTM International: West Conshohocken, PA, USA, 2011.

31. Huheey, J.E.; Keiter, E.A.K.; Keiter, R.L.; Medhi, O.K. *Inorganic Chemistry: Principles of Structure and Reactivity*; Harper & Row: New York, NY, USA, 1983.
32. Joint Committee on Powder Diffraction Standards (JCPDS)-International Centre for Diffraction Data (ICDD). *Card No. 44-1294*; Joint Committee on Powder Diffraction Standards (JCPDS)-International Centre for Diffraction Data (ICDD): Swarthmore, PA, USA, 1995.

The Effect of Mg Adding Order on the Liquid Structure and Solidified Microstructure of the Al-Si-Mg-P Alloy: An Experiment and *ab Initio* Study

Xiangzhen Zhu, Wei Jiang, Mingrui Li, Huan Qiao, Yuying Wu, Jingyu Qin * and Xiangfa Liu *

Key Laboratory for Liquid-Solid Structural Evolution and Processing of Materials, Ministry of Education, Shandong University, 17923 Jingshi Road, Jinan 250061, China; E-Mails: zhuxiangzhen8705@163.com (X.Z.); jiangwei0806@126.com (W.J.); miraclelisdu@sina.com (M.L.); qiaohuan1124@126.com (H.Q.); wuyuying@sdu.edu.cn (Y.W.)

* Authors to whom correspondence should be addressed; E-Mails: qinjy@sdu.edu.cn (J.Q.); xfliu@sdu.edu.cn (X.L.)

Academic Editor: Enrique Louis

Abstract: In this paper, the relationship between the liquid structure and the corresponding solidified microstructure of an Al-Si-Mg-P alloy was studied. Experimental results show that Mg can reduce the phosphorous-modification effect if it was added after adding Al-P alloy. However, when it is added before adding Al-P alloy, Mg has no effect on the phosphorous-modification. It is considered that the difference in liquid structure induced by changing the adding order of Mg should be responsible for the above phenomenon, and was investigated by *ab initio* molecular dynamics simulation (AIMD). It was believed that the high-active Mg atoms could bond P atoms to form P-Mg clusters and then reduce the modification effect of AlP, when pure Mg was added into the prepared Al-Si-P melt. When the pure Mg was added into Al-Si melt before adding Al-P alloy, the Mg atoms would be first occupied by Si atoms to form Mg-Si clusters, and thus lose the ability to eliminate P-Al clusters which dissolve into melt later, leading to a good phosphorous-modification effect.

Keywords: AlP; Mg; modification; Al-Si alloy; liquid structure; evolution; simulation

---

1. Introduction

The structural relationship between the liquid melt and the corresponding solidified alloy has been a long-standing concern and is not very clear now. In 1927, Levi focused on the correlation between the difference in the liquid melt and its resulting varying performance of solid alloys which have the same composition and underwent the same processing [1]. Since then, many researchers have made attempts to clarify this problem [2–5]. Currently, increasingly strict performance requirements of casting are put forward; therefore, the study of controlling liquid metal state, which is the base to produce an excellent casting, is receiving increasing attention.

Al-Si based alloys are important materials which have been widely used in piston production and other automotive applications. To improve the properties of near-eutectic and hypereutectic Al-Si alloys, it is imperative to modify primary Si grains and add alloying elements. In 1933, researchers found that AlP is a very effective compound to modify primary Si phase [6]. Due to the similar crystal parameters, AlP phase could act as high-efficiency heterogeneous nucleation substrate for the primary Si, resulting in an increase in the number of small Si particles [6–10]. Since then, many modifiers contained P have been invented, such as Cu-P alloy, Si-P alloy and Al-P alloy. Alloying is also an effective method to strengthen the Al-Si based alloys. Mg is an important alloying element to enhance the elevated-temperature properties of Al-Si alloys [11,12]. However, due to its high reactivity, the burning loss of Mg is not a negligible problem. Hence, a secondary addition of Mg is a very common process in casting shop for returning to a normal Mg level. As a high-active element, Mg has the potential to react with P element. This reaction may reduce the modification effect of AlP. However, no study about this problem has been carried out.

In this paper, the fading phosphrous-modification of Si phase induced by changing order of adding Mg and Al-P alloy was revealed by our experiments. To understand the mechanism behind this phenomenon, the liquid structures of the corresponding Al-Si-Mg-P melt and their evolution paths should be investigated. The *ab initio* molecular dynamics simulation (AIMD), which is a computational chemistry method based on quantum chemistry, was adopted to calculate the information of clusters in the Al-Si-Mg-P system.

2. Experimental and Computational Methods

The eutectic Al-12.6Si alloy (all compositions quoted in this work are in wt.% unless otherwise stated) was used to research the effect of adding Mg on the phosphrous-modification of Si phase. The adoptive modifier is Al-3P alloy rod, which was supplied by Shandong Al & Mg Melt Technology Co. Ltd. (Jinan, Shandong, China).

In this study, the Al-12.6Si alloy was remelted in a clay-graphite crucible by an induction furnace. Then, the melt was transferred into a resistance furnace at 780 °C. The following holding process and addition of Mg and Al-P alloy were all operated in the resistance furnace. The addition level of Al-3P alloy (1.0%) and pure Mg (1.5%) is same for two phosphorous-modified samples. The phosphorous-modification parameters are also the same: holding 40 min at 780 °C after adding Al-3P alloy. The only difference in the preparation process is the order of adding Mg. For one Al-12.6Si alloy (marked as “B alloy”), the addition of Mg occurs before adding Al-3P alloy. While for another alloy (marked as “C alloy”), the addition of Mg occurs after adding Al-3P alloy. The specific operating procedures are shown in Figure 1. The melts were poured into a cast iron mold with a size of 100 mm × 40 mm × 30 mm. The solidification rate of the alloys is about 14 K/s. Specimens with the size of 15 mm × 10 mm × 10 mm were obtained from the center of the entire casting alloys and then mechanically ground and polished using standard routines. The microstructures of the alloys were characterized and analyzed by optical microscope and field emission scanning electron microscope (FESEM, model Hitachi-SU-70, Tokyo, Japan).

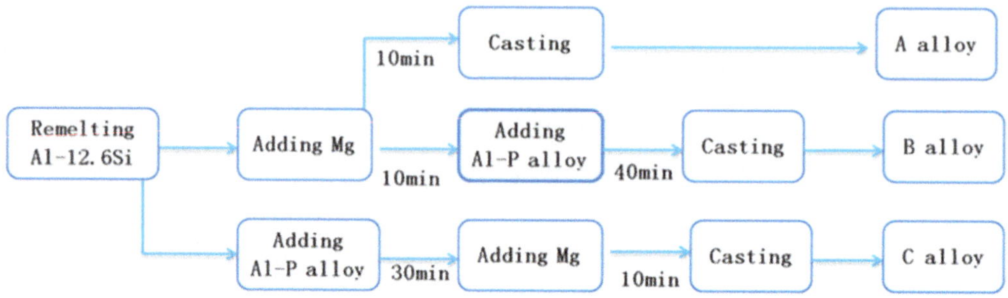

Figure 1. The preparing processes of three samples.

In the experiment, the content of P element in Al-Si melt is only 300 ppm and Mg is 1.5%. As for AIMD, if one kind of atom in an alloy is too small, the statistical error of the structure functions will be large. So, the concentrations of P and Mg were set higher than these in the experimental melt. The AIMD simulations were executed using the Vienna *ab initio* simulation package (VASP) [13] by implementing the projector augmented-wave method [14]. A gradient corrected energy function [15] was adopted. The plane-wave basis set contained components with energies up to 270 eV. Generally, only the Γ-point was used to sample the supercell Brillouin zone. All the simulations were carried out in the canonical ensemble (NVT) through a Nosé [16] thermostat with a characteristic frequency equal to 52 ps^{-1}. Newton's equations of motion were integrated using the Verlet algorithm with a time step of 3 fs. The system of $Al_{179}Mg_9Si_6P_6$ with 200 atoms was equilibrated at T = 1173 K (1373 K) for 9 ps with 2000 configurations from the last 6 ps periods being collected for structural analysis.

3. Results and Discussion

Figure 2 shows the microstructures of three Al-12.6Si-1.5Mg alloys. It is clear that the Si phase in the A alloy without P exists in the form of eutectic Si phase, and the Mg_2Si phase also exists in the matrix, as shown in Figure 2a,b. There is no proper Al-Si-Mg diagram that can describe in detail all the phases (containing α-Al, eutectic Mg_2Si, primary Si and eutectic Si) in Al-12.6Si-1.5Mg alloys, but we can speculate on these phases based on the binary Al-Si and Mg-Si diagrams. As it is well known, Al-12.6% Si is just the eutectic Al-Si alloy. However, since part of Si atoms was bonded by Mg atoms to form Mg_2Si phase, the content of Si which would crystallize as Si phase is less than 12.6%. Therefore, the microstructure of Al-12.6Si-1.5Mg alloy possesses the characteristics of hypoeutectic Al-Si alloys, and no primary Si particles were found. While, when Al-P alloy was added into the melt after adding Mg, some small primary Si particles appear, as shown in Figure 2c.

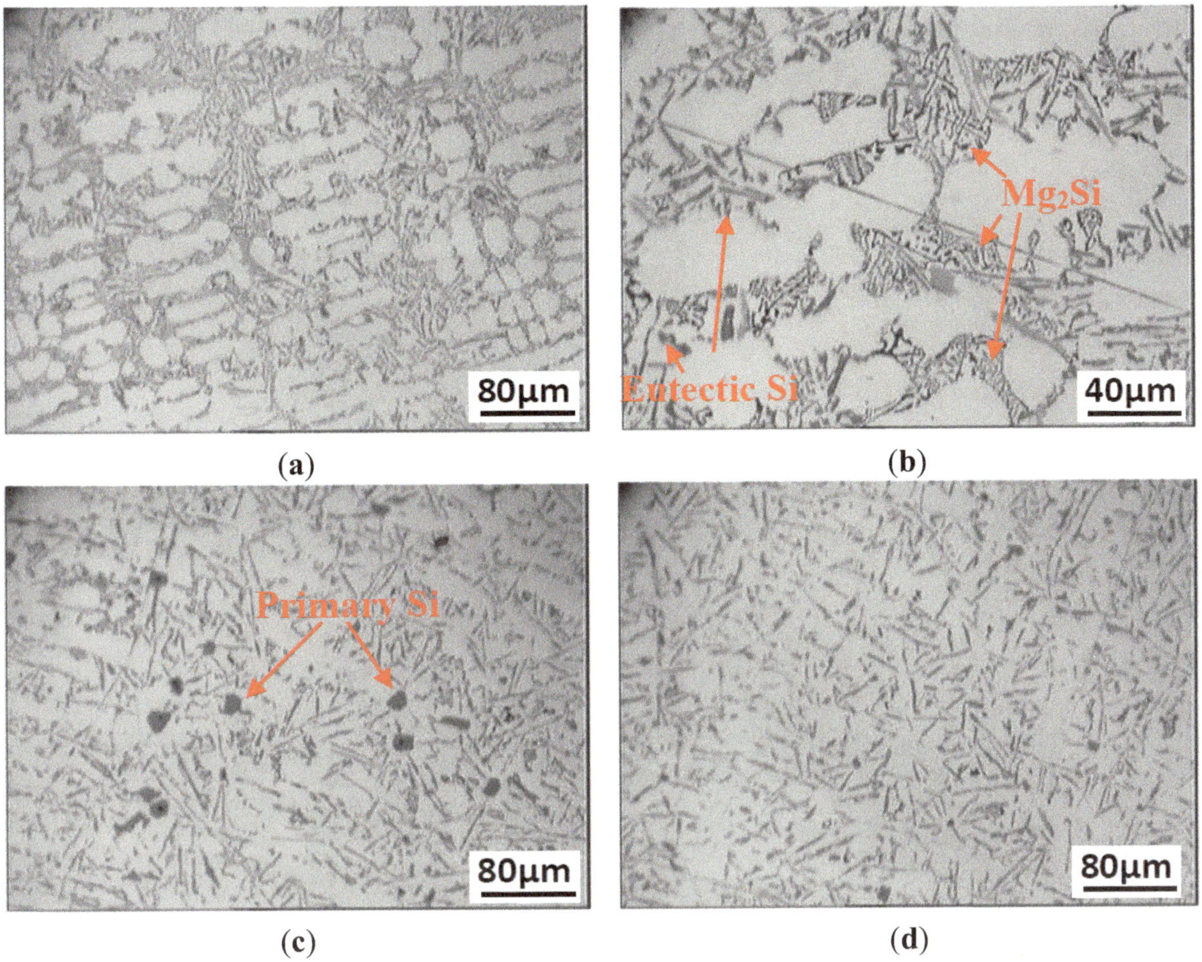

Figure 2. The microstructures of Al-12.6Si-1.5Mg alloys: (**a**,**b**) A alloy; (**c**) B alloy with some primary Si grains, into which Al-P alloy was added after adding Mg; (**d**) C alloy with few primary Si grains, into which Al-P alloy was added before adding Mg.

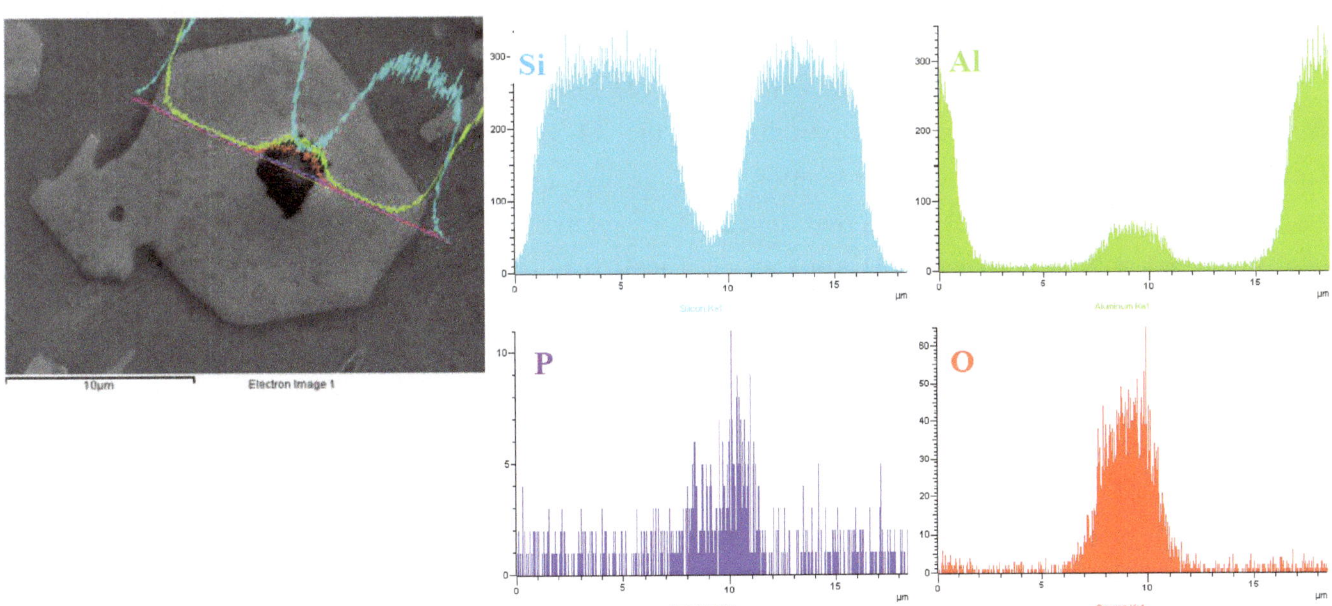

Figure 3. The EDS analysis of the black phase in the center of a primary Si particle. The black phase contains Al, P and O, and is considered as AlP.

Figure 3 shows the EDS analysis of a typical primary Si particle. It was found that a black phase exists in the center of primary Si phase, and is rich in Al, P and O elements. It is considered that this black phase is AlP, and the existence of O element is due to the oxidation and hydrolysis of AlP during specimen preparation [7]. AlP phase and Si phase have the same face-centered cubic (fcc) crystal structure and very similar lattice parameters (a_{AlP} = 5.45 Å, a_{Si} = 5.42 Å), as shown in Figure 4. So, AlP can act as the highly effective heterogeneous nucleation substrate for primary Si phase. Due to the existence of AlP, many Si atoms in this eutectic melt could be absorbed and bonded by AlP during solidification, and then precipitated as primary Si phase, rather than eutectic Si phase. This transformation from the eutectic Si phase to the primary Si phase is the aim of the phosphorous-modification.

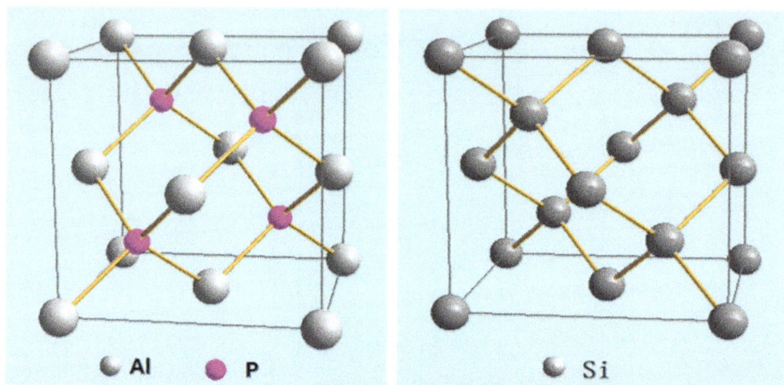

(a) (b)

Figure 4. The crystal structures of (**a**) AlP; (**b**) Si. They both have face-centered cubic structure, and a_{AlP} = 5.45 Å, a_{Si} = 5.42 Å.

Figure 2d displays the microstructure of C alloy. Though the C alloy was also supplied with Al-3P alloy, it has few primary Si particles. It indicates the AlP did not play its role of heterogeneous nucleation substrate. Compared with B alloy, the only difference in the preparation process is the order of adding Mg element. It is supposed that Mg plays a pivotal role in reducing the modification effect of AlP.

It is well known that the liquid metal is the source of casting, and the solid microstructure originates from the liquid structure. The liquid structure of the Al-Si-Mg-P alloy was analyzed using AIMD simulations to investigate the solidification mechanisms caused by particular Mg alloying orders. As discussed above, it is considered that the reactions related with Mg are the key point. It seems that Mg may react with AlP, according to our experimental results. In general, Mg tends to strongly react with the Si element to form Mg_2Si. Hence, the competition between the above two reactions should be investigated, and the affinities of P-Mg and Si-Mg were analysed in this study. For the convenience of comparing these two affinities, the atomic percents of Si and P were all set to the same value (3%). The atomic percent of Mg was set as 4.5%. So, the ration of Mg to P is just that in Mg_3P_2 compound, and the ration of Mg to Si is less than that in Mg_2Si compound. In other words, the content of Si is in excess, which could enhance the affinity of Si-Mg and is a little similar with the conditions in the experimental melt.

Figure 5 shows the partial correlation functions (PCFs) of $Al_{179}Mg_9Si_6P_6$ alloy at 1173 K. In order to describe the liquid structure of this Al-Si-Mg-P alloy comprehensively, there needs to be 10 PCFs, namely, $g_{PAl}(r)$, $g_{PMg}(r)$, $g_{PSi}(r)$, $g_{PP}(r)$, $g_{MgAl}(r)$, $g_{MgMg}(r)$, $g_{MgSi}(r)$, $g_{SiAl}(r)$, $g_{SiSi}(r)$ and $g_{AlAl}(r)$. Since the

$g_{AlAl}(r)$ is rather normal, we only present the other nine PCFs in this investigation. To facilitate the comparison and analysis, three figures, which are related with P, Mg and Si, respectively, are used to show these nine PCFs.

Four P-related g(r) curves are plotted in Figure 5a. It is observed clearly that the first peak of the $g_{PMg}(r)$ curve is much higher than those of other three *P*-related g(r) curves, indicating that the affinity of P-Mg pairs is strongest. The P-Al pairs have a second strong affinity. Now, we tend to $g_{PSi}(r)$ and $g_{PP}(r)$ curves. Since the second peak of $g_{PSi}(r)$ curve is more obvious than the first peak, the Si atoms tend to exist at the second coordination shell rather than at the first coordination shell of P atoms, avoiding meet P atoms directly. The atomic radius of P atom is 0.11 nm. So, if there were P-P pairs, a pronounced first peak should be at 0.22 nm. However, the $g_{PP}(r)$ curve in Figure 1a does not demonstrate the first peak at 0.22 nm, meaning the P-P pairs are extremely scarce. The structural information from $g_{PSi}(r)$ and $g_{PP}(r)$ curves is in good accordance with our previous study [17]. On the basis of above analysis, it can be concluded that P atoms tend to combine with Mg atoms preferentially and then with Al atoms, but would not bond with Si atoms or other P atoms. It also can be shown that Mg element has the potential to eliminate P-Al clusters and reduce the modification effect of AlP.

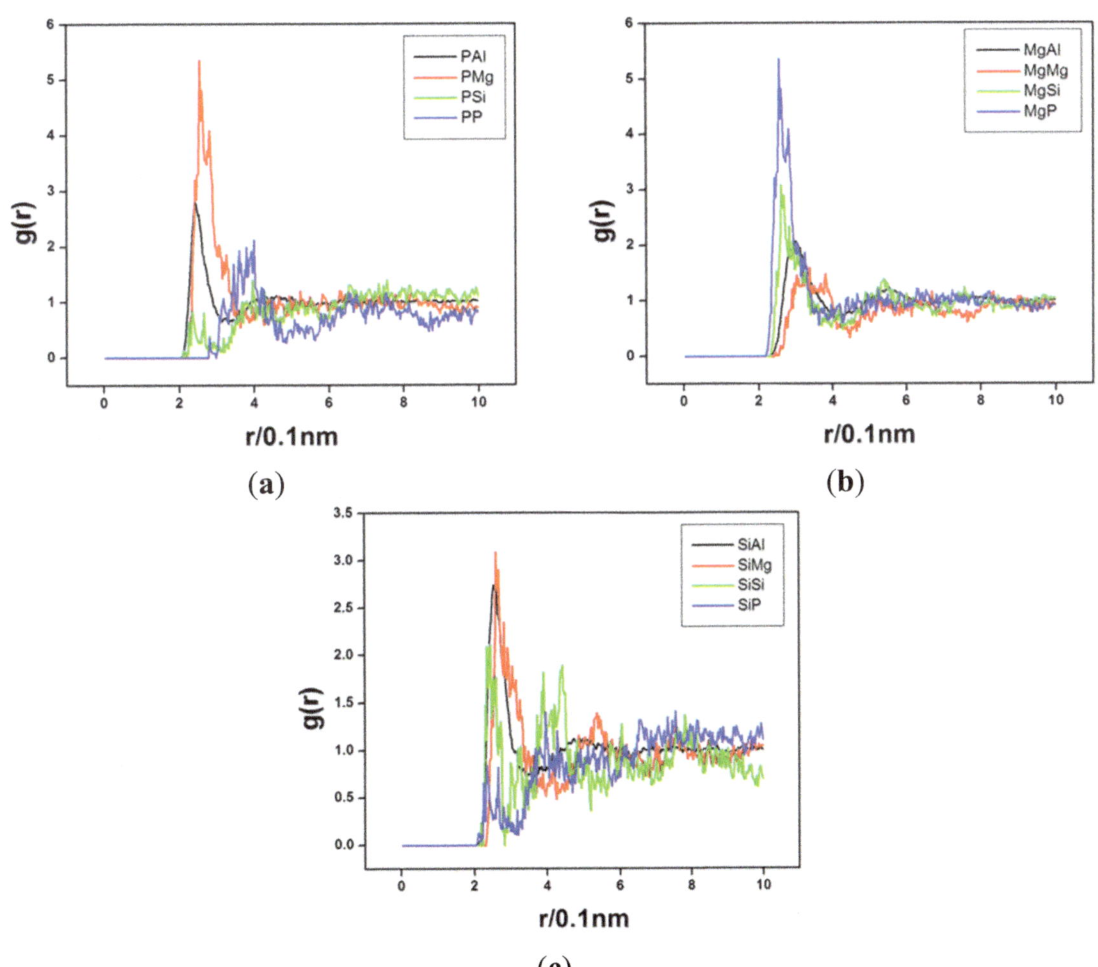

Figure 5. The partial pair correlation functions of $Al_{179}Mg_9Si_6P_6$ system at 1173K. (**a**) g(r) curves related with P element; (**b**) g(r) curves related with Mg element; (**c**) g(r) curves related with Si element. The g(r) curves show that the affinity of P-Mg is strongest, followed by Mg-Si and P-Al.

The chemical environment around P atoms is shown in Table 1. In the table, N_{ij} is the partial coordination number around a P atom, R_{ij} is the percentage of element i in the partial coordination number of element j. The parameter D_{ij} is the deviation degree between R_{ij} and its nominal value (random distribution), reflecting the chemical short-range order (CSRO) around P atoms. D_{ij} is defined as:

$$D_{ij} = (R_{ij} - \text{Nominal}) / \text{Nominal} \times 100\% \tag{1}$$

Table 1. Chemical environment around P atoms.

Terms	N_{ij}		R_{ij} (%)		Nominal	D_{ij} (%)	
	1173 K	1373 K	1173 K	1373 K		1173 K	1373 K
PAl	5.91	5.95	88.70	92.15	89.50	−0.89	2.96
PMg	0.72	0.46	10.75	7.16	4.50	138.89	59.11
PSi	0.04	0.05	0.54	0.69	3.00	−82.00	−77.00
PP	0.00	0.00	0.00	0.00	3.00	−100	−100

Both at 1373 K and 1173 K, the partial coordination numbers of Al atoms around P atoms (5.95 and 5.91, respectively) are very close to 6, which is similar with that in the Al-Si-P system [17]. While, the number of Mg atoms in the first coordination shell is less than 1. It should be noted the number of Al atoms is high, mainly because the solvent Al atoms are extremely plentiful in this Al-rich melt, and not because the P-Al cluster is the strongest cluster. To measure the strength or importance of the clusters, R_{ij} and D_{ij} should be investigated.

At 1373 K, the R_{PAl} (92.15) and R_{PMg} (7.16) are all greater than their nominal values (89.50 and 4.50). However, the D_{PAl} is only 2.96, meaning that the Al content in the first coordination shell of P atoms is just a little higher than the nominal value (random distribution). In other words, P-Al cluster is not a very strong cluster in this system. For P-Mg pairs, the value of D_{PMg} is up to 59.11. It means P atoms have a higher tendency to combine with Mg atoms.

Moreover, when the temperature is cooled down to 1173 K, the partial coordination number of Mg atoms increases and the value of D_{PMg} reaches 138.89. While, the three parameters of P-Al pairs (N_{PAl}, R_{PAl} and D_{PAl}) all reduce as the temperature drops. These changes of parameters indicate that the P-Mg cluster would be enhanced and P-Al cluster would be weakened during the solidification process. What should be paid particular attention is that the D_{PAl} is −0.89. This negative value is of much significance: when the temperature is below 1173 K, the CRSO between P atoms and Al atoms is very weak, and the P-Mg cluster is the most pronounced cluster in this alloy.

The structural information of P-Si pairs and P-P pairs is also shown in Table 1. The parameters N_{PSi}, which are only 0.05 at 1373 K and 0.04 at 1173 K, demonstrate that the partial coordination number of Si atoms is very little, and becomes lower as the temperature decreases. As for P-P pairs, the N_{PP} value of 0 and D_{PP} value of −100 declare that in fact there are no P-P pairs in the melt.

Now, based on the above integrated information from the g(r) curves and chemical environment, the behavior rule of P atoms is revealed. When P atoms enter the melt, they react with Mg atoms preferentially, followed by Al atoms. Furthermore, P atoms do not combine with Si atoms and other P atoms.

According to the behavior rule of P atoms, it seems that the addition of Mg would eliminate part of P-Al cluster and reduce the modification effect of AlP, according to the conditions of C alloy. However, when the Mg was added to Al-Si melt before adding Al-P alloy, the modification effect was positive, just as in the case of B alloy. So, the behavior rule of Mg atoms should also be studied carefully.

Figure 5b shows the four Mg-related g(r) curves. It is obvious the first peak of $g_{MgP}(r)$ curve is much higher than those of other three g(r), implying that Mg atoms tend to combine with P atoms, preferentially. It should be known that $g_{ij}(r)$ is equal to $g_{ji}(r)$ in simulation, so $g_{MgP}(r)$ is the same with $g_{PMg}(r)$. The next most important curve is the $g_{MgSi}(r)$ curve, which has the second-ranked peak. Hence, for Mg atoms, it is the second-favorite choice to combine with Si atoms. Then, Al atoms also have a certain degree of potential to bond with Mg atoms, but Mg-Mg pairs are few. The Si-related g(r) curves are also considered, as shown in Figure 5c. With regards to Si atoms, combining with Mg atoms is the preferred choice. It means that the Mg atoms would form Mg-Si clusters in Al-Si melt.

One arbitrary configuration out of 2000 is visualized in Figure 6. By inspection of the visualized configurations, it is found that P atoms tend to be near Mg atoms, and so do Si atoms, agreeing with the information from g(r) curves.

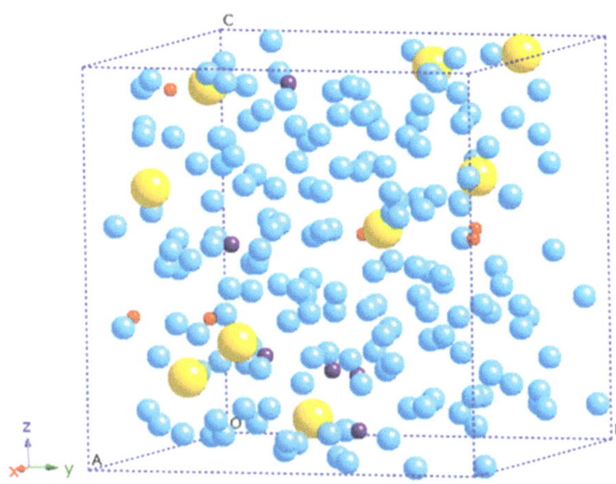

Figure 6. Visualization of an arbitrary configuration of liquid $Al_{179}Mg_9Si_6P_6$ system at 1173 K (Al: blue ball; Mg: yellow ball; Si: violet ball; and P: red ball).

Based on the above analysis, the three strongest clusters are listed as follows: (1) P-Mg cluster > (2) Mg-Si cluster > (3) P-Al cluster.

Among them, the P-Mg cluster is much stronger than the other two clusters. However, the fact that the first peak of $g_{MgSi}(r)$ curve is just a little higher than that of $g_{PAl}(r)$ shows that the Mg-Si cluster is just a little stronger than P-Al cluster in this $Al_{179}Mg_9Si_6P_6$ system.

It should be specially mentioned that the above analysis reflects the conditions of the melt in equilibrium. While, in our experiments, the raw materials are not added to the Al-Si melt at the same time. So, some pre-existing clusters would come into being when the first addition has been completed but the second addition has not been done. Even after the second addition, those pre-existing can also survive for a certain time. They need some time to evolve to equilibrated clusters. However, if the melt was casted before it could reach the balanced or near-balanced state, the survived pre-existing clusters would play a role, making the solid microstructure different from the equilibrated microstructure.

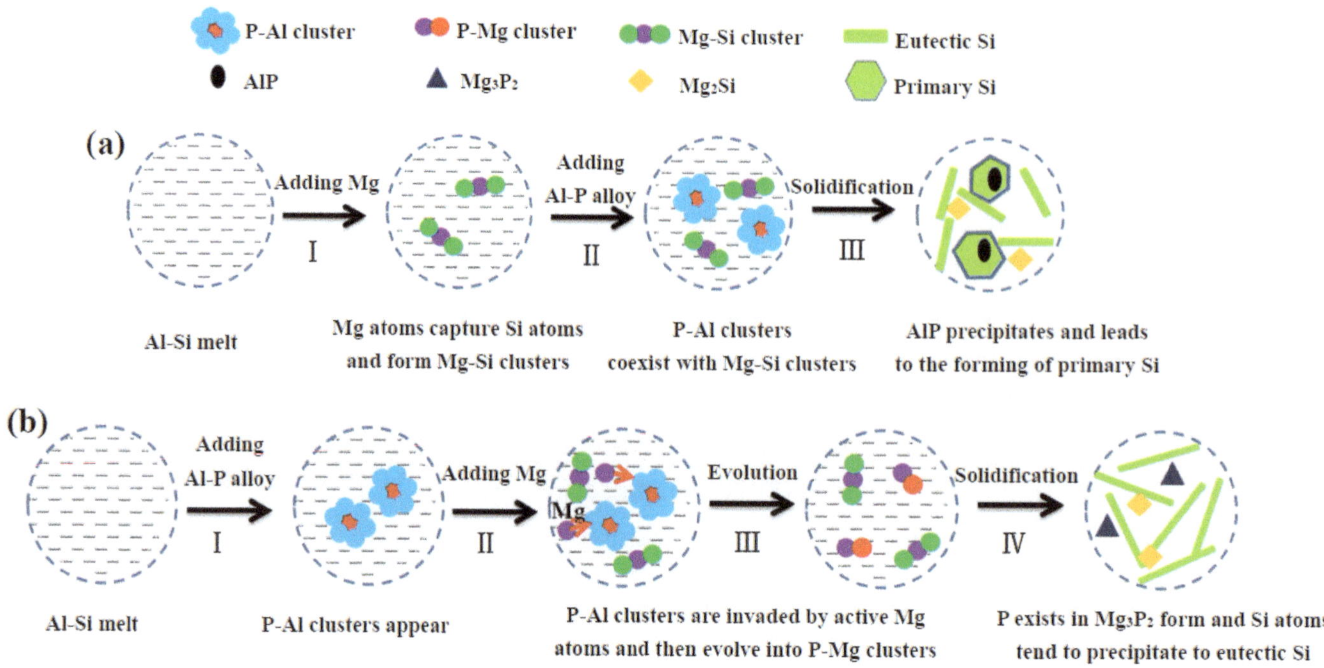

Figure 7. The schematics of the two evolution paths of the Al-Si-Mg-P liquid melt structure induced by changing order of adding Mg and its resulting solidification microstructure. (**a**) when pure Mg was added into Al-Si melt before addition of Al-P alloy, Mg-Si clusters formed firstly and can coexist with P-Al clusters. In the solidification process, the P-Al clusters can precipitate as AlP and then lead to the forming of primary Si; (**b**) when pure was added in to melt after the addition of Al-P alloy, P-Al clusters were invaded by active Mg atoms dissolved from pure Mg and then evolved into P-Mg clusters. In the solidification process, P precipitated as Mg_3P_2, rather than AlP. So, no primary Si particles formed due to the absence of AlP.

Figure 7 depicts the two evolution paths of the Al-Si-Mg-P liquid melt structure induced by implementing different adding orders of Mg. For B alloy, its evolution path is relatively simple and can be divided into three stages, as shown in Figure 7a. In stage I, pure Mg was added into the binary Al-Si melt. In this phosphorous-free melt, the high-active Mg atoms would bond Si atoms and form Mg-Si clusters. The Al-P alloy was added into the melt in the following stage II. Then, AlP compound, which preexists in the Al-P alloy, would dissolve in the melt and evolves into P-Al clusters. That is to say the Mg-Si clusters co-existed with P-Al clusters at the beginning of the stage II. According to the above AIMD results of the melt in equilibrium, the Mg-Si clusters should react with P-Al clusters and evolve to P-Mg clusters. While, considering the experimental result of alloy B, it is speculated that this evolution in the experimental melt could not be finished in a short time: (1) In the experimental Al-12.6Si-1.5Mg-0.03P melt, the amounts of Al atoms and Si atoms are much more than these of Mg atoms and P atoms. So, The P-Al clusters are the P-centered clusters [17], and Mg-Si clusters are the Mg-centered clusters. If P atoms could combine with Mg atoms to form P-Mg clusters, P atoms should get close to and go through the Si atoms zone surrounding the centered Mg atoms in Mg-Si clusters. While, as discussed above, there exists a large tendency to repel each other between P atoms and Si atoms, which blocks the formation of P-Mg clusters. (2) The electronegativity, which was introduced by Pauling in 1932 [18], is a chemical property that describes the tendency of an atom or a functional group to attract electrons (or electron density) towards itself. The electronegativity difference actually reflects the bonding nature of atomic pairs in alloys. The lager electronegativity difference between constituent elements helps the formation of clusters and compounds [19]. While, according to the electronegativity equalization principle, once Mg atoms combine with Si atoms to form Mg-Si clusters, the electronegativity of Mg would increase, decreasing the electronegativity difference between Mg and P atoms. So, the tendency of Mg to combine P becomes weak. In other words, the Mg atoms in Mg-Si clusters have a lower activity, losing the ability to eliminate P-Al clusters quickly. Therefore, P-Al clusters could coexist with Mg-Si clusters for a long time. However, after only 40 min, the melt was poured into a mould and solidification (stage III) took place. With the decrease in the melt temperature, the co-existing Mg-Si clusters and P-Al clusters grew into crystalline phases correspondingly: Mg_2Si and AlP phases. As the highly effective heterogeneous substrate for primary Si phase, AlP could absorb and bond many Si atoms, and then make them precipitate as primary Si phase, resulting in a microstructure like that of B alloy.

As discussed above, Si plays a suppressive role in the elimination of part of P-Al clusters induced by Mg and ensures a good phosphorous-modification effect (*i.e.*, ensure the transformation from part of eutectic Si plates to some small and distributed primary Si grains), by combining with Mg atoms and reducing the activity of Mg. This phenomenon also appeared in Li's experiment [20]. In his experiment, the prepared Al-20Mg_2Si-3Si alloy could also be modified by AlP. The content of Mg in the prepared Al-20Mg_2Si-3Si alloy is up to 12.7%, but all Mg atoms were bonded with Si atoms and existed in the Mg-Si clusters before adding Al-P alloy. So, Mg had no effect on the phosphorous-modification

As for the C alloy, its evolution path can be divided into four stages, as shown in Figure 7b. The Al-P alloy was added firstly in stage I, resulting in many P-Al clusters appearing in the Al-Si melt. In the next stage II, pure Mg with high activity was added and dissolved into the melt. As discussed above, elemental Mg tends to combine with P atoms preferentially. So, the pre-existing P-Al clusters were invaded by these active Mg atoms and then evolved into P-Mg clusters in a short period of time (stage III). Of course, Mg atoms are excessive, and most of them also combined with Si atoms, forming Mg-Si clusters. The next stage IV is the solidification. P would precipitate in the form of Mg_3P_2, which could not induce the formation of primary Si. So, the microstructure of C alloy has few primary Si grains.

4. Conclusions

In this paper, the reduction of a phosphorus-modification effect, induced by changing the adding order of Mg and an Al-P alloy, was investigated. The AIMD simulation was used to investigate the liquid structure of Al-Si-Mg-P system to reveal the potential mechanism.

(1) Elemental Mg has the high potential to combine with P atoms. If pure Mg was added into the melt which contains many P atoms, Mg can eliminate P-Al clusters quickly and reduce the phosphorous-modification effect.

(2) Elemental Si can prevent P-Al clusters from the elimination induced by Mg. If the pure Mg was added into Al-Si melt before adding Al-P alloy, Si atoms can combine with Mg atoms and reduce the activity of Mg, then make them lose the ability to eliminate P-Al clusters and ensure a good phosphorous-modification effect.

Acknowledgments

The authors wish to thank the financial support of the National Natural Science Foundation of China (No.50971082), the National Natural Science Foundation of China (No.51001065) and the National Basic Research Program of China (973 Program, No.2012CB825702).

Author Contributions

Xiangzhen Zhu, Jingyu Qin, Yuying Wu and Xiangfa Liu conceived and designed the experiments and simulations; Wei Jiang, Mingrui Li and Huan Qiao performed the experiments; Jingyu Qin performed the simulations; Xiangzhen Zhu, Yuying Wu, Jingyu Qin and Xiangfa Liu analyzed the data; Xiangzhen Zhu wrote the paper.

Conflicts of Interest

The authors declare no conflict of interest.

References

1. Levi, A. Heredity in cast iron. *The Iron Age* **1927**, *6*, 960–965.
2. Frank, F. Supercooling of liquids. *Proc. R. Soc. London Ser. A* **1952**, *215*, 43–46.
3. Vertman, A.; Samarin, A.; Yakobson, A. Structure of liquid eutectics. *Izv. Akad. Nauk SSSR, Otd. Tekhn. Nauk., Met. i Toplivo* **1960**, *3*, 17–21.
4. Li, P.-J.; Zhang, Y.F.; Nikitin, V.; Kandalova, E.; Nikitin, K. Hereditary effect of al-based modifiers and grain refiners on structure and properties of A356. 2 alloys. *Trans. Nonferrous Met. Soc. China* **2002**, *12*, 233–237.
5. Liu, X.-F.; Bian, X.F.; Qi, X.G.; Ma, J. Structural heredity between Al5Ti1B and AlTi, AlB master alloys. *Trans. Nonferrous Met. Soc. China* **1999**, *9*, 806–809.
6. Sterner, R.; Rainer, C. Aluminum silicon alloy with a phosphorus content of 0.001%–0.1%. U.S. Patent 1940922, 26 December 1933.
7. Zhu, X.; Wu, Y.; Li, C.; Li, P.; Qiao, H.; Liu, X. The dispersive orientated-precipitation of AlP on alumina film and its effect on the primary Si gathering behavior in the Al–Si alloy surface layer. *CrystEngComm* **2014**, *16*, 5583–5590.
8. Ho, C.; Cantor, B. Heterogeneous nucleation of solidification of Si in Al-Si and Al-Si-P alloys. *Acta Metall. Mater.* **1995**, *43*, 3231–3246.
9. Faraji, M.; Todd, I.; Jones, H. Effect of solidification cooling rate and phosphorus inoculation on number per unit volume of primary silicon particles in hypereutectic aluminium—Silicon alloys. *J. Mater. Sci.* **2005**, *40*, 6363–6365.
10. Liu, X.; Qiao, J.; Liu, Y.; Li, S.; Bian, X. Modification performance of the Al-P master alloy for eutectic and hypereutectic Al-Si alloys. *Acta Metall. Sin.* **2004**, *40*, 471–476.
11. Samuel, F.H.; Samuel, A.M.; Liu, H. Effect of magnesium content on the ageing behaviour of water-chilled Al-Si-Cu-Mg-Fe-Mn (380) alloy castings. *J. Mater. Sci.* **1995**, *30*, 2531–2540.
12. Caceres, C.H.; Davidson, C.J.; Griffiths, J.R.; Wang, Q.G. The effect of Mg on the microstructure and mechanical behavior of Al-Si-Mg casting alloys. *Metall. Mater. Trans. A* **1999**, *30*, 2611–2618.

13. Kresse, G.; Furthmüller, J. Efficiency of *ab initio* total energy calculations for metals and semiconductors using a plane-wave basis set. *Comput. Mater. Sci.* **1996**, *6*, 15–50.
14. Kresse, G.; Joubert, D. From ultrasoft pseudopotentials to the projector augmented-wave method. *Phys. Rev. B* **1999**, *59*, 1758.
15. Perdew, J.P.; Wang, Y. Pair-distribution function and its coupling-constant average for the spin-polarized electron gas. *Phys. Rev. B* **1992**, *46*, 12947.
16. Nosé, S. A unified formulation of the constant temperature molecular dynamics methods. *J. Chem. Phys.* **1984**, *81*, 511–519.
17. Qin, J.; Zuo, M.; Gu, T.; Liu, X. The featured local structural units in liquid $Al_{80}Si_{15}P_5$ alloy and their relationship with si modification. *J. Alloy. Compd.* **2010**, *492*, 525–528.
18. Pauling, L. The nature of the chemical bond. IV. The energy of single bonds and the relative electronegativity of atoms. *J. Am. Chem. Soc.* **1932**, *54*, 3570–3582.
19. Lu, P.; Liu, C.; Dong, D. Effects of atomic bonding nature and size mismatch on thermal stability and glass-forming ability of bulk metallic glasses. *J. Non-Cryst. Solids* **2004**, *341*, 93–100.
20. Li, C.; Liu, X.; Wu, Y. Refinement and modification performance of Al–P master alloy on primary Mg_2Si in Al–Mg–Si alloys. *J. Alloy. Compd.* **2008**, *465*, 145–150.

Permissions

All chapters in this book were first published in Metals, by MDPI; hereby published with permission under the Creative Commons Attribution License or equivalent. Every chapter published in this book has been scrutinized by our experts. Their significance has been extensively debated. The topics covered herein carry significant findings which will fuel the growth of the discipline. They may even be implemented as practical applications or may be referred to as a beginning point for another development.

The contributors of this book come from diverse backgrounds, making this book a truly international effort. This book will bring forth new frontiers with its revolutionizing research information and detailed analysis of the nascent developments around the world.

We would like to thank all the contributing authors for lending their expertise to make the book truly unique. They have played a crucial role in the development of this book. Without their invaluable contributions this book wouldn't have been possible. They have made vital efforts to compile up to date information on the varied aspects of this subject to make this book a valuable addition to the collection of many professionals and students.

This book was conceptualized with the vision of imparting up-to-date information and advanced data in this field. To ensure the same, a matchless editorial board was set up. Every individual on the board went through rigorous rounds of assessment to prove their worth. After which they invested a large part of their time researching and compiling the most relevant data for our readers.

The editorial board has been involved in producing this book since its inception. They have spent rigorous hours researching and exploring the diverse topics which have resulted in the successful publishing of this book. They have passed on their knowledge of decades through this book. To expedite this challenging task, the publisher supported the team at every step. A small team of assistant editors was also appointed to further simplify the editing procedure and attain best results for the readers.

Apart from the editorial board, the designing team has also invested a significant amount of their time in understanding the subject and creating the most relevant covers. They scrutinized every image to scout for the most suitable representation of the subject and create an appropriate cover for the book.

The publishing team has been an ardent support to the editorial, designing and production team. Their endless efforts to recruit the best for this project, has resulted in the accomplishment of this book. They are a veteran in the field of academics and their pool of knowledge is as vast as their experience in printing. Their expertise and guidance has proved useful at every step. Their uncompromising quality standards have made this book an exceptional effort. Their encouragement from time to time has been an inspiration for everyone.

The publisher and the editorial board hope that this book will prove to be a valuable piece of knowledge for researchers, students, practitioners and scholars across the globe.

List of Contributors

Sravya Tekumalla, Sankaranarayanan Seetharaman and Manoj Gupta
Department of Mechanical Engineering, National University of Singapore, 9 Engineering Drive 1, Singapore 117576, Singapore

Abdulhakim Almajid
Mechanical Engineering Department, College of Engineering, King Saud University, Riyadh 11421, Saudi Arabia

Zhaoxin Du, Jingshun Liu and Shufeng Lv
School of Materials Science and Engineering, Inner Mongolia University of Technology, Hohhot 010051, China

Shulong Xiao, Lijuan Xu, Fantao Kong and Yuyong Chen
National Key Laboratory of Science and Technology on Precision Heat Processing of Metals, Harbin Institute of Technology, Harbin 150001, China

Fang-Qiu Zu
Liquid/Solid Metal Processing Institute, School of Materials Science & Engineering, Hefei University of Technology, Hefei 230009, China

Daniela Nickel, Dagmar Dietrich, Thomas Mehner, Philipp Frint, Dagobert Spieler and Thomas Lampke
Technische Universität Chemnitz, Professur Oberflächentechnik/Funktionswerkstoffe, 09107 Chemnitz, Germany

Holger Schwab and Uta Kühn
IFW Dresden, Institut für Komplexe Materialien, Postfach 270116, D-01171 Dresden, Germany

Konda Gokuldoss Prashanth
IFW Dresden, Institut für Komplexe Materialien, Postfach 270116, D-01171 Dresden, Germany
R & D Engineer, Additive manufacturing Center, Sandvik AB, 81181 Sandviken, Sweden

Lukas Löber
IFW Dresden, Institut für Komplexe Materialien, Postfach 270116, D-01171 Dresden, Germany
R & D Engineer, Sauer GmbH–DMG MORI SEIKI, D-87459 Pfronten, Germany

Jürgen Eckert
IFW Dresden, Institut für Komplexe Materialien, Postfach 270116, D-01171 Dresden, Germany
TU Dresden, Institut für Werkstoffwissenschaft, D-01062 Dresden, Germany

Shujun Qiu, Hailiang Chu, Yongjin Zou, Cuili Xiang, Huanzhi Zhang, Fen Xu, Lixian Sun and Huaiying Zhou
Guangxi Key Laboratory of Information Materials, Guilin University of Electronic Technology, Guilin 541004, China
School of Materials Science and Engineering, Guilin University of Electronic Technology, Guilin 541004, China

Jianling Huang
School of Materials Science and Engineering, Guilin University of Electronic Technology, Guilin 541004, China

Liuzhang Ouyang
School of Materials Science and Engineering, South China University of Technology, Guangzhou 510641, China

Pubo Li, Tijun Chen and Suqing Zhang
State Key Laboratory of Advanced Processing and Recycling of Nonferrous Metals, Lanzhou University of Technology, Lanzhou 730050, China

Renguo Guan
School of Materials and Metallurgy, Northeastern University, Shenyang 110004, China

Tina Fichtner, ChanghaiWang, Aleksandr A. Levin, Guido Kreiner, Catalina Salazar Mejia and Claudia Felser
Max-Planck-Institute for Chemical Physics of Solids (MPI CPFS), Noethnitzer Str. 40, DE-01187 Dresden, Germany

Simone Fabbrici
Institute of Materials for Electronics and Magnetism, National Research Council (IMEM-CNR), Parco Area delle Scienze 37/A, I-43124 Parma, Italy
Laboratory of Micro and Submicro Enabling Technologies for Emilia-Romagna (MIST E-R), via Gobetti 101, I 40129 Bologna, Italy

Franca Albertini
Institute of Materials for Electronics and Magnetism, National Research Council (IMEM-CNR), Parco Area delle Scienze 37/A, I-43124 Parma, Italy

Lisa Köhler, Kristin Hockauf and Thomas Lampke
Institute of Materials Science and Engineering, Technische Universität Chemnitz, Erfenschlager Str. 73, 09125 Chemnitz, Germany

Maxim Murashkin and Ruslan Valiev
Institute for Physics of Advanced Materials, Ufa State Aviation Technical University, Ufa 450000, Russia
Laboratory for Mechanics of Bulk Nanostructured Materials, Saint Petersburg State University, Saint-Petersburg 199034 Russia

Ilchat Sabirov
IMDEA Materials Institute, Getafe, Madrid 28906, Spain

Dmitriy Prosvirnin, Vladimir Terentiev and Sergey Dobatkin
Baikov Institute for Metallurgy and Materials Science of the Russian Academy of Sciences, Moscow 119991, Russia

Ilya Ovid'ko
Research Laboratory for Mechanics of New Nanomaterials, Saint-Petersburg State Polytechnical University, Saint-Petersburg 195251, Russia

Yibo Ai, Chang Sun and Weidong Zhang
National Center for Materials Service Safety, University of Science and Technology Beijing, Beijing 100083, China

Hongbo Que
Qishuyan Institute Co., Ltd., China South Locomotive & Rolling Stock Corporation Limited, Changzhou 213011, China

Soo Yeol Lee
Department of Materials Science and Engineering, Chungnam National University, Daejeon 305-764, Korea

Huamiao Wang
Materials Science and Technology, Los Alamos National Laboratory, Los Alamos, NM 87544, USA

Michael A. Gharghouri
Canadian Neutron Beam Centre, Canadian Nuclear Laboratories, Chalk River, ON K0J 1J0, Canada

Halil Ibrahim Kurt and Murat Oduncuoglu
Technical Sciences, University of Gaziantep, Gaziantep 27310, Turkey

Mehmet Kurt
Department of Metallurgical and Materials Engineering, Engineering Faculty, Fırat University, Elazığ 23119, Turkey

Iman S. El-Mahallawi
Mining Petroleum and Metallurgical Engineering Department, Cairo University, Giza 12613, Egypt

Ahmed Yehia Shash
Mechanical Design and Production Department, Cairo University, Giza 12613, Egypt

Amer Eid Amer
Mechanical Design and Production Engineering Department, Faculty of Engineering, Beni-Suef University, Beni-Suef 62111, Egypt

Mi-Kyung Han, Jae-Bong Im, Moon-Jin Hwang, Bong-Jun Kim, Hae-Young Kim and Yeong-Joon Park
Department of Dental Materials and Medical Research Center for Biomineralization Disorders, School of Dentistry, Chonnam National University, Gwangju 500-757, Korea

Xiangzhen Zhu, Wei Jiang, Mingrui Li, Huan Qiao, Yuying Wu, Jingyu Qin and Xiangfa Liu
Key Laboratory for Liquid-Solid Structural Evolution and Processing of Materials, Ministry of Education, Shandong University, 17923 Jingshi Road, Jinan 250061, China